Handbook of Medicinal Plants of the World for Aging

Handbook of Medicinal Plants of the World for Aging: Botany, Ethnopharmacology, Natural Products, and Molecular Pathways provides an unprecedented comprehensive overview of more than 100 plants used globally as medicine with the potential to prevent premature aging. This handbook covers the pathophysiology of aging from the molecular and cellular to the organ levels, as well as the current state of knowledge about the modes of action of natural products from plants on the pathophysiological pathways related to the (i) cardiovascular system and metabolism, (ii) central nervous system, (iii) kidneys, (iv) bones, (v) skin and hair, and (vi) immune system.

Medicinal plants are presented alphabetically. For each plant is indicated the botanical family, synonyms, and common names in English, French, German, Portuguese, Russian, and Spanish. For each plant, the reader will also find the part used, active principles, medical history, contemporary medicinal uses, as well as pharmacological, clinical, and toxicological studies. The bibliographical references have been carefully selected for their relevance. This handbook is intended for medical doctors, nurses, pharmacists, dieticians, and nutritionists, as well as readers with interest in health food and herbs.

FEATURES

- Alphabetical presentation of over 100 medicinal plants and the pharmacological rationales for their uses for aging
- Discusses the medical history, current medicinal uses, and potential candidates for the prevention of premature aging
- Introduces the molecular mechanism of natural products on the pathophysiology of aging
- Contains a selection of bibliographic references
- A useful research tool for postgraduates, academics, and the pharmaceutical, herbal, or nutrition industries

Handbook of Medicinal Plants of the World for Aging: Botany, Ethnopharmacology, Natural Products, and Molecular Pathways presents comment sections that invite further research and reflection on the fascinating and timely subject of herbals for healthy aging. This is an ideal reference text for medicinal plant enthusiasts.

Handbook of Medicinal Plants of the World for Aging

Botany, Ethnopharmacology, Natural Products, and Molecular Pathways

Christophe Wiart Pharm.D., Ph.D.

Institute of Tropical Biology and Conservation, University Malaysia Sabah

CRC Press
Taylor & Francis Group
Boca Raton London New York

CRC Press is an imprint of the
Taylor & Francis Group, an **informa** business

Cover image: © shutterstock

First edition published 2024
by CRC Press
2385 Executive Center Drive, Suite 320, Boca Raton, FL 33431

and by CRC Press
4 Park Square, Milton Park, Abingdon, Oxon, OX14 4RN

© 2024 Christophe Wiart

CRC Press is an imprint of Taylor & Francis Group, LLC

Library of Congress Cataloging-in-Publication Data
Names: Wiart, Christophe, author.
Title: Handbook of medicinal plants of the world for aging : botany, ethnopharmacology, natural products, and molecular pathways / authored by Christophe Wiart, Associate Professor, School of Pharmacy, University of Nottingham Malaysia Campus.
Description: First edition. | Boca Raton : CRC Press, 2024. | Includes bibliographical references and index.
Identifiers: LCCN 2023031384 (print) | LCCN 2023031385 (ebook) | ISBN 9781032293974 (hbk) | ISBN 9781032293981 (pbk) | ISBN 9781003301455 (ebk)
Subjects: LCSH: Botany, Medical—Handbooks, manuals, etc. | Aging—Prevention—Handbooks, manuals, etc.
Classification: LCC QK99.A1 W527 2024 (print) | LCC QK99.A1 (ebook) | DDC 581.6/34—dc23/eng/20231017
LC record available at https://lccn.loc.gov/2023031384
LC ebook record available at https://lccn.loc.gov/2023031385

ISBN: 978-1-032-29397-4 (hbk)
ISBN: 978-1-032-29398-1 (pbk)
ISBN: 978-1-003-30145-5 (ebk)

DOI: 10.1201/9781003301455

Typeset in Times
by Apex CoVantage, LLC

A ma grand-mère, Madame Renée Monllor,
A ma mère, Madame Flora Monllor,
A ma famille,
A mes amis et maîtres, les arbres

Contents

Preface

People around the world have been observed living healthily to very old ages without medical care by mainly relying on medicinal plants. In our Western civilization, where the medical systems offer great assistance in maintaining good health, we still observe a recrudescence of pathologies and long-term treatments, which raises the question of whether the medicinal plants and food plants used globally could delay the physiological process of aging and related pathologies. The present handbook covers the physiology of aging from the molecular and cellular to organs levels, as well as the current state of knowledge about the mode of action of natural products from plants on the pathophysiological pathways related to the aging of the (i) cardiovascular system and metabolism, (ii) central nervous system, (iii) kidneys, (iv) bones, (v) skin and hair, and (vi) immune system.

The purpose of this handbook is to provide readers with a compendium of botanical, phytochemical, historical, ethnopharmacological, and pharmacological information on over 100 medicinal plants that can be found around the world and have the potential to prevent premature aging. The plants are presented alphabetically. For each plant is indicated the botanical family; synonyms; and common names in English, French, German, Portuguese, Russian, and Spanish. The reader will also find the parts used, active principles, medical history, contemporary medicinal uses, and the results of relevant pharmacological and clinical studies. The bibliographical references have been carefully selected for their relevance.

This book has been deliberately written in a very simple and accessible way to give students a clear introduction to the subject. This work is also intended for university lecturers and researchers working on anti-aging. Finally, reading this book is recommended for doctors, pharmacists, nurses, dieticians, researchers in the private sector, and anyone who wishes to learn more about the virtues of these plants against premature aging.

There is a global keen interest about medicinal plants, yet the very teaching of medicinal plants has been removed from most medical and pharmacy schools. It is clear that pharmacy students need a full course in medicinal plants, at least six semesters, in order to be able to prevent poisoning and frauds involving the use of adulterated products. The manufacture, prescription, and delivery of medicinal plants must be the responsibility of well-trained pharmacists, if not graduates from national schools of herbalism under pharmaceutical control could do the job. Furthermore, none of the plants listed in this handbook is intended for pediatric or veterinary use.

I have no doubt that the time will come when humans will be able to live much longer because of medicines that will be able to slow the physiological process of aging, and it is my hope that this book will assist in some way in the discovery of such "immortality pills".

I wish to express my gratitude to CRC Press, especially Hilary Lafoe and Sukirti Singh, as well as Dr. David J. Newman for his excellent foreword, and University Malaysia Sabah for providing me with the conditions favorable to the writing of this book.

Christophe Wiart Pharm.D., Ph.D.
Kota Kinabalu,
June 16, 2023

Foreword

It gives me significant pleasure to write a foreword to Christophe Wiart's latest compendium covering plant-related treatments for "premature ageing'. This is an area that has not been covered to any significant extent in translated treatises based on TCM (Traditional Chinese Medicine) or Ayurvedic-derived treatments (Indian subcontinent and contiguous areas). Ageing includes loss of cognition and includes dementia, which are frequently obvious to relatives of the "patient" and have been treated for centuries in many parts of the world.

This treatise covers the areas that "modern-day" natural product chemists and pharmacognosists would search in order to identify the active principles involved in the "treatment modalities" covered in the text. He has used indigenous lore to identify the plant(s) and/or their components and subsequent treatment, covering over 100 medicinal plants. What is of definite utility is the listing of the different names ascribed to a plant depending upon the sources of the information. This is a point that is frequently overlooked when describing a medicinal plant, since cross-referencing the name(s) used in different languages permits one to search current databases for information.

Another "plus" is his linkage of identified plant entities to information as to their "formal toxicities" which in some cases link-back to identified chemical entities. This part of the overall dataset should not stop any scientific work on the components that are described, since today, methods of delivery of toxic agents are well defined and used.

Therefore, I definitely recommend this handbook to anyone who is interested in potential treatments for premature ageing, be they scientific or medical professionals or people who are interested in the topic for general interest.

David J. Newman, DPhil
Retired Chief, Natural Products Branch,
National Cancer Institute, NIH, USA

Author Biography

Christophe Wiart was born on 12th of August 1967 in Saint Malo, France. After his A-levels, he completed his Pharm.D. at the Facultée des Sciences Pharmaceutiques et Biologiques, Université Rennes 2 (France) and earned his Ph.D. in Natural Products Chemistry at the Universiti Pertanian Malaysia. He has taught pharmacognosy at the University of Malaya, and elsewhere. Dr. Wiart is the author of *Medicinal Plants of the Asia-Pacific: Drugs from the Future?* (2006), *Medicinal Plants of Asia and the Pacific* (2006), *Ethnopharmacology of Medicinal Plants: Asia and the Pacific* (2006), *Medicinal Plants from the East* (2010), *Medicinal Plants from China, Korea and Japan: Bioresource for Tomorrow's Drug and Cosmetic Discovery* (2012), *Lead Compounds from Medicinal Plants for the Treatment of Cancer* (2012), *Lead Compounds from Medicinal Plants for the Treatment of Neurodegenerative Diseases* (2013), *Medicinal Plants in Asia for Metabolic Syndrome* (2018), *Medicinal Plants from West Bengal and Bangladesh* (2019), *Medicinal Plants in Asia and Pacific for Parasitic Infections: Botany, Ethnopharmacology, Molecular Basis, and Future Prospect* (2020), *Medicinal Plants in Asia and the Pacific for Zoonotic Pandemics* (2021). He has published numerous articles. Dr. Wiart is presently completing a book on the medicinal plants of North Borneo. Other current research interests include the ethnopharmacological study of the medicinal plants of Southeast Asia for the development of herbals and lead therapeutic compounds.

1 Yarrow (*Achillea millefolium* L.)

Etymology: After *Achilleos* (Achilleus), an ancient Greek mythological war-
rior of the Trojan War, and from the Latin *millefolium* = thousands of leaves
Family: Asteraceae
Synonyms: *Achillea californica* Pollard; *Achillea lanulosa* Nutt.; *Achillea pec-
ten-veneris* Pollard
Common names: Yarrow; achillée millefeuille (Fr.); gemeine schafgarbe (Ger.);
planta milefólio (Port.); тысячелистник (Rus.); planta de milenrama (Spa.)
Part used: Leaf
Constituents: Sesquiterpene lactones (leucodin, achillin) (Li et al., 2021),
flavones (artemetin) (Falk et al., 1975).
Medical history: Dioscorides and Galen call yarrow *millefolium* and recommend it for wounds and internal injuries. The plant was used as a styptic in
Middle Ages France and was called "*herbe aux charpentiers*". In Scotland,
a decoction of leaves was used for cold and to treat common ailments in
children. In England, it was vulnerary and was used to stop internal bleeding and profuse mucous discharges. In France, an infusion was used as an
emmenagogue, for fevers, and for lochia. In Sweden, it was employed to
make beer (Guibourt, 1836).
Medicinal uses: Period-related painful spasms and to facilitate digestion,
2–4 g of plant powder in 250 mL boiling water 3 or 4 times daily between
meals (European Union); diuretic, inflammations (Turkey); rheumatoid
arthritis, gout, angina pectoris (Iran); kidney stones, hypertension (Afghanistan); diuretic, dysentery (Pakistan); toothache (India)

Blood pressure: Aqueous methanol extract of aerial parts at concentrations
ranging from about 2 to 4 mg/mL inhibited phenylephrine- and potassium
ions- induced contraction of aortic rings, calcium-related vascular smooth
muscle cells, and endothelium-dependent relaxant effects. This extract
administered intravenously to rats at a dose of 10 mg/kg evoked about 20%
fall in blood pressure (Khan & Gilani, 2011).
Hydroalcoholic extract of aerial parts (containing the methoxyflavone artemetin) given at the single oral dose of 100 mg/kg to rats evoked after 3 hours
a transient decrease in blood pressure from 116.4 to 100 mmHg (de Souza
et al., 2011). Leucodin and achillin given orally at 50 mg/kg decreased
systolic and diastolic blood pressure in spontaneously hypertensive rat.
In vitro, these sesquiterpene lactones evoked vasorelaxant effects that were
inhibited by N(ω)-nitro-*L-arginine* methyl ester (L-NAME) and potassium

DOI: 10.1201/9781003301455-1

1

chloride, suggesting nitric oxide (NO) and calcium ion channel blockage (Arias-Durán et al., 2021).

Plasma lipids and glucose: Hydroalcoholic extract given orally at 100 mg/kg for 28 days to streptozotocin-induced diabetic rats decreased glycemia from about 450 to 250 mg/dL, total cholesterol from about 120 to 75 mg/dL, and plasma triglycerides from about 80 to 40 mg/dL (Rezaei et al., 2020).

Kidneys: A single dose of 300 mg/kg hydroethanolic extract of aerial parts given to rats orally increased urine excretion by about 60% after 8 hours, as well as urinary sodium and potassium (De Souza et al., 2013).

Immune system: Methanol extract of aerial parts given intraperitoneally to mice at 100 mg/kg/day evoked an increase in leukocytes count (Al-Ezzy et al., 2018).

Skin and hair: A 2-month treatment with an extract at 2% reduced the appearance of wrinkles and pores (Pain et al., 2011). Aqueous extract of flowers promoted the growth of hair (Grollier & Rosenbaum, 1990).

Brain: Aqueous extract given orally at 2.8 mg/kg/day for 14 days protected rats against 6-hydroxydopamine-induced Parkinson's disease (Akramian et al., 2015).

Comments: (i) Infusions of leaves could be of value in delaying cardiovascular aging.

(ii) Most of the plants with the potential to delay aging were known to Greek and Roman physicians more than 2000 years ago. Nero's Greek physician, Pedanus Dioscorides, in the first century described thousands of plants and their uses, including yarrow, in his monumental book titled *De Materia Medica*. The Greek physician Claudius Galenus (129–216 AD), or Galen, is the author of another monumental book: *de Alimentorum Facultatibus*. The works of both Dioscorides and Galen were used as absolute medical texts in Europe until the end of the 19th century. Pharmacy students and even medical students need to read their work.

(iii) L-NAME is an inhibitor of NO synthetase, the enzyme responsible for the production of NO by vascular endothelial cells that relaxes vascular smooth muscle cells. This synthetic molecule is used to induce models of hypertension.

(iv) Streptozotocin is a bacterial toxin that destroys pancreatic β-cells and is used to mimic models of diabetes in rodents.

REFERENCES

Akramian Fard, M., Moghaddam Ahmadi, A., Ayyobi, F., Nakisa, H., Hadadian, Z., Shabani, M., Tavakoli, M. and Shamsi Zadeh, A., 2015. Effects of Achillea millefolium aqueous extract in a Parkinsons disease model induced by intra-cerebral-ventricular injection of 6-hydroxydopamine in male rats. *Journal of Sabzevar University of Medical Sciences*, 22(1), pp. 36–44.

Al-Ezzy, R.M., Al Anee, R.S. and Ibrahim, N.A., 2018. Assessments of immunological activity of Achillea millefolium methanolic extract on albino male mice. *Journal of Pharmacy and Pharmacology*, 6, pp. 563–569.

Arias-Durán, L., Estrada-Soto, S., Hernández-Morales, M., Millán-Pacheco, C., Navarrete-Vázquez, G., Villalobos-Molina, R., Ibarra-Barajas, M. and Almanza-Pérez, J.C., 2021. Antihypertensive and vasorelaxant effect of leucodin and achillin isolated from Achillea millefolium through calcium channel blockade and NO production: In vivo, functional ex vivo and in silico studies. *Journal of Ethnopharmacology*, 273, p. 113948.

De Souza, P., Crestani, S., da Silva, R.D.C.V., Gasparotto, F., Kassuya, C.A.L., da Silva-Santos, J.E. and Junior, A.G., 2013. Involvement of bradykinin and prostaglandins in the diuretic effects of Achillea millefolium L. (Asteraceae). *Journal of Ethnopharmacology*, 149(1), pp. 157–161.

de Souza, P., Gasparotto Jr, A., Crestani, S., Stefanello, M.É.A., Marques, M.C.A., da Silva-Santos, J.E. and Kassuya, C.A.L., 2011. Hypotensive mechanism of the extracts and artemetin isolated from Achillea millefolium L. (Asteraceae) in rats. *Phytomedicine*, 18(10), pp. 819–825.

Falk, A.J., Smolenski, S.J., Bauer, L. and Bell, C.L., 1975. Isolation and identification of three new flavones from Achillea millefolium L. *Journal of Pharmaceutical Sciences*, 64(11), pp. 1838–1842.

Grollier, J.F., Rosenbaum, G. and L'Oreal, S.A., 1990. *Cosmetic composition for the treatment of hair, particularly oily hair, based on an extract of yarrow (Achillea millefolium L)*. U.S. Patent 4,948,583.

Guibourt, N.J.B.G., 1836. *Histoire abrégée des drogues simples*. Méquignon-Marvis Père et fils.

Khan, A.U. and Gilani, A.H., 2011. Blood pressure lowering, cardiovascular inhibitory and bronchodilatory actions of Achillea millefolium. *Phytotherapy Research*, 25(4), pp. 577–583.

Li, H., Li, J., Liu, M., Xie, R., Zang, Y., Li, J. and Aisa, H.A., 2021. Guaianolide sesquiterpene lactones from Achillea millefolium L. *Phytochemistry*, 186, p. 112733.

Pain, S., Altobelli, C., Boher, A., Cittadini, L., Favre-Mercuret, M., Gaillard, C., Sohm, B., Vogelgesang, B. and André-Frei, V., 2011. Surface rejuvenating effect of Achillea millefolium extract. *International Journal of Cosmetic Science*, 33(6), pp. 535–542.

Rezaei, S., Ashkar, F., Koohpeyma, F., Mahmoodi, M., Gholamalizadeh, M., Mazloom, Z. and Doaei, S., 2020. Hydroalcoholic extract of Achillea millefolium improved blood glucose, liver enzymes and lipid profile compared to metformin in streptozotocin-induced diabetic rats. *Lipids in Health and Disease*, 19(1), pp. 1–7.

2 Venus's Hair Fern (*Adiantum capillus-veneris* L.)

Etymology: From the Greek *Adianton* = does not take or retain water and the Latin *capillus-veneris* = hair of Venus

Family: Pteridaceae

Synonyms: *Adiantum michelii* Christ; *Adiantum subemarginatum* Christ

Common names: Venus's hair fern, maiden hair fern; capillaire de Montpellier (Fr.); Venus-Frauenhaarfarn (Ger.); culantrillo de pozo (Spa.); capilária (Port.); девичий папоротник (Rus.)

Part used: Whole plant

Constituents: Flavone glycosides (kaempferol and quercetin glycosides) (Akabori & Hasegawa, 1969), oleanane triterpenes (Nakane et al., 2002).

Medical history: Diuretic, pectoral, expel urinary stones (Dioscorides). The plant was used by Romans to dye hair and to promote the growth of hair. Treated cough and cold in 17th-century France according to Lémery (1716).

DOI: 10.1201/9781003301455-2

Good for cold and diseases of lungs and kidneys in the lectures of Alston (1770). It was also used to prepare a *"sirop de capillaire"* in 19th-century France for cough.

Medicinal uses: Uric acid, prostate enlargement (Turkey); cough, difficulty breathing, nephritis, renal stones, hypertension (Iraq); jaundice, cough, hemorrhoid, fever (Iran); cough, fever, diuretic, cold, hepatitis (Pakistan); fever, cough, diuretic, piles, wounds, bronchial problems (India); cystitis, fever (China)

Blood pressure: Aqueous extract given orally to rats induced with hypertension using L-NAME reduced blood pressure from 156 to 121 mmHg (Aziz & Dizaye., 2019).

Plasma lipids and glucose: Aqueous extract given orally at 400 mg/kg/day for 21 days decreased blood glucose from 242 to 174 mg/dL (metformin at 50 mg/kg/day: 102 mg/dL) in rats induced with diabetes via streptozotocin injection (Ranjan et al., 2014). Aqueous extract of leaves delivered orally at 100 mg/kg/day for 8 weeks decreased plasma cholesterol from 5.3 to 3.1 mmol/L and low-density lipoprotein (LDL)-cholesterol from 3.6 to 1.4 mmol/L in rats on a high-cholesterol diet (Mutar et al., 2021).

Kidneys: Hydroalcoholic extract given orally at 255.2 mg/kg/day for 7 days to rats prevented the formation of kidney stones, damage to the medulla, glomeruli, tubules, and interstitial spaces, kidney blood vessel proliferation, and inflammatory cell infiltration induced by ethylene glycol and ammonium chloride (Ahmed et al., 2013).

Aqueous extract given orally to rats increased urine secretion from 0.6 to 1.5 mL/hr/kg and glomerular filtration rate from 193.4 to 286.3 mL/hr/kg, as well as urinary sodium ion excretion (Aziz & Dizaye., 2019).

Immune system: Ethanol extract given to mice poisoned with lipopolysaccharides decreased the pancreatic expression of nuclear factor kappa-light-chain-enhancer of activated B cells (NF-kB) (Yuan et al., 2013).

Skin and hair: Ethanol extract at 1% part of a preparation for 21 days prevented testosterone-induced hair fall in mice (Noubarani et al., 2014).

Comments: (i) Infusions of this plant could potentially be of value in delaying cardiovascular aging.

(ii) Nicolas Lémery (1645–1715) was a French physician and chemist who wrote the *Dictionnaire Universel des Drogues Simples*.

(iii) Charles Alston (1683–1760) was a professor of materia medica at the University of Edinburgh, and many of the plants examined here are discussed in his lectures.

(iv) Urinary stones (uric acid, calcium, struvite) tend to appear in middle-aged and elderly men and after menopause in women from a mix of genetic predisposition, bacterial infection, and dietary issues such as excess red meat and alcohol and dehydration. Magnesium, citric acid, and potassium decrease the formation of urinary stones (Yasui et al., 2017; Wang et al., 2021). The chronic presence of stones in the kidneys, as well as the alteration of kidney structure from hyperglycemia (diabetes), results in local inflammation, decreased glomerular filtration rate (Chou et al., 2011), high blood pressure (Wardle et al., 1996), and stroke (Lee et al., 2010).

REFERENCES

Ahmed, A., Wadud, A., Jahan, N., Bilal, A. and Hajera, S., 2013. Efficacy of Adiantum capillus veneris Linn in chemically induced urolithiasis in rats. *Journal of Ethnopharmacology*, *146*(1), pp. 411–416.

Akabori, Y. and Hasegawa, M., 1969. Flavonoid Pattern in the Pteridaceae II Flavonoid Constituents in the Fronds of Adiantum capillus-veneris and A. cuneatum. *Shokubutsugaku Zasshi*, *82*(973), pp. 294–297.

Alston, C., 1770. *Lectures on the Materia Medica: Containing the Natural History of Drugs, their Virtues and Doses: Also Directions for the Study of the Materia Medica; and an Appendix on the Method of Prescribing*. Edward and Charles Dilly.

Aziz, R.S. and Dizaye, K.A.W.A., 2019. Diuretic effect of Adiantum capillus and its chemical constituents in hypertensive rats. *International Journal of Pharmaceutical Research*, *11*(3).

Chou, Y.H., Li, C.C., Hsu, H., Chang, W.C., Liu, C.C., Li, W.M., Ke, H.L., Lee, M.H., Liu, M.E., Pan, S.C. and Wang, H.S., 2011. Renal function in patients with urinary stones of varying compositions. *The Kaohsiung Journal of Medical Sciences*, *27*(7), pp. 264–267.

Lee, M., Saver, J.L., Chang, K.H., Liao, H.W., Chang, S.C. and Ovbiagele, B., 2010. Low glomerular filtration rate and risk of stroke: Meta-analysis. *BMJ*, *341*.

Lémery, N., 1716. *Traité universel des drogues simples, mises en ordre alphabétique. Où l'on trouve leurs différens noms . . . et tout ce qu'il y a de particulier dans les animaux, dans les végétaux, et dans les minéraux*. Au dépend de la Companie.

Mutar, W.M., Al-Hayawi, A.Y. and Asaad, M.M., 2021. Reduction of cholesterol and estimation of genotoxic effects of an aqueous infusion of adiantum capillus-veneris leaves in male white rats' testicular tissue. *Systematic Reviews in Pharmacy*, *12*(2), pp. 266–272.

Nakane, T., Maeda, Y., Ebihara, H., Arai, Y., Masuda, K., Takano, A., Ageta, H., Shiojima, K., Cai, S.Q. and Abdel-Halim, O.B., 2002. Fern constituents: Triterpenoids from Adiantum capillus-veneris. *Chemical and Pharmaceutical Bulletin*, *50*(9), pp. 1273–1275.

Noubarani, M., Rostamkhani, H., Erfan, M., Kamalinejad, M., Eskandari, M.R., Babaeian, M. and Salamzadeh, J., 2014. Effect of Adiantum capillus veneris linn on an animal model of testosterone-induced hair loss. *Iranian Journal of Pharmaceutical Research: IJPR*, *13*(Suppl), p. 113.

Ranjan, V., Vats, M., Gupta, N. and Sardana, S., 2014. Antidiabetic potential of whole plant of Adiantum capillus veneris linn. In streptozotocin induced diabetic rats. *International Journal of Pharmaceutical and Clinical Research*, *6*(4), pp. 341–347.

Wang, H., Fan, J., Yu, C., Guo, Y., Pei, P., Yang, L., Chen, Y., Du, H., Meng, F., Chen, J. and Chen, Z., 2021. Consumption of tea, alcohol, and fruits and risk of kidney stones: A prospective cohort study in 0.5 million Chinese adults. *Nutrients*, *13*(4), p. 1119.

Wardle, E.N., Kurihara, I., Saito, T., Obara, K., Shoji, Y., Hirai, M., Soma, J., Sato, H., Imai, Y., Abe, K. and Yuasa, S., 1996. Hypertension in kidney stone patients. *Nephron*, *73*(4), pp. 569–572.

Yasui, T., Okada, A., Hamamoto, S., Ando, R., Taguchi, K., Tozawa, K. and Kohri, K., 2017. Pathophysiology-based treatment of urolithiasis. *International Journal of Urology*, *24*(1), pp. 32–38.

Yuan, Q., Zhang, X., Liu, Z., Song, S., Xue, P., Wang, J. and Ruan, J., 2013. Ethanol extract of Adiantum capillus-veneris L. suppresses the production of inflammatory mediators by inhibiting NF-κB activation. *Journal of Ethnopharmacology*, *147*(3), pp. 603–611.

3 Bael (*Aegle marmelos* (L.) Corrêa)

Aegle marmelos (L.) Corrêa

Etymology: From the Greek *aigle* = radiance and Portuguese *marmelo* = quince

Family: Rutaceae

Synonyms: *Bilacus marmelos* (L.) Kuntze; *Crateva marmelos* L.

Common names: Bael, Bengal quince, stone apple; le fruit de bael (Fr.); belbaum (Ger.); membrillo de bengala (Spa.); marmelo de bengala (Port.)

Parts used: Fruit, leaf

Constituents: Coumarins (auraptene) (Bhardwaj & Nandal, 2015), fibers in fruit pulp (Charoensiddhi & Anprung, 2008).

Medical history: In Ayurvedic medicine, Bengal quince has been used as heart medicine from the dawn of time. The pulp was used as a mild astringent in 19th-century USA and for dysentery in 19th-century England.

Medicinal uses: Dysentery, constipation, heat stroke (Bangladesh); cooling, jaundice, asthma, diuretic, laxative, diabetes, urinary trouble, heart palpitation, arthritis, trouble (India); sunstroke (Myanmar)

Blood pressure: Juice of leaves given to type-2 diabetic patients orally at 20 g per day for 8 weeks decreased blood pressure (Nigam & Nambiar, 2019).

Plasma lipids and cholesterol: Aqueous extract of fruits given orally to streptozotocin-induced diabetic rats at 250 mg/kg twice a day for 30 days decreased blood glucose from 336.8 to 96.2 mg/dL (normal group: 85.8 ng/dL) and increased plasma insulin from 10.7 to 17.9 μU/mL (normal group: 19.5 μU/mL) (Kamalakkannan & Prince, 2004). Juice of leaves given to type-2 diabetic patients orally at 20 g per day for 8 weeks decreased glycemia, plasma cholesterol, and plasma antioxidant capacity (Nigam & Nambiar, 2019).

Heart: Aqueous extract of leaves given orally at 200 mg/kg/day for 35 days to rodents prevented isoprenaline-induced cardiac insults as indicated by the near normalization of myocardial lactate dehydrogenase (a marker of cellular injuries), heart weight, and heart and aortic cholesterol (Prince & Rajadurai, 2005). Auraptene from the root bark decreased the beating rate of mouse myocardial cells *in vitro* with an IC_{50} of 0.6 μg/mL (Kakiuchi et al., 1991).

Kidneys: Aqueous extract of fruits given orally to streptozotocin-induced diabetic rats at 250 mg/kg twice a day for 30 days brought kidney malondialdehyde and glutathione close to normal values (Kamalakkannan & Prince, 2004).

DOI: 10.1201/9781003301455-3

Immune system: Methanol extract given orally at 100 mg/kg boosted the immune system of mice (Patel & Asdaq, 2010).

Brain: Ethanol extract of leaves given orally at 200 mg/kg/day for 2 weeks protected rats against dementia induced by intracerebroventricular injection of streptozotocin. This regimen decreased hippocampal malondialdehyde (MDA) and increased cholinergic function (Raheja et al., 2019).

Comments: (i) The fruit pulp of Bengal quince has soluble and insoluble dietary fiber contents of 11.2 and 8.6 g/100 g dry weight, respectively (Charoensiddhi & Anprung, 2008) and as such could inhibit intestinal cholesterol absorption and help control diabetes. The leaves, if not toxic, could be developed as over-the-counter treatments for heart and vascular aging.

(ii) It is currently generally accepted that aging involves oxidative stress, a continuous and increasing exposure of cells to reactive oxygen species (ROS), which alter the chemical structure of lipids, proteins, RNA, and DNA via oxidation and peroxidation. MDA is an indicator of lipid peroxidation, and it increases with aging (Mutlu-Türkoğlu et al., 2003). Concurrently, aging is synonymous with decreased levels of glutathione, a peptide whose physiological role is to reduce hydrogen peroxide (coming from the superoxide anion (O^{2-}) via superoxide dismutase (SOD) into water (Yang et al., 1995). It can be seen as a spiral of oxidative events of increasing amplitude whereby with time mitochondria exposed to ROS generate more ROS and, as such, are exposed to more ROS, etc. We will one day be able to live much longer with the help of drugs that will stop this oxidative spiral. Diabetes intensifies oxidative stress, and bael fruit pulp is interesting in this context.

REFERENCES

Bhardwaj, R.L. and Nandal, U., 2015. Nutritional and therapeutic potential of bael (*Aegle marmelos* Corr.) fruit juice: A review. *Nutrition & Food Science*, *45*(6), pp. 895–919.

Charoensiddhi, S. and Anprung, P., 2008. Bioactive compounds and volatile compounds of Thai bael fruit (Aegle marmelos (L.) Correa) as a valuable source for functional food ingredients. *International Food Research Journal*, *15*(3), pp. 287–295.

Kakiuchi, N., Senaratne, L.R., Huang, S.L., Yang, X.W., Hattori, M., Pilapitiya, U. and Namba, T., 1991. Effects of constituents of Beli (Aegle marmelos) on spontaneous beating and calcium-paradox of myocardial cells. *Planta Medica*, *57*(01), pp. 43–46.

Kamalakkannan, N. and Prince, P., 2004. Antidiabetic and anti-oxidant activity of Aegle marmelos extract in streptozotocin-induced diabetic rats. *Pharmaceutical Biology*, *42*(2), pp. 125–130.

Mutlu-Türkoğlu, Ü., İlhan, E., Öztezcan, S., Kuru, A., Aykaç-Toker, G. and Uysal, M., 2003. Age-related increases in plasma malondialdehyde and protein carbonyl levels and lymphocyte DNA damage in elderly subjects. *Clinical Biochemistry*, *36*(5), pp. 397–400.

Nigam, V. and Nambiar, V.S., 2019. Aegle marmelos leaf juice as a complementary therapy to control type 2 diabetes–Randomised controlled trial in Gujarat, India. *Advances in Integrative Medicine*, *6*(1), pp. 11–22.

Patel, P. and Asdaq, S.M.B., 2010. Immunomodulatory activity of methanolic fruit extract of Aegle marmelos in experimental animals. *Saudi Pharmaceutical Journal*, *18*(3), pp. 161–165.

Prince, P.S.M. and Rajadurai, M., 2005. Preventive effect of Aegle marmelos leaf extract on isoprenaline-induced myocardial infarction in rats: Biochemical evidence. *Journal of Pharmacy and Pharmacology*, *57*(10), pp. 1353–1357.

Raheja, S., Girdhar, A., Kamboj, A., Lather, V. and Pandita, D., 2019. Aegle marmelos leaf extract ameliorates the cognitive impairment and oxidative stress induced by intracerebroventricular streptozotocin in male rats. *Life Sciences*, *221*, pp. 196–203.

Yang, C.S., Chou, S.T., Liu, L., Tsai, P.J. and Kuo, J.S., 1995. Effect of ageing on human plasma glutathione concentrations as determined by high-performance liquid chromatography with fluorimetric detection. *Journal of Chromatography B: Biomedical Sciences and Applications*, *674*(1), pp. 23–30.

4 Grains of Paradise (*Aframomum melegueta* (Roscoe) K. Schum.)

Aframomum melegueta (Roscoe) K. Schum.

Etymology: From the Hebrew *aphrah* = dust related to Africa and the Latin *amomum* = cardamon and from the old French name of the plant, *maliguette*
Family: Zingiberaceae
Synonym: *Amomum melegueta* Roscoe
Common names: Alligator pepper, grains of paradise; maniguette, poivre du paradis (Fr.); paradieskörner (Ger.); granos del paraíso (Spa.); grãos-do-paraíso (Port.)
Part used: Seed
Constituents: Essential oil (humulene and caryophyllene) (Ajaiyeoba & Ekundayo, 1999), alkylphenols (6-gingerol, 6-paradol, 6-shogaol) (Sugita et al., 2013).
Medical history: The seeds were sold in the markets of Lyon in France in 1245. It is the *Cardamomum majus* of Matthioli (1572), and it was used in the 19th century to give taste to liquors as well as for perfumes (Guibourt, 1836).
Medicinal uses: Aphrodisiac (Africa)

Plasma lipids and glucose: Methanol extract of seeds given orally to rats at 400 mg/kg for 21 days decreased plasma cholesterol from 95.3 to 67.2 mg/dL, triglycerides from 50.9 to 34.3 mg/dL, and LDL-cholesterol from 36.9 to 12.7 mg/dL and increased HDL-cholesterol from 48.1 to 54.4 mg/dL (Onoja et al., 2015).

Aqueous extract of seeds given orally at 400 mg/kg to alloxan-induced diabetic rats for 14 days decreased fasting glycemia from 25 to 4.9 mmol/L (Adesokan et al., 2010).

Adipose tissues: Ethanol extract of seeds given to healthy men at 40 mg after breakfast, 30 mg after lunch, and 30 mg after dinner for 4 weeks increased energy expenditure from brown adipose tissues (Sugita et al., 2013).

Bones and cartilages: 6-Paradol (also known as gingerone) at the concentration of 1 µM enhanced the activity of alkaline phosphatase (ossification marker), increased the cellular content of 1,25-dihydroxy calciferol, and decreased the activity of acid phosphatase (bone resorption marker) in SAOS-2 cells (Abdel-Naim et al., 2017).

DOI: 10.1201/9781003301455-4

Brain: Aqueous extract given orally at 25 mg/kg/day for 3 days prevented dementia induced by scopolamine in mice. This regimen decreased MDA and increased cerebral levels of glutathione and SOD (Ishola et al., 2016).

Comments: (i) Pietro Andrea Matthioli (1501–1577) was an Italian botanist and personal physician to the Emperor Maximilian II and the author of *"Commentaires sur les Six Livres de Pedacius Dioscorides Anazarbeen de la matière medicinale"* (1572).

(ii) The French pharmacist Nicolas Jean Baptiste Gaston Guibourt (1790–1867) was a professor of pharmaceutical natural history at the École de pharmacie de Paris and author of *Histoire abrégée des drogues simples* (1836).

(iii) With aging, the ability to gain energy from adipose tissue declines (Bartke et al., 2021), and in mice, increased metabolic rate translates into longevity (Westbrook et al., 2009). The seeds of alligator pepper might potentially boost metabolism in elderly persons.

REFERENCES

Abdel-Naim, A.B., Alghamdi, A.A., Algandaby, M.M., Al-Abbasi, F.A., Al-Abd, A.M., Abdallah, H.M., El-Halawany, A.M. and Hattori, M., 2017. Phenolics isolated from Aframomum meleguta enhance proliferation and ossification markers in bone cells. *Molecules*, 22(9), p. 1467.

Adesokan, A.A., Akanji, M.A. and Adewara, G.S., 2010. Evaluation of hypoglycaemic efficacy of aqueous seed extract of Aframomum melegueta in alloxan-induced diabetic rats. *Sierra Leone Journal of Biomedical Research*, 2(2).

Ajaiyeoba, E.O. and Ekundayo, O., 1999. Essential oil constituents of *Aframomum melegueta* (Roscoe) K. Schum. seeds (alligator pepper) from Nigeria. *Flavour and Fragrance Journal*, 14(2), pp. 109–111.

Bartke, A., Brannan, S., Hascup, E., Hascup, K. and Darcy, J., 2021. Energy metabolism and aging. *The World Journal of Men's Health*, 39(2), p. 222.

Guibourt, N.J.B.G., 1836. *Histoire abrégée des drogues simples*. Méquignon-Marvis Père et fils.

Ishola, I.O., Awoyemi, A.A. and Afolayan, G.O., 2016. Involvement of antioxidant system in the amelioration of scopolamine-induced memory impairment by grains of paradise (*Aframomum melegueta* K. Schum.) extract. *Drug Research*, 66(9), pp. 455–463.

Matthioli, P.A., 1572. *Commentaires sur les Six Livres de Pedacius Dioscorides Anazarbeen de la matière medicinale*. A l'Escue de Milan.Onoja, S.O., Ezeja, M.I., Omeh, Y.N. and Emeh, E.C., 2015. Phytochemical and hypolipidemic effects of methanolic extract of Aframomum melegueta seed. *European Journal of Medicinal Plants*, 5(4), p. 377.

Sugita, J., Yoneshiro, T., Hatano, T., Aita, S., Ikemoto, T., Uchiwa, H., Iwanaga, T., Kameya, T., Kawai, Y. and Saito, M., 2013. Grains of paradise (Aframomum melegueta) extract activates brown adipose tissue and increases whole-body energy expenditure in men. *British Journal of Nutrition*, 110(4), pp. 733–738.

Westbrook, R., Bonkowski, M.S., Strader, A.D. and Bartke, A., 2009. Alterations in oxygen consumption, respiratory quotient, and heat production in long-lived GHRKO and Ames dwarf mice, and short-lived bGH transgenic mice. *Journals of Gerontology Series A: Biological Sciences and Medical Sciences*, 64, pp. 443–451.

5 Common Agrimony (*Agrimonia eupatoria* L.)

Agrimonia eupatoria L.

Etymology: From the Latin *ager* = field as Dioscorides recorded that it grew near cultivated fields ("*agrimonia, quod in agris abundat*") and *eupatorium* in honor of King Eupator, who first found the plant. Also called *hepatorium* because the plant is good for the liver.

Family: Rosaceae

Common names: Common agrimony; aigremoine eupatoire (Fr.); kleiner odermennig (Ger.); agrimonia común (Spa.); agrimônia comum (Port.); репешок обыкновенный (Rus.)

Part used: Leaf

Constituents: Flavones (quercitrin and its glycosides) (Tomlison et al., 2003), ellagitannin (agrimoniin) (Grochowski et al., 2017).

Medical history: Common agrimony was used by Roman physicians for ulcers (Dioscorides) and was good for the liver (Galen). In the 12th century, Arab physicians like Serapion the younger considered it hot and dry to the third degree. Alston (1770) in his lectures described common agrimony as aperitive, cooling, and vulnerary. The plant was used as an astringent, a tonic, and a febrifuge in 19th-century Europe as well as a substitute for tea and a treatment for diarrhea and for inflammation of the mouth and throat (Guibourt, 1836).

Medicinal uses: Diuretic, laxative, goiter, hernia (Turkey); diuretic, tonic, astringent (India); diuretic, astringent (Myanmar)

Kidneys: Infusion given orally at 3 g/kg induced urination in rats (Giachetti et al., 1989).

Plasma lipids and glucose: Rats fed a diet containing 62.5 g/kg of agrimony were protected against diabetes induced by streptozotocin. *In vitro*, aqueous extract of agrimony at the concentration of 1 mg/mL increased the secretion of insulin by pancreatic β-cells by about fourfold (Gray & Flatt, 1998).

Kidneys: Infusion given orally at 3 g/kg induced urination in rats (Giachetti et al., 1989).

Brain: Methanol extract protected mouse hippocampal neurons against glutamate-induced oxidative stress (Lee et al., 2010).

Warning: Intake of the plant (dose unknown) has been associated with tubulointerstitial nephritis, proteinuria, hypertension due to lower filtration function of the kidney, and renal tubular dysfunction (Batyushin et al., 2013).

DOI: 10.1201/9781003301455-5

Comments: (i) Infusion of leaves could potentially be of value as a diuretic, in controlling glycemia in diabetes, and in attenuating age-related oxidation. (ii) Patients with steatosis (a liver saturated with fats) are often obese, with impaired flow-mediated vasodilatation, arterial stiffness, and atherosclerosis, and they are prone to cardiovascular diseases (Targher et al., 2010). Steatosis accounts for increased inflammatory markers in the plasma such as tumor necrosis factor-α (TNF α) and interleukin-6, which contribute to insulin resistance and the development of type-2 diabetes (Shoelson et al., 2007). Healthy volunteers taking an infusion of leaves (1 g in 200 mL, 10 mins) twice a day for 30 days had increased plasma antioxidant capacity and decreased interleukin-6 (from about 10 to 5 pg/mL) (Ivanova et al., 2013). Aqueous extract of aerial parts given orally at 300 mg/kg/day to rats poisoned with ethanol for 8 weeks prevented the elevation of serum aminotransferase activities, pro-inflammatory cytokines, and lipid peroxidation and decreased glutathione (Yoon et al., 2012).

(iii) Serapion the younger was a 12th-century Arab physician and the author of *Liber de Simplicibus Medicamentis.*

(iv) The Greek physician Hippocrates (460–370 BC) asserts that hot, dry, cold, and humid are the four qualities that describe humans' physical conditions and that diseases result from an imbalance between 4 humors: blood (hot and wet), yellow bile (hot and dry), black bile (cold and dry), and phlegm (cold and wet). He also believes that medicinal plants could restore balance; for instance, that applying a cold plant could correct an overheated condition (Jouanna, 2012). It may sound strange and quite unscientific, but closer investigation and study of medicinal plants in daily use make it clear that there is some truth to explore there. In Pakistan and India, for instance, Unani practitioners still use the Hippocratic system (as well as Galen's teachings) and have managed to improve the well-being of many in destitute areas where doctors do not open their clinics. I was myself diagnosed as hot and dry to the third degree by hakeems in Karachi and advised to take tea, and it worked out to be very useful.

REFERENCES

Alston, C., 1770. *Lectures on the Materia Medica: Containing the Natural History of Drugs, their Virtues and Doses: Also Directions for the Study of the Materia Medica; and an Appendix on the Method of Prescribing.* Edward and Charles Dilly.

Batyushin, M.M., Sadovnichaya, N.A., Rudenko, L.I. and Povilaitite, P.E., 2013. Clinical case of nephropathy caused by the reception of the Agrimonia eupatoria. *Vestnik Urologii* (3), pp. 30–38.

Giachetti, D., Taddei, I., Cenni, A. and Taddei, E., 1989. Diuresis from distilled water compared with that from vegetable drugs. *Planta Medica*, *55*(01), pp. 97–97.

Gray, A.M. and Flatt, P.R., 1998. Actions of the traditional anti-diabetic plant, Agrimony eupatoria (agrimony): Effects on hyperglycaemia, cellular glucose metabolism and insulin secretion. *British Journal of Nutrition*, *80*(1), pp. 109–114.

Grochowski, D.M., Skalicka-Woźniak, K., Orhan, I.E., Xiao, J., Locatelli, M., Piwowarski, J.P., Granica, S. and Tomczyk, M., 2017. A comprehensive review of agrimoniin. *Annals of the New York Academy of Sciences, 1401*(1), pp. 166–180.

Guibourt, N.J.B.G., 1836. *Histoire abrégée des drogues simples*. Méquignon-Marvis Père et fils.

Ivanova, D., Vankova, D. and Nashar, M., 2013. Agrimonia eupatoria tea consumption in relation to markers of inflammation, oxidative status and lipid metabolism in healthy subjects. *Archives of Physiology and Biochemistry, 119*(1), pp. 32–37.

Jouanna, J., 2012. The legacy of the Hippocratic treatise the nature of man: The theory of the four humours. In *Greek Medicine from Hippocrates to Galen* (pp. 335–359). Brill.

Lee, K.Y., Hwang, L., Jeong, E.J., Kim, S.H., Kim, Y.C. and Sung, S.H., 2010. Effect of neuroprotective flavonoids of Agrimonia Eupatoria on glutamate-induced oxidative injury to HT22 hippocampal cells. *Bioscience, Biotechnology, and Biochemistry, 74*(8), pp. 1704–1706.

Shoelson, S.E., Herrero, L. and Naaz, A., 2007. Obesity, inflammation, and insulin resistance. *Gastroenterology, 132*(6), pp. 2169–2180.

Targher, G., Day, C.P. and Bonora, E., 2010. Risk of cardiovascular disease in patients with non-alcoholic fatty liver disease. *New England Journal of Medicine, 363*(14), pp. 1341–1350.

Tomlinson, C.T., Nahar, L., Copland, A., Kumarasamy, Y., Mir-Babayev, N.F., Middleton, M., Reid, R.G. and Sarker, S.D., 2003. Flavonol glycosides from the seeds of *Agrimonia eupatoria* (Rosaceae). *Biochemical Systematics and Ecology, 31*(4), pp. 439–441.

Yoon, S.J., Koh, E.J., Kim, C.S., Zee, O.P., Kwak, J.H., Jeong, W.J., Kim, J.H. and Lee, S.M., 2012. Agrimonia Eupatoria protects against chronic ethanol-induced liver injury in rats. *Food and Chemical Toxicology, 50*(7), pp. 2335–2341.

6 Shallot (*Allium ascalonicum* L.)

Allium ascalonicum L.

Etymology: From the Latin *allium* = garlic and Ascalon = an ancient city in Israel

Family: Amaryllidaceae

Common names: Shallot; échalote (Fr.); schalotte (Ger.); chalote (Spa.); chalota (Port.); шалот (Rus.)

Part used: Bulb

Constituents: Essential oil (diallylsulfides) (Rattanachaikunsopon & Phumkhachor, 2009), flavones (quercetin, isorhamnetin, and their glycosides), furostane-type saponins (Fattorusso et al., 2002).

Medical history: Called *cepa ascalonica* by Matthioli (1572). Lémery (1716) praises the merits of shallot for urinary stones and as a diuretic.

Medicinal uses: Antidote for dog and snake bites and after childbirth (India); earache (Pakistan, the Philippines)

Blood pressure: After 2 months, shallot mixed at 6.2% to pellets and administered to streptozotocin-induced diabetic mice protected thoracic aortic rings against decreased contractile responsiveness to potassium chloride and noradrenaline (Fallahi et al., 2009).

Plasma lipids and glucose: Shallot mixed at 6.2% to pellets decreased glycemia in streptozotocin-induced diabetic mice after 2 months (Fallahi et al., 2009). Ethanol extract of shallot given at 5 mg/kg/day to rabbits for 2 weeks evoked anticoagulant effects and decreased serum cholesterol by about 40% (Singh & Chaturvedi, 2016). Two doses of bulb juice given orally to streptozotocin-induced diabetic mice (1 g/100 g body weight) for 14 days decreased glycemia by about 50% (Luangpirom et al., 2013).

Immune system: Intake of shallot boosts the immune system (Mirabeau & Samson, 2012

Warning: Shallot eaten in excess could interfere with blood coagulation (Spolarich & Andrews, 2007)

Comment: Adding raw shallot to the diet in moderation could help control glycemia and maintain reactivity of vascular smooth muscle cells.

REFERENCES

Fallahi, F., Roghani, M. and Bagheri, A., 2009. The effect of oral feeding of Allium ascalonicum L. on thoracic aorta contractile response in diabetic rats. *Koomesh*, *10*(3), pp. 213–218.

DOI: 10.1201/9781003301455-6

15

Fattorusso, E., Iorizzi, M., Lanzotti, V. and Taglialatela-Scafati, O., 2002. Chemical composition of shallot (Allium ascalonicum Hort.). *Journal of Agricultural and Food Chemistry*, *50*(20), pp. 5686–5690.

Lémery, N., 1716. *Traité universel des drogues simples, mises en ordre alphabétique. Où l'on trouve leurs différens noms . . . et tout ce qu'il y a de particulier dans les animaux, dans les végétaux, et dans les minéraux.* Au dépend de la Companie.

Luangpirom, A., Kourchampa, W., Junaimuang, T., Somsapt, P. and Sritragool, O., 2013. Effect of shallot (Allium ascalonicum L.) bulb juice on hypoglycemia and sperm quality in streptozotocin induced diabetic mice. *Animal Biology & Animal Husbandry*, *5*(1).

Matthioli, P.A., 1572. *Commentaires sur les Six Livres de Pedacius Dioscorides Anazarbeen de la matière medicinale.* A l'Escue de Milan.

Mirabeau, T.Y. and Samson, E.S., 2012. Effect of Allium cepa and Allium sativum on some immunological cells in rats. *African Journal of Traditional, Complementary and Alternative Medicines*, *9*(3), pp. 374–379.

Rattanachaikunsopon, P. and Phumkhachorn, P., 2009. Shallot (Allium ascalonicum L.) oil: Diallyl sulfide content and antimicrobial activity against food-borne pathogenic bacteria. *African Journal of Microbiology Research*, *3*(11), pp. 747–750.

Singh, G. and Chaturvedi, G.N., 2016. Experimental study on anticoagulant and fibrinolysis activities of single clove garlic (Allium ascalonicum). *Journal of Ayurveda Physicians & Surgeons (JAPS)(EISSN 2394–6350)*, *3*(2), pp. 32–35.

Spolarich, A.E. and Andrews, L., 2007. An examination of the bleeding complications associated with herbal supplements, antiplatelet and anticoagulant medications. *American Dental Hygienists' Association*, *81*(3), pp. 67–67.

7 Leek (*Allium porrum* L.)

Allium porrum L.

Etymology: From the Latin *allium* = garlic and *porrum* = leek

Family: Amaryllidaceae

Common names: Leek; poirreaux (Fr.); lauch (Ger.); puerro (Spa.); alho-poró (Port.); порей (Rus.)

Parts used: Bulb, stem

Constituents: Essential oil (alkylsulfur) (Casella et al., 2013), spirostane-type saponins (Harmatha et al., 1987), flavone glycosides (Fattorusso et al., 2001), fructan-type polysaccharides (inulin) (Causey et al., 2000).

Medical history: According to Dioscorides, leek is a diuretic, it promotes women's fertility, and it purges arteries. According to Pliny the Elder, the Roman emperor Nero ate leek monthly to have a better voice, and he recommends it for nose bleed. Dioscorides and Pliny write that leek is an aphrodisiac. Simeon Seth describes leek as hot and dry to the second degree. The Medical School of Salerno (10th century) recommends the smoke of leek seeds for toothache and says that it makes women sterile and can treat nosebleed.

Medicinal uses: Earache (Turkey); cough (India)

Blood pressure: Addition of lyophilized leek (1%) into a high-cholesterol diet for 8 weeks decreased blood pressure in rats (Ahn et al., 1991). In rats, the oral administration of alcoholic extract of bulbs and leaves at 500 mg/kg/day for 8 weeks prior to L-NAME-induced hypertension decreased systolic blood pressure and normalized plasma MDA and NO levels (Badary et al., 2013).

Plasma lipids and glucose: Addition of lyophilized leek (1%) into a high-cholesterol diet for 8 weeks caused a decrease of plasma triglycerides in rats (Ahn et al., 1991). In a subsequent study, intake of dried powdered leek for 14 weeks in rats poisoned with a high-cholesterol diet decreased plasma triglycerides from 144.7 to 128.2 mg/dL (Fatoorechi et al., 2016). The soluble fiber inulin, known to occur in leek, given at 20g/day to volunteers decreased plasma triglycerides to 40 mg/dL (Causey et al., 2000).

Immune system: Intake of leek boosts the immune system (Mirabeau & Samson, 2012

DOI: 10.1201/9781003301455-7

Comments: (i) Addition of leek in the diet could be beneficial for controlling blood pressure and plasma triglycerides.

(ii) Pliny the Elder, or Gaius Plinius Secundus (23 to 79 AD), was a Roman naturalist and the author of the book *Naturalis Historia*, which was used until the end of the 19th century as a reference for physicians and apothecaries.

(iii) The Medical School of Salerno (10th century) brought the medical knowledge of Arabian physicians into medieval Europe and authored the *Regimen sanitatis Salernitanum*, used as a medical reference text until the end of the 19th century. Salerno's hospital was settled to heal wounded Crusaders.

(iv) Simeon Seth (1035–1100) was a Jewish physician in Byzantium.

(v) Non-soluble fibers in leek and other vegetables alter the absorption of cholesterol and triglycerides as well as the reabsorption of bile salts that are in turn excreted with fecal matter (García-Herrera et al., 2014; Pasquier et al., 1996; Vahouny et al., 1980).

(vi) It is clear that the intake of leeks is beneficial for aging. In France, recipes for delicious leek soups are available.

REFERENCES

Ahn, R.M., Go, G.S. and Hwang, S.H., 1991. Protective effect of leek (Allium odorum L.) on the cholesterol fed rats poisoning in rats. *Journal of the Korean Applied Science and Technology*, 8(2), pp. 183–189.

Badary, O.A., Yassin, N.A., El-Shenawy, S., El-Moneem, M.A. and Al-Shafeiy, H.M., 2013. Study of the effect of Allium porrum on hypertension induced in rats. *Revista latino-americana de química*, 41(3), pp. 149–160.

Casella, S., Leonardi, M., Melai, B., Fratini, F. and Pistelli, L., 2013. The role of diallyl sulfides and dipropyl sulfides in the in vitro antimicrobial activity of the essential oil of garlic, Allium sativum L., and leek, Allium porrum L. *Phytotherapy Research*, 27(3), pp. 380–383.

Causey, J.L., Feirtag, J.M., Gallaher, D.D., Tungland, B.C. and Slavin, J.L., 2000. Effects of dietary inulin on serum lipids, blood glucose and the gastrointestinal environment in hypercholesterolemic men. *Nutrition Research*, 20(2), pp. 191–201.

Fatoorechi, V., Rismanchi, M. and Nasrollahzadeh, J., 2016. Effects of Persian leek (Allium ampeloprasum) on hepatic lipids and the expression of proinflammatory gene in hamsters fed a high-fat/high-cholesterol diet. *Avicenna Journal of Phytomedicine*, 6(4), p. 418.

Fattorusso, E., Lanzotti, V., Taglialatela-Scafati, O. and Cicala, C., 2001. The flavonoids of leek, Allium porrum. *Phytochemistry*, 57(4), pp. 565–569.

García-Herrera, P., Morales, P., Fernández-Ruiz, V., Sánchez-Mata, M.C., Cámara, M., Carvalho, A.M., Ferreira, I.C., Pardo-de-Santayana, M., Molina, M. and Tardío, J., 2014. Nutrients, phytochemicals and antioxidant activity in wild populations of Allium ampeloprasum L., a valuable underutilized vegetable. *Food Research International*, 62, pp. 272–279.

Harmatha, J., Mauchamp, B., Arnault, C. and Sláma, K., 1987. Identification of a spirostane-type saponin in the flowers of leek with inhibitory effects on growth of leek-moth larvae. *Biochemical Systematics and Ecology*, 15(1), pp. 113–116.

Mirabeau, T.Y. and Samson, E.S., 2012. Effect of Allium cepa and Allium sativum on some immunological cells in rats. *African Journal of Traditional, Complementary and Alternative Medicines*, 9(3), pp. 374–379.

Mousavian, S.Z., Eidi, A. and Zarringhalam, M.J., 2013. Evaluation of the diuretic activity of aerial parts of allium porrum in rats.

Pasquier, B., Armand, M., Guillon, F., Castelain, C., Borel, P., Barry, J.L., Pieroni, G. and Lairon, D., 1996. Viscous soluble dietary fibers alter emulsification and lipolysis of triacylglycerols in duodenal medium in vitro. *The Journal of Nutritional Biochemistry, 7*(5), pp. 293–302.

Vahouny, G.V., Tombes, R., Cassidy, M.M., Kritchevsky, D. and Gallo, L.L., 1980. Dietary fibers: V. Binding of bile salts, phospholipids and cholesterol from mixed micelles by bile acid sequestrants and dietary fibers. *Lipids, 15*(12), pp. 1012–1018.

8 Garlic (*Allium sativum* L.)

Allium sativum L.

Etymology: From the Latin *allium* = garlic and *sativum* = cultivated

Family: Amaryllidaceae

Synonym: *Allium pekinense* Prokhanov

Common names: Garlic; ail (Fr.); knoblauch (Ger.); ajo (Spa.); alho (Port.); чеснок (Rus.)

Part used: Bulb

Constituents: Essential oil (thiosulfinates including allicin) (Harris et al., 2001), flavones (quercetin, myricetin, and apigenin) (Azzini et al., 2014).

Medicinal history: Garlic was known to Roman physicians as a diuretic, anthelminthic, anti-inflammatory, and antidote for snake bites and rabies, and Dioscorides used it for hemorrhoids, angina pectoris, and impotence. Galen describes garlic as hot and dry to the fourth degree. The Medical School of Salerno (10th century) says that garlic is antidotal but detrimental to the eye. Matthioli (1572) indicates that garlic gives wind; is laxative, diuretic, and anthelmintic; and unclogs arteries when taken with black olives. Daléchamps (1615) asserts that garlic is good to give strength to farmers. Garlic was still used in 19th-century Western Europe clinical practice. Guibourt (1836) writes that garlic was used in 19th-century France to make an antiseptic medicine called "*vinaigre des 4 voleurs*" and called garlic "*thériaque du pauvre*", that could be translated as the poor man's theriac." In 19th-century North America, Pereira (1843) attests that garlic was taken internally for cough, to stimulate urination, for intestinal worms, and externally for tumors. Frederick Porter Smith (1871) notes that garlic was used as a tonic and to prevent goiter and pestilential diseases in 19th-century China.

Medicinal use: Hypertension (eaten raw) in Turkey, Afghanistan, Iran, Iraq, India, Bangladesh, the Philippines; induce urination (Bangladesh); cardiac disorder (Afghanistan); longevity and virility (Myanmar); headache (Thailand)

Blood pressure: Intake of garlic causes systolic and diastolic blood pressure to decrease (Wang et al., 2015).

Plasma lipids and glucose: About one half to one clove per day decreases total serum cholesterol by about 9% (Warshafsky et al., 1993). Garlic powder given for 48 months to volunteers (50 to 80 years old) decreased the formation of atherosclerotic plaques in carotid and femoral arteries by 6–13% (Koscielny et al., 1999).

DOI: 10.1201/9781003301455-8

Heart: Isolated hearts of rat fed garlic powder (1% of standard chow) for 10 weeks were resistant to ventricular tachycardia and fibrillation caused by ischemia and reperfusion (Isensee et al., 1993). Intake of garlic at a daily dose of about 1.2 g/day decreases plasma high-sensitivity C-reactive protein (hs-CRP) (Taghizadeh et al., 2019).

Gallbladder: Garlic powder at 0.6% of diet for 10 weeks prevented the formation of gallstones in mice on a cholesterol and bile salt-enriched diet (Vidyashankar et al., 2008).

Kidneys: In obese rats, garlic extract at 500 mg/kg/day for 4 weeks mitigated renal injuries induced by ischemia and reperfusion, decreased serum creatinine from 1.7 to 1.5 mg/dL and decreased renal MDA, and increased renal glutathione and SOD (Ali et al., 2016).

Bones and cartilages: Essential oil given orally to ovariectomized rats attenuated the development of osteoporosis (Mukherjee et al., 2004).

Immune system: Aqueous extract given orally at 250 mg/kg to rats increased white blood cells count (Mirabeau & Samson, 2012).

Brain: In a transgenic model of mice with Alzheimer's, intake of 40 mg/kg/day of an extract for 4 weeks increased soluble amyloid precursor protein α (a neurogenic and neuroprotective protein) by 25% and decreased amyloid β proteins 40 and 42 by about 30% (Chauhan, 2003).

Warnings: Excessive intake of garlic poses the risk of bleeding. Garlic oil given orally to rats at 50 mg/kg/day evoked anticoagulant effects mediated by thrombin inhibition (Chan et al., 2007). Excessive ingestion of garlic evokes very unpleasant side effects, including anemia, and in high doses, the juice is said to be fatal (Yamato et al., 2005).

Comments: (i) Jacques Daléchamps (1513–1588) was a French physician and botanist.

(ii) Jonathan Pereira (1804–1853) was a British professor of materia medica at the Aldersgate Medical School.

(iii) Including one or two fresh cloves daily in the diet could participate in delaying cardiovascular aging. LDL infiltrates into the arterial intima, and with aging, the accumulation of cholesterol triggers an inflammatory response (foam cells, calcification, migration of smooth muscle cells, deposit of fibers) that leads to the formation of atheroma (a groat-like mass) which attracts circulating immune cells. Atheromatous plaques stiffen arteries, interfere with NO's ability to relax vascular smooth muscle cells, and with time expose the sub-endothelial collagen that causes platelet aggregation, the formation of a thrombus in the circulation, and stroke or cardiac infarction (Gargiulo et al., 2016). For systems with low tolerance, it is worth investigating if boiled garlic has similar results.

(iv) Frederick Porter Smith (1833–1888) was a British medical missionary in China.

(v) Garlic is an anticoagulant and must be avoided in patients on warfarin (Kansara & Jani, 2017).

REFERENCES

Ali, S.I., Alhusseini, N.F., Atteia, H.H., Idris, R.A.E.S. and Hasan, R.A., 2016. Renoprotective effect of a combination of garlic and telmisartan against ischemia/reperfusion-induced kidney injury in obese rats. *Free Radical Research*, *50*(9), pp. 966–986.

Azzini, E., Durazzo, A., Foddai, M.S., Temperini, O., Venneria, E., Valentini, S. and Maiani, G., 2014. Phytochemicals content in Italian garlic bulb (Allium sativum L.) varieties. *Journal of Food Research*, *3*(4), p. 26.

Chan, K.C., Yin, M.C. and Chao, W.J., 2007. Effect of diallyl trisulfide-rich garlic oil on blood coagulation and plasma activity of anticoagulation factors in rats. *Food and Chemical Toxicology*, *45*(3), pp. 502–507.

Chauhan, N.B., 2003. Anti-amyloidogenic effect of Allium sativum in Alzheimer's transgenic model Tg2576. *Journal of Herbal Pharmacotherapy*, *3*(1), pp. 95–107.

Daléchamps, 1615. *De l' histoire generale des plantes simples*. Chez Heritier Guillaume Rouille.

Gargiulo, S., Gramanzini, M. and Mancini, M., 2016. Molecular imaging of vulnerable atherosclerotic plaques in animal models. *International Journal of Molecular Sciences*, *17*(9), p. 1511.

Guibourt, N.J.B.G., 1836. *Histoire abrégée des drogues simples*. Méquignon-Marvis Père et fils.

Harris, J.C., Cottrell, S.L., Plummer, S. and Lloyd, D., 2001. Antimicrobial properties of Allium sativum (garlic). *Applied Microbiology and Biotechnology*, *57*(3), pp. 282–286.

Isensee, H., Rietz, B. and Jacob, R., 1993. Cardioprotective actions of garlic (Allium sativum). *Arzneimittel-Forschung*, *43*(2), pp. 94–98.

Kansara, M.B. and Jani, A.J., 2017. Possible interactions between garlic and conventional drugs: A review. *PBE*, *4*(2), p. 7.

Koscielny, J., Klüssendorf, D., Latza, R., Schmitt, R., Radtke, H., Siegel, G. and Kiesewetter, H., 1999. The antiatherosclerotic effect of Allium sativum. *Atherosclerosis*, *144*(1), pp. 237–249.

Matthioli, P.A., 1572. *Commentaires sur les Six Livres de Pedacius Dioscorides Anazarbeen de la matière medicinale*. A l'Escue de Milan.

Mirabeau, T.Y. and Samson, E.S., 2012. Effect of Allium cepa and Allium sativum on some immunological cells in rats. *African Journal of Traditional, Complementary and Alternative Medicines*, *9*(3), pp. 374–379.

Mukherjee, M., Das, A.S., Mitra, S. and Mitra, C., 2004. Prevention of bone loss by oil extract of garlic (Allium sativum Linn.) in an ovariectomized rat model of osteoporosis. *Phytotherapy Research: An International Journal Devoted to Pharmacological and Toxicological Evaluation of Natural Product Derivatives*, *18*(5), pp. 389–394.

Pereira, J., 1843. *The Elements of Materia Medica and Therapeutics*. Lea and Blanchard.

Taghizadeh, M., Hamedifard, Z. and Jafarnejad, S., 2019. Effect of garlic supplementation on serum C-reactive protein level: A systematic review and meta-analysis of randomized controlled trials. *Phytotherapy Research*, *33*(2), pp. 243–252.

Vidyashankar, S., Sambaiah, K. and Srinivasan, K., 2008. Dietary garlic and onion reduce the incidence of atherogenic diet-induced cholesterol gallstones in experimental mice. *British Journal of Nutrition*, *101*(11), pp. 1621–1629.

Wang, H.P., Yang, J., Qin, L.Q. and Yang, X.J., 2015. Effect of garlic on blood pressure: A meta-analysis. *The Journal of Clinical Hypertension*, *17*(3), pp. 223–231.

Warshafsky, S., Kamer, R.S. and Sivak, S.L., 1993. Effect of garlic on total serum cholesterol: A meta-analysis. *Annals of Internal Medicine*, *119*(7_Part_1), pp. 599–605.

Yamato, O., Kasai, E., Katsura, T., Takahashi, S., Shiota, T., Tajima, M., Yamasaki, M. and Maede, Y., 2005. Heinz body hemolytic anemia with eccentrocytosis from ingestion of Chinese chive (Allium tuberosum) and garlic (Allium sativum) in a dog. *Journal of the American Animal Hospital Association*, *41*(1), pp. 68–73.

9 Onions (*Allium cepa* L.)

Allium cepa L.

Etymology: From the Latin *allium* = garlic and *cepa* = onion
Family: Amaryllidaceae
Common names: Onions; zwiebeln (Ger.); cebolas (Port.); cebollas (Spa.)
Part used: Bulb
Constituents: Syn-propanethial-S-oxide (Stice et al., 2020), flavones (quercetin, kaempferol) (Bilyk et al., 1984).
Medical history: Raw onions were recommended for cough and externally for tumors and earaches in 19th-century Europe (Pereira, 1843). Onion tea was given to persons suffering from catarrh, fever, headache, cholera, diarrhea, dysentery, urinary disorders, and rheumatic affections in 19th-century China (Porter Smith, 1871).
Medicinal uses: Sprain, edema, cold flu, and headache (Turkey); dysentery, aphrodisiac, prevention of heart diseases, high blood pressure (Pakistan); fever, nausea, vomiting, abscess, dysentery, bleeding piles (India); high blood pressure, headache, cold, constipation (Bangladesh); cough, diuretic (Myanmar); boils, measles (the Philippines); post-partum (Korea)

Blood pressure: In rats poisoned with L-NAME, onion powder at 5% of diet for 4 weeks increased NO synthetase activity in aorta and kidneys, increased content of NO in urine, decreased systolic blood pressure from about 180 to 160 mmHg, and brought to almost normal values ROS in the plasma (Sakai et al., 2003). Juice of onion leaves given to rats at 2 g/kg/day for 4 weeks decreased systolic blood pressure from 116 to 106 mmHg and inhibited platelet adhesion, which increased of tail bleeding time (Chen et al., 2000).
Plasma lipids and glucose: Aqueous extract given orally to rats at 1.5 mg/kg/day for 28 days decreased cholesterol from 314.2 to 119.8 mg/dL, triglycerides from 270.6 to 209.4 mg/dL, and LDL from 197.8 to 178.6 mg/dL and increased HDL-cholesterol from 38,4 to 65.3 mg/dL (Emmanuel & James, 2011).
Gallbadder: Onion powder at 2% of diet for 10 weeks prevented the formation of gallstones in mice on a cholesterol and bile salt-enriched diet (Vidyashankar et_al., 2008).
Immune system: Aqueous extract given orally at 250 mg/kg to rats increased white blood cells count (Mirabeau & Samson, 2012).

DOI: 10.1201/9781003301455-9

Brain: Methanol extract given orally at 168 mg/kg/day for 13 days protected rats against dementia induced by intracerebrovascular injection of streptozotocin (Kaur et al., 2020).

REFERENCES

Bilyk, A., Cooper, P.L. and Sapers, G.M., 1984. Varietal differences in distribution of quercetin and kaempferol in onion (Allium cepa L.) tissue. *Journal of Agricultural and Food Chemistry*, *32*(2), pp. 274–276.

Chen, J.H., Chen, H.I., Tsai, S.J. and Jen, C.J., 2000. Chronic consumption of raw but not boiled Welsh onion juice inhibits rat platelet function. *The Journal of Nutrition*, *130*(1), pp. 34–37.

Emmanuel, U.C. and James, O., 2011. Comparative effects of aqueous garlic (Allium sativum) and onion (Allium cepa) extracts on some haematological and lipid indices of rats. *Annual Research & Review in Biology*, pp. 37–44.

Kaur, R., Randhawa, K., Kaur, S. and Shri, R., 2020. Allium cepa fraction attenuates STZ-induced dementia via cholinesterase inhibition and amelioration of oxidative stress in mice. *Journal of Basic and Clinical Physiology and Pharmacology*, *31*(3).

Mirabeau, T.Y. and Samson, E.S., 2012. Effect of Allium cepa and Allium sativum on some immunological cells in rats. *African Journal of Traditional, Complementary and Alternative Medicines*, *9*(3), pp. 374–379.

Pereira, J., 1843. *The Elements of Materia Medica and Therapeutics*. Lea and Blanchard.

Porter Smith, F., 1871. *Contributions Towards the Materia Medica and Natural History of China for the Use of Medical Missionaries and Students*. American Presbytarian Mission Press.

Sakai, Y., Murakami, T. and Yamamoto, Y., 2003. Antihypertensive effects of onion on NO synthase inhibitor-induced hypertensive rats and spontaneously hypertensive rats. *Bioscience, Biotechnology, and Biochemistry*, *67*(6), pp. 1305–1311.

Stice, S.P., Thao, K.K., Khang, C.H., Baltrus, D.A., Dutta, B. and Kvitko, B.H., 2020. Thiosulfinate tolerance is a virulence strategy of an atypical bacterial pathogen of onion. *Current Biology*, *30*(16), pp. 3130–3140.

Vidyashankar, S., Sambaiah, K. and Srinivasan, K., 2008. Dietary garlic and onion reduce the incidence of atherogenic diet-induced cholesterol gallstones in experimental mice. *British Journal of Nutrition*, *101*(11), pp. 1621–1629.

10 Indian Aloe (*Aloe vera* L.)

Aloe vera L.

Etymology: From the Syriac word *Alwai* and from the Latin *vera* = true

Family: Asphodelaceae

Synonym: *Aloe barbadensis* Mill.

Common names: Barbados aloe, Indian aloe; aloes (Fr.); áloe (Spa.); babosa (Port.); алоэ вера (Rus.)

Constituents: Anthraquinones, chromones, polysaccharides (acemannans) (Hamman, 2008)

Medical history: Dioscorides recommends the dry latex of Indian aloe for wounds, jaundice, and sore eyes and as a cathartic. Fusch (1555) defines the latex as hot to the first degree and dry to the third degree.

Medicinal uses: Laxative (Iraq); diabetes (Pakistan); spleen and liver diseases, enlarged lymphatic glands (India); diuretic, edema, liver diseases, asthma, jaundice, bladder stones, high blood pressure (Nepal); alopecia (the Philippines; Papua New Guinea)

Blood pressure: Aqueous solution of gel (30%) given to rats orally at 200 mg/kg/day for 20 days decreased blood pressure and plasmatic glutathione induced by streptozotocin poisoning (Jain et al., 2010).

Heart: Aqueous solution of gel (30%) given to streptozotocin-induced diabetic rats orally at 200 mg/kg/day for 20 days brought close to normal catalase (CAT) and markers of lipid peroxidation and mitigated myocardium structural insults (Jain et al., 2010).

Plasma lipids and glucose: Ethanol extract of gel given orally at 300 mg/kg/day for 21 days to streptozotocin-induced diabetic rats decreased glycemia from 332.2 to 96.8 mg/dL (normal value: 85.8 mg/dL), cholesterol from 228.3 to 98.3 mg/dL (normal: 92.6 mg/dL), triglycerides from 229.3 to 79.2 mg/dL (normal value: 73.5 mg/dL), and LDL-cholesterol from 139.2 to 48.5 mg/dL (normal value: 45.1 mg/dL) (Rajasekaran et al., 2006).

Aqueous solution of gel (30%) given to rats orally at 200 mg/kg/day for 20 days decreased blood glucose from 324.1 to 81.7 mg/dL (normal value: 69.3 mg/dL) (Jain et al., 2010).

Bones and cartilages: Polysaccharides given orally at 300 mg/kg/day for 12 weeks to ovariectomized rats increased trabecular bones density (Yao et al., 2022).

DOI: 10.1201/9781003301455-10

Skin and hair: Intake of gel at 1.2 g/day for 90 days by women aged above 45 increased the expression of type I procollagen mRNA in the dermis, improved facial skin elasticity, and decreased the formation of wrinkles (Cho et al., 2009).

Brain: The gel given at 200 mg/kg/days for 10 days prevented nitrite-induced memory impairments in rats (Kaithwas et al., 2007).

Warnings: There are reports of liver poisoning associated with over-the-counter *Aloe vera* L. products (Lee et al., 2014; Parlati et al., 2017), but in Bangladesh, the gel of *Aloe vera* L. is used to make a refreshing drink used daily by many of people without side effects (Mohammed Rahmatullah, personal communication).

Use cautiously in patients with oral hypoglycemic agents, renal insufficiency, cardiac disease, or electrolyte abnormalities (Posadzki et al., 2013)

Comments: (i) It is clear that intake of gel could delay cardiovascular aging.

(ii) The German physician and botanist Leonhart Fusch (1501–1566) was a professor of medicine at the University of Tünbingen and author of the book *De Historia Stirpium Commetarii Insignes* (1542).

(iii) With aging, intake of saturated fats, processed food, and glucose-enriched food cause insulin resistance and exhaust the ability of pancreatic β-cells to secrete insulin, resulting in type 2 diabetes especially in genetically predisposed patients (López-Otín et al., 2016).

(iv) Increased glycemia activates endothelial NAPH oxidase and consequently generates ROS that interfere with the physiological vasodilatating effect of NO, resulting in hypertension (Dhananjayan et al., 2016; Triggle et al., 2020).

(v) With insulin deprivation, or insulin resistance, hormone-sensitive lipase (physiologically inhibited by insulin) in adipose tissues is overactive, resulting in the release in the plasma of fatty acids (Holm, 2003) and at the hepatic level in the increased production of plasmatic triglycerides and cholesterol via acetyl-CoA (Bai & Li, 2019).

(vi) The progressive removal of medicinal plant teachings from the curricula of schools of pharmacy for the last 50 years has sinister consequences for the public's craving for safe herbal remedies. The manufacture and control of herbal products must be the responsibility of properly trained pharmacists, meaning five full years including botany, toxicology, pharmacology, and clinical application of medicinal plants.

(vii) The second outermost layer of the skin or dermis is extracellularly made of collagen (type I and III) and procollagen. Decreasing collagen content in the dermis from aging results in the formation of wrinkles.

REFERENCES

Bai, L. and Li, H., 2019. Innate immune regulatory networks in hepatic lipid metabolism. *Journal of Molecular Medicine*, 97(5), pp. 593–604.

Cho, S., Lee, S., Lee, M.J., Lee, D.H., Won, C.H., Kim, S.M. and Chung, J.H., 2009. Dietary Aloe vera supplementation improves facial wrinkles and elasticity and it increases the type I procollagen gene expression in human skin in vivo. *Annals of Dermatology*, 21(1), pp. 6–11.

Dhananjayan, R., Koundinya, K.S., Malati, T. and Kutala, V.K., 2016. Endothelial dysfunction in type 2 diabetes mellitus. *Indian Journal of Clinical Biochemistry*, *31*, pp. 372–379.

Fusch, L., 1555. *De Historia Stirpium Commetarii Insignes*. Lugduni Apud Ioan Tornaesium.

Hamman, J.H., 2008. Composition and applications of Aloe vera leaf gel. *Molecules*, *13*(8), pp. 1599–1616.

Holm, C., 2003. Molecular mechanisms regulating hormone-sensitive lipase and lipolysis. *Biochemical Society Transactions*, *31*(6), pp. 1120–1124.

Jain, N., Vijayaraghavan, R., Pant, S.C., Lomash, V. and Ali, M., 2010. Aloe vera gel alleviates cardiotoxicity in streptozocin-induced diabetes in rats. *Journal of Pharmacy and Pharmacology*, *62*(1), pp. 115–123.

Kaithwas, G., Dubey, K., Bhtia, D., Sharma, A.D. and Pillai, K., 2007. Reversal of sodium nitrite induced impairment of spontaneous alteration by Aloe vera gel: Involvement of cholinergic system. *Pharmacologyonline*, *3*, pp. 428–437.

Lee, J., Lee, M.S. and Nam, K.W., 2014. Acute toxic hepatitis caused by an aloe vera preparation in a young patient: A case report with a literature review. *The Korean Journal of Gastroenterology*, *64*(1), pp. 54–58.

López-Otín, C., Galluzzi, L., Freije, J.M., Madeo, F. and Kroemer, G., 2016. Metabolic control of longevity. *Cell*, *166*(4), pp. 802–821.

Parlati, L., Voican, C.S., Perlemuter, K. and Perlemuter, G., 2017. Aloe vera-induced acute liver injury: A case report and literature review. *Clinics and Research in Hepatology and Gastroenterology*, *41*(4), pp. e39–e42.

Posadzki, P., Watson, L. and Ernst, E., 2013. Contamination and adulteration of herbal medicinal products (HMPs): An overview of systematic reviews. *European Journal of Clinical Pharmacology*, *69*, pp. 295–307.

Rajasekaran, S., Ravi, K., Sivagnanam, K. and Subramanian, S., 2006. Beneficial effects of Aloe vera leaf gel extract on lipid profile status in rats with streptozotocin diabetes. *Clinical and Experimental Pharmacology and Physiology*, *33*(3), pp. 232–237.

Triggle, C.R., Ding, H., Marei, I., Anderson, T.J. and Hollenberg, M.D., 2020. Why the endothelium? The endothelium as a target to reduce diabetes-associated vascular disease. *Canadian Journal of Physiology and Pharmacology*, *98*(7), pp. 415–430.

Yao, X.W., Liu, H.D., Ren, M.X., Li, T.L., Jiang, W.K., Zhou, Z., Liu, Z.Y. and Yang, M., 2022. Aloe polysaccharide promotes osteogenesis potential of adipose-derived stromal cells via BMP-2/Smads and prevents ovariectomized-induced osteoporosis. *Molecular Biology Reports*, *49*(12), pp. 11913–11924.

11 Citron Verbena (*Aloysia citrodora* Paláu)

Aloysia citrodora Paláu

Etymology: After the Spanish queen Maria Luisa of Parma (1751–1819) and from the Latin *citrodora* = with lemon fragrance

Family: Verbenaceae

Synonyms: *Aloysia triphylla* Royle; *Lippia critriodora* Humb. Bonpl. et Kth.; *Lippia critrodora* Kunth.; *Lippia triphylla* (L'Hér.) Kuntz

Common names: Citron verbena, lemon verbena; verveine citronnée (Fr.); zitronenverbene (Ger.); hierba luisa (Spa.); verbena de limão (Port.); вербены лимонной (Rus.)

Constituents: Essential oil (neral, geranial), flavone glycosides (Skaltsa & Shammas, G., 1988), hydroxycinnamic derivatives (verbascoside, up to about 3.5% dry weight of leaves) (Etemad et al., 2016; Corbi et al., 2018).

Medical history: In Chile, the plant has been used for fever (Anales de la Universidad de Chile, 1861).

Medicinal uses: Mental stress, flatulence (5 g leaves in 100 mL of boiling water as a decoction, 3 times daily; Europe); carminative, dizziness (Argentina); cough, insomnia (Cuba); tachycardia, anxiety, headache (Paraguay)

Plasma lipids and glucose: Extract of the plant (containing verbascoside 0.5%) mixed in food (5 mg/kg) fed to rabbits for 80 days improved lipid profile and glycemia and boosted sirtuin 1 (SIRT1) activity in the heart (Corbi et al., 2018). In rats poisoned by a high-fat diet for 3 months, verbascoside at 2 mg/kg/day for 6 weeks decreased total cholesterol from 13.7 to 3.9 mmol/L (normal: 1.4 mmol/L) and triglycerides from 2.1 to 0.9 mmol/L (normal: 0.6 mmol/L) and corrected hr-CRP and interleukin-6 to close to normal (Rifai & Ridker, 2001).

Brain: An extract containing verbascoside given for 28 days to patients with multiple sclerosis decreased plasmatic interferon γ and interleukin-12 while increasing interleukin-4 and -10 (Mauriz et al., 2015).

Warning: Verbascoside mut be avoided at all cost at all stages of pregnancy (Etemad et al., 2016).

Comments: (i) Mononuclear cells, macrophages, and T lymphocytes activated in atheromatous plaque (cardiovascular inflammation) secrete cytokines and the downstream synthesis of hr-CRP, which are used as clinical markers of coronary disease risks (Rifai & Ridker, 2001).

DOI: 10.1201/9781003301455-11

(ii) Sirtuins (SIRTs) are enzymes that are activated by calorie restriction and that command cell cycle arrest and DNA repair and regulate the expression of antioxidant enzymes (Pardo & Boriek, 2020). SIRT1 inhibits NADPH oxidase activation and protects endothelial function in the rat aorta (Zarzuelo et al., 2013). Mitigation of NADPH oxidase 2 activity inhibits peroxynitrite formation (Zielonka et al., 2016).

(iii) Mental stress contributes to hypertension and the development of cardiovascular diseases (Pickering, 2001). As a mild anxiolytic and by decreasing hrCRP, lemon verbena (in tea form) could potentially contribute to protecting the cardiovascular system against premature aging.

(iv) Interleukin-10 and interleukin-4 are down-regulatory cytokines produced by Th2 cells that inhibit the production of other cytokines. IFN-γ accounts for demyelination via the activation of mononuclear cells (Mauriz et al., 2015).

REFERENCES

Corbi, G., Conti, V., Komici, K., Manzo, V., Filippelli, A., Palazzo, M., Vizzari, F., Davinelli, S., Di Costanzo, A., Scapagnini, G. and Ferrara, N., 2018. Phenolic plant extracts induce Sirt1 activity and increase antioxidant levels in the rabbit's heart and liver. *Oxidative Medicine and Cellular Longevity*, *2018*.

Etemad, L., Zafari, R., Moallem, S.A., Vahdati-Mashhadian, N., Shirvan, Z.S. and Hosseinzadeh, H., 2016. Teratogenic effect of verbascoside, main constituent of Lippia citriodora leaves, in mice. *Iranian Journal of Pharmaceutical Research: IJPR*, *15*(2), p. 521.

Mauriz, E., Vallejo, D., Tuñón, M.J., Rodriguez-López, J.M., Rodríguez-Pérez, R., Sanz-Gómez, J. and del Camino García-Fernández, M., 2015. Effects of dietary supplementation with lemon verbena extracts on serum inflammatory markers of multiple sclerosis patients. *Nutricion hospitalaria*, *31*(2), pp. 764–771.

Pardo, P.S. and Boriek, A.M., 2020. SIRT1 regulation in ageing and obesity. *Mechanisms of Ageing and Development*, *188*, p. 111249.

Pickering, T.G., 2001. Mental stress as a causal factor in the development of hypertension and cardiovascular disease. *Current Hypertension Reports*, *3*(3), p. 249.

Rifai, N. and Ridker, P.M., 2001. High-sensitivity C-reactive protein: A novel and promising marker of coronary heart disease. *Clinical Chemistry*, *47*(3), pp. 403–411.

Skaltsa, H. and Shammas, G., 1988. Flavonoids from Lippia citriodora. *Planta Medica*, *54*(5), pp. 465–465.

Zarzuelo, M.J., López-Sepúlveda, R., Sánchez, M., Romero, M., Gómez-Guzmán, M., Ungvary, Z., Pérez-Vizcaíno, F., Jiménez, R. and Duarte, J., 2013. SIRT1 inhibits NADPH oxidase activation and protects endothelial function in the rat aorta: Implications for vascular aging. *Biochemical Pharmacology*, *85*(9), pp. 1288–1296.

Zielonka, J., Zielonka, M., VerPlank, L., Cheng, G., Hardy, M., Ouari, O., Ayhan, M.M., Podsiadły, R., Sikora, A., Lambeth, J.D. and Kalyanaraman, B., 2016. Mitigation of NADPH oxidase 2 activity as a strategy to inhibit peroxynitrite formation. *Journal of Biological Chemistry*, *291*(13), pp. 7029–7044.

12 Greater Galangal (*Alpinia galanga* (L.) Willd.)

Alpinia galanga (L.) Willd.

Etymology: After the Italian physician Prospero Alpini (1553–1617) and from the Arabic name of the plant, *caluegia*

Family: Zingiberaceae

Synonyms: *Amomum galanga* (L.) Lour.; *Languas galanga* (L.) Stuntz; *Maranta galanga* L.; *Zingiber galanga* (L.) Stokes

Common names: Greater galangal; Java galangal root; galanga majeur (Fr.); galgantwurzel (Ger.); galanga maior (Port.); galanga mayor (Spa.)

Part used: Rhizome

Constituents: Essential oil (1,8-cineole) (Raina et al., 2014), *p*-hydroxycinnamaldehyde (Barik et al., 1987), flavones (galangin) (Tungmunnithum et al., 2020).

Medical history: According to Daléchamps (1615), greater galangal is hot and dry to the third degree and is good for the stomach, the heart, flu, and as a male aphrodisiac. The Portuguese physician Da Orta described the plant in 1563 as tall like a spear. Pierre Pomet (1694) states that greater galangal was used to make vinegar. The plant was used as a tonic in 19th-century France. According to George Watt (1889), hakeems in India used it for impotence, bronchitis, diabetes mellitus, and dyspepsia and as a mouth antiseptic.

Medicinal uses: Fever, rheumatism (India); piles, cough (Bangladesh); fever, inflammation, carminative, asthma, heart disease, diuretic, chest pain (Myanmar)

Blood pressure: Galangin given at 50 mg/kg/day orally for 4 weeks to rats caused a decrease in blood pressure (Prasatthong et al., 2021).

Plasma lipids and glucose: Alcoholic extract of rhizomes given orally at 200 mg/kg/day for 40 days to streptozotocin-induced diabetic rats decreased glycemia from 524.2 to 203.6 mg/dL (normal value 103.4 mg/dL) and plasma cholesterol from 86.5 to 50.5 mg/dL (normal value: 49.2 mg/dL) (Kaushik et al., 2013).

Galangin given at 50 mg/kg/day orally for 4 weeks to rats on a high-fat diet decreased fasting serum insulin, fasting glucose from 105.4 to 100.4 mg/dL, total cholesterol from 1.3 to 1 mmol/L, and triglycerides from

DOI: 10.1201/9781003301455-12

0.7 to 0.5 mmol/L and increased HDL-cholesterol from 0.7 to 1.3 mmol/L (Prasatthong et al., 2021).

Heart: Galangin given orally at 1 mg/kg/day for 14 days protected rats against isoprotenelol-induced hypotension, tachycardia, cardiac oxidative insults, and cardiac fibrosis (Thangaiyan et al., 2020). Galangin given at 50 mg/kg/day orally for 4 weeks to rats on a high-fat diet decreased heart beat from 378.5 to 336 bpm and prevented cardiac hypertrophy and inflammation (Prasatthong et al., 2021).

Adipose tissues: Galangin given at 50 mg/kg/day orally for 4 weeks to rats on a high-fat diet decreased epidydimal pad fat (Prasatthong et al., 2021).

Kidneys: Alcoholic extract of rhizomes given orally at 200 mg/kg/day for 40 days to streptozotocin-induced diabetic rats decreased urinary albumin from 0.3 to 0.09 g/dL and prevented kidney cytoarchitecture modification as well as kidney oxidative insults, as evidenced by an almost normalization of MDA, glutathione, SOD, and CAT (Kaushik et al., 2013).

Brain: Chloroform extract given orally at 400 mg/kg/day for 20 days protected mice against dementia induced by the intracerebrovascular injection of amyloid β proteins (JC et al., 2011).

Warning: Greater galangal stimulates thirst, a feeling of warmth, and insomnia.

Comment: (i) The Jewish Portuguese physician Garcia da Orta (1501–1568) lived in India and wrote his observations of local medicinal plants in the form of dialogues in a book titled *Colóquios dos simples e drogas da India*.

(ii) Pierre Pomet (1658–1699) was a French pharmacist, the druggist of King Louis XIV, and the author of *Histoire Générale des Drogues* (1694).

(iii) George Watt was a Scottish botanist (1851–1930) based in India and the author of a book titled *A Dictionary of the Economic Products of India* (1889).

(iv) Obesity is correlated with a low-grade but constant inflammatory state where the plasma levels of pro-inflammatory interleukin-6, interleukin-1β, and TNF α increases. TNF α blocks insulin signaling through the phosphorylation of insulin receptors, contributing to insulin resistance and inducing mitochondrial dysfunction (Ferrucci & Fabbri, 2018).

(v) The rhizomes could potentially have some usefulness in protecting kidneys in diabetes.

REFERENCES

Barik, B.R., Kundu, A.B. and Dey, A.K., 1987. Two phenolic constituents from Alpinia galanga rhizomes. *Phytochemistry*, 26(7), pp. 2126–2127.

Daléchamps, 1615. *De l' histoire generale des plantes simples*. Chez Heritier Guillaume Rouille.

Ferrucci, L. and Fabbri, E., 2018. Inflammageing: Chronic inflammation in ageing, cardiovascular disease, and frailty. *Nature Reviews Cardiology*, 15(9), pp. 505–522.

JC, H.S., Alagarsamy, V., Diwan, P.V., Kumar, S., Nisha, J.C. and Reddy, N., 2011. Neuro-protective effect of Alpinia galanga (L.) fractions on Aβ (25–35) induced amnesia in mice. *Journal of Ethnopharmacology*, *138*(1), pp. 85–91.

Kaushik, P., Kaushik, D., Yadav, J. and Pahwa, P., 2013. Protective effect of Alpinia galanga in STZ induced diabetic nephropathy. *Pakistan Journal of Biological Sciences*, *16*(16), pp. 804–811.

Pomet, P., 1694. *Histoire Générale des Drogues*. Chez Jean Baptiste Loyson et Auguste Pillon.

Prasatthong, P., Meephat, S., Rattanakanokchai, S., Khamseekaew, J., Bunbupha, S., Prachaney, P., Maneesai, P. and Pakdeechote, P., 2021. Galangin resolves cardiometabolic disorders through modulation of AdipoR1, COX-2, and NF-κB expression in rats fed a high-fat diet. *Antioxidants*, *10*(5), p. 769.

Raina, A.P., Verma, S.K. and Abraham, Z., 2014. Volatile constituents of essential oils isolated from Alpinia galanga Willd. (L.) and A. officinarum Hance rhizomes from North East India. *Journal of Essential Oil Research*, *26*(1), pp. 24–28.

Thangaiyan, R., Arjunan, S., Govindasamy, K., Khan, H.A., Alhomida, A.S. and Prasad, N.R., 2020. Galangin attenuates isoproterenol-induced inflammation and fibrosis in the cardiac tissue of albino wistar rats. *Frontiers in Pharmacology*, *11*, p. 585163.

Tungmunnithum, D., Tanaka, N., Uehara, A. and Iwashina, T., 2020. Flavonoids Profile, Taxonomic Data, History of Cosmetic Uses, Anti-Oxidant and Anti-Aging Potential of Alpinia galanga (L.) Willd. *Cosmetics*, *7*(4), p. 89.

Watt, G., 1889. *A Dictionary of the Economic Products of India*. Printed by the Superintendent of Government Printing, India.

13 Lesser Galangal (*Alpinia officinarum* Hance)

Alpinia officinarum Hance

Etymology: After the Italian physician Prospero Alpini (1553–1617) and from the Latin *officinarum* = of medicinal value

Family: Zingiberaceae

Synonyms: *Languas officinarum* (Hance) Farw.; *Languas officinarum* (Hance) P.H. Hô

Common names: Lesser galangal; petit galanga (Fr.); galgant (Ger.); galanga menor (Spa.); galanga menor (Port.)

Part used: Rhizome

Constituents: Essential oil (1,8-cineole) (Zhang et al., 2020), diarylheptanoids (5-hydroxy-7-(4′-hydroxy-3′-methoxyphenyl)-1-phenyl-3-heptanone) (Shin et al., 2004; Ling et al., 2010), flavones (galangin, 3-methylethergalangin) (Shin et al., 2003; Verza et al., 2011).

Medical history: Arab physicians of the Middle Ages, including Avicenna, were familiar with lesser galangal. In 16th-century Italy, physicians including Fioravanti held lesser galangal as a useful ingredient in a formulation against plagues and syphilis. Matthioli (1572) defines it as hot and dry to the third degree. In 19th-century European medical practice, it was used as an aromatic stimulant, while in India, it was used as a carminative tonic to reduce the quantity of urine in diabetes and correct foul breath when chewed. The juice was used for sore throat, as a tonic, and an aphrodisiac (Watt, 1889).

Medicinal uses: Nausea, indigestion (Turkey); carminative, rheumatism (Iraq, Iran); tonic (Myanmar)

Blood pressure: Ethanol extract of rhizomes given at 500 mg/kg/day for 4 weeks to rats poisoned with a diet high in fat, sugar, and salt normalized systolic blood pressure from 141.4 to 126.5 mmHg and diastolic blood pressure from 90.8 to 80.1 mmHg (Javaid et al., 2021). *In vitro*, ethanol extract of rhizomes at the concentration of 30 µg/mL evoked relaxation of rat arteries contracted by phenylephrine, and this effect was endothelium independent and dependent on calcium ion influx in smooth muscle cells (Haam et al., 2022).

1,8-Cineole relaxed rat isolated aorta (IC_{50}: 663.2 µg/mL) via an endothelial-dependent mechanism (Pinto et al., 2009).

DOI: 10.1201/9781003301455-13

Heart: Ethanol extract of rhizomes given at 500 mg/kg/day for 4 weeks to rats poisoned with a diet high in high fat, sugar, and salt decreased heart rate from 466.2 to 370.4 bpm (normal: 355.2 bpm) (Javaid et al., 2021).

Plasma lipids and glucose: Rats fed a cholesterol-enriched chow containing 5% of ethanol extract of rhizomes for 6 weeks had decreases in total cholesterol from 3.5 to 2.5 mmol/L and triglycerides from 1.8 to 1.2 mmol/L (Xia et al., 2010). Ethanol extract of rhizomes given at 500 mg/kg/day for 4 weeks to rats poisoned with a diet high in fat, sugar, and salt decreased total cholesterol from 154.2 to 89.2 mg/dL (Javaid et al., 2021).

Adipose tissues: Rats fed a cholesterol-enriched chow containing 5% of ethanol extract of rhizomes for 6 weeks decreased the mass of adipose tissues from 53.3 to 40.5 g/g body weight, normalized total cholesterol from 3.5 to 2.5 mmol/L, and triglycerides from 1.8 to 1.2 mmol/L (Xia et al., 2010).

Kidneys: Ethanol extract of rhizomes given at 500 mg/kg/day for 4 weeks to rats poisoned with a diet high in fat, sugar, and salt evoked diuretic effects (Javaid et al., 2021).

Bones and cartilages: Ethanol extract given orally at 300 mg/kg/day for 12 weeks enhanced bone strength and prevented the deterioration of trabecular microarchitecture in ovariectomized rats (Su et al., 2016).

Immune system: Aqueous extract given to mice orally at 100 mg/kg increased circulating phagocytic activity (Bendjeddou et al., 2003).

Skin and hair: Aqueous extract given topically (50 µg/mL) to mice for 10 weeks prevented the formation of wrinkles induced by ultraviolet B irradiation (Jung et al., 2022).

Brain: 7-(4-Hydroxyphenyl)-1-phenyl-4E-hepten-3-one at the concentration of 0.5 µM prevented amyloid β-induced decrease of dendritic branches in hippocampal neurons (Huang et al., 2016).

Warnings: Chicken fed a diet containing 2, 5, or 10% for 4 weeks had altered blood parameters including increased uric acid (Seddeag et al., 2010). Ethanol extract of rhizomes given orally to rats at 100 mg/kg/day induced heart and lung weight changes and increased red blood cells count (Qureshi et al., 1992).

Comments: (i) Avicenna (Ibn Sina) (980–1037) was a Persian physician and author of the monumental medical book *Al-Qanun fit-Tibb*, used as a reference in European medical practice. Arabs had a monopoly on the spice supply to Europe before being supplanted by Portuguese, Dutch, and British companies. Spices were sought for the tables and pride of the wealthy, at exorbitant prices, and were considered "black gold" from the Roman empire to the end of the 18th century.

(ii) Leonardo Fioravanti (1517–1588) was an Italian physician.

(iii) Pancreatic lipases are enzymes that catalyze the release of fatty acids from dietary triglycerides. Lesser galangal yields diarylheptanoids that inhibit pancreatic lipase such as 5-hydroxy-7-(4'-hydroxy-3'-methoxyphenyl)-1-phenyl-3-heptanone (Shin et al., 2004) and flavones including 3-methylethergalangin (Shin et al., 2003).

(iv) Plants classified as "hot" by our ancient peers tend to decrease adipose mass.

(v) Lesser galangal could potentially be used as a functional food in small dietary amounts and occasionally to prevent cardiovascular aging.

REFERENCES

Bendjeddou, D., Lalaoui, K. and Satta, D., 2003. Immunostimulating activity of the hot water-soluble polysaccharide extracts of Anacyclus pyrethrum, Alpinia galanga and Citrullus colocynthis. *Journal of ethnopharmacology*, *88*(2–3), pp. 155–160

Haam, C.E., Byeon, S., Choi, S.J., Lim, S., Choi, S.K. and Lee, Y.H., 2022. Vasodilatory effect of Alpinia officinarum extract in rat mesenteric arteries. *Molecules*, *27*(9), p. 2711.

Huang, X., Tang, G., Liao, Y., Zhuang, X., Dong, X., Liu, H., Huang, X.J., Ye, W.C., Wang, Y. and Shi, L., 2016.7-(4-Hydroxyphenyl)-1-phenyl-4E-hepten-3-one, a Diarylheptanoid from Alpinia officinarum, Protects Neurons against Amyloid-β Induced Toxicity. *Biological and Pharmaceutical Bulletin*, *39*(12), pp. 1961–1967.

Javaid, F., Mehmood, M.H. and Shaukat, B., 2021. Hydroethanolic extract of A. officinarum hance ameliorates hypertension and causes diuresis in obesogenic feed-fed rat model. *Frontiers in Pharmacology*, p. 1437.

Jung, J.M., Kwon, O.Y., Choi, J.K. and Lee, S.H., 2022. Alpinia officinarum Rhizome ameliorates the UVB induced photoaging through attenuating the phosphorylation of AKT and ERK. *BMC Complementary Medicine and Therapies*, *22*(1), p. 232.

Ling, Z.H.A.O., Wei, Q.U., Ju-Qin, F.U. and LIANG, J.Y., 2010. A new diarylheptanoid from the rhizomes of Alpinia officinarum. *Chinese Journal of Natural Medicines*, *8*(4), pp. 241–243.

Pinto, N.V., Assreuy, A.M.S., Coelho-de-Souza, A.N., Ceccatto, V.M., Magalhães, P.J.C., Lahlou, S. and Leal-Cardoso, J.H., 2009. Endothelium-dependent vasorelaxant effects of the essential oil from aerial parts of Alpinia zerumbet and its main constituent 1, 8-cineole in rats. *Phytomedicine*, *16*(12), pp. 1151–1155.

Qureshi, S., Shah, A.H. and Ageel, A.M., 1992. Toxicity studies on Alpinia galanga and Curcuma longa. *Planta Medica*, *58*(2), pp. 124–127.

Seddeag, M., Madawe, G., El Badwi, S. and Bakhiet, A., 2010. The effect of dietary Alpinia officinarum (Hance) supplementation in Bovns-type chicks. *International Journal of Poultry Science*, *9*(5), pp. 499–502.

Shin, J.E., Han, M.J. and Kim, D.H., 2003.3-Methylethergalangin isolated from Alpinia officinarum inhibits pancreatic lipase. *Biological and Pharmaceutical Bulletin*, *26*(6), pp. 854–857.

Shin, J.E., Han, M.J., Song, M.C., Baek, N.I. and Kim, D.H., 2004.5-Hydroxy-7-(4'-hydroxy-3'-methoxyphenyl)-1-phenyl-3-heptanone: A pancreatic lipase inhibitor isolated from Alpinia officinarum. *Biological and Pharmaceutical Bulletin*, *27*(1), pp. 138–140.

Su, Y., Chen, Y., Liu, Y., Yang, Y., Deng, Y., Gong, Z., Chen, J., Wu, T., Lin, S. and Cui, L., 2016. Antiosteoporotic effects of Alpinia officinarum Hance through stimulation of osteoblasts associated with antioxidant effects. *Journal of Orthopedic Translation*, *4*, pp. 75–91.

Verza, S.G., Pavei, C., Borré, G.L., Silva, A.P., González Ortega, G. and Mayorga, P., 2011. Determination of Galangin in commercial extracts of Alpinia officinarum by RP-HPLC-DAD. *Latin American Journal of Pharmacy*, *30*.

Xia, D.Z., Yu, X.F., Wang, H.M., Ren, Q.Y. and Chen, B.M., 2010. Anti-obesity and hypolipidemic effects of ethanolic extract from Alpinia officinarum Hance (Zingiberaceae) in rats fed high-fat diet. *Journal of Medicinal Food*, *13*(4), pp. 785–791.

Zhang, L., Pan, C., Ou, Z., Liang, X., Shi, Y., Chi, L., Zhang, Z., Zheng, X., Li, C. and Xiang, H., 2020. Chemical profiling and bioactivity of essential oils from Alpinia officinarum Hance from ten localities in China. *Industrial Crops and Products*, *153*, p. 112583.

14 Marshmallow (*Althaea officinalis* L.)

Althaea officinalis L.

Etymology: From the Greek *althain* = to cure and the Latin *officinalis* = of medicinal value

Family: Malvaceae

Synonyms: *Althaea micrantha* Wiesb. ex Borbás; *Althaea taurinensis* DC.; *Althaea vulgaris* Bubani; *Malva althaea* E.H.L. Krause; *Malva officinalis* (L.) Schimp. & Spenn.

Common names: Marshmallow; guimauve (Fr.); mäusespeck (Ger.); malvavisco (Spa.); проскурняк (Rus.)

Parts used: Flower, root

Constituents: Flavone glycosides (hypolaetin-8-glucoside), coumarins, phenolic acids (Gudej, 1991).

Medical history: Dioscorides used marshmallow as an anti-inflammatory, vulnerary, and diuretic, for kidney stones, and dysentery. Galen adds that it is good for bleeding, and Pliny the Elder recommends its use for asthma. It was cultivated in medieval Europe under the directive of the emperor Charlemagne (10th century). According to Fusch (1555), the plant is hot and dry to the first degree. It was used as a demulcent in 19th-century Europe. In 19th-century India, the root was used for cough and for irritation of the intestines and bladder (Watt, 1889).

Medicinal uses: Dry cough, carminative, 0.5–3 g in 150 mL of boiling water, several times daily (European Union) diuretic, kidney stones (Turkey); asthma, urinary retention, hypertension (Iran); asthma, kidney pain, hypertension (Afghanistan)

Plasma glucose and lipids: Aqueous extract of flowers added to the drinking water (corresponding to 50 mg/kg) of rats on a high-fat diet for 30 days resulted in an increase of HDL-cholesterol from 37.5 to 47.4 mg/dL and a decrease of plasma glucose from 110.5 to 92.5 mg/dL (Hage-Sleiman et al., 2011).

Hydroalcoholic extract of roots given orally to streptozotocin-induced diabetic rats at 200 mg/kg/day for 4 weeks decreased glycemia and plasma lipids (Ashtiyani et al., 2015).

Comment: Infusions of roots could potentially be beneficial for preventing or controlling excessive plasma lipid and glucose, which help accelerate cardiovascular aging.

DOI: 10.1201/9781003301455-14

REFERENCES

Ashtiyani, C., Yarmohammady, P., Hosseini, N. and Ramazani, M., 2015. The effect of Althaea officinalis. L root alcoholic extract on blood sugar level and lipid profiles of streptozotocin induced-diabetic rats. *Iranian Journal of Endocrinology and Metabolism*, *17*(3), pp. 238–250.

Fioravanti, L., 1571. *Il Reggimento Della Peste, Dell'Eccellente Dottore Et Cavaliero M. Leonardo Fioravanti Bolognese: Nel quale si tratta che cosa sia la peste, & da che procede, & quello che doveriano fare i Prencipi per conservar' i suoi popoli da essa; & ultimamente, si mostrano mirabili secreti da curarla, cosa non mai piu scritta da niuno in questo modo.* Sessa.

Fusch, L., 1555. *De Historia Stirpium Commetarii Insignes.* Lugduni Apud Ioan Tornaesium.

Gudej, J., 1991. Flavonoids, phenolic acids and coumarins from the roots of Althaea officinalis. *Planta Medica*, *57*(03), pp. 284–285.

Hage-Sleiman, R., Mroueh, M. and Daher, C.F., 2011. Pharmacological evaluation of aqueous extract of Althaea officinalis flower grown in Lebanon. *Pharmaceutical Biology*, *49*(3), pp. 327–333.

Matthioli, P.A., 1572. *Commentaires sur les Six Livres de Pedacius Dioscorides Anazarbeen de la matière medicinale.* A l'Escue de Milan.

Watt, G., 1889. *A Dictionary of the Economic Products of India.* Printed by the Superintendentof Government Printing, India.

15 Khella
(*Ammi visnaga* (L.) Lam.)

Ammi visnaga (L.) Lam.

Etymology: From the Greek *ammi* = name of a plant in the family Apiaceae and the Spanish *bisnaga* = pick-tooth
Family: Apiaceae
Synonym: *Daucus visnaga* L.
Common names: Khella, toothpickweed; herbe aux cure-dents, herbe aux gencives (Fr.); zahnstocherkraut (Ger.)
Part used: Seed
Constituent: Furanocoumarins (khellin, visnagin) (Rauwald et al., 1994).
Medical history: Khella was known to Dioscorides as *keraskomen* and to Theophrastus as *raphanon agrian*. It was a diuretic and antispasmodic medicine for the Arab physicians of the Middle Ages.
Medicinal uses: Gall inflammation and bile stone (Turkey); Unani practitioners use the plant for spasms in renal colic, bronchial asthma, whooping cough, and angina pectoris (India)

Blood pressure: Khellin given orally at 120 mg improved the condition of patients with angina pectoris (Anrep et al., 1946). This furanocoumarin enhances calcium ion extrusion or sequestration in vascular smooth muscle cells (Ubeda & Villar, 1989).

Visnagin relaxed mesenteric arteries contracted by noradrenaline and this effect was not affected by endothelium removal (Duarte et al., 2000). Plasma glucose and lipids: Aqueous extract at 20 mg/kg reduced blood glucose in normal rats 6 hours after a single oral administration and 9 days after repeated oral administration, and these effects were more pronounced in streptozotocin-induced diabetic rats (Jouad et al., 2002).

Kidneys: Aqueous decoction of seeds given to rats for 4 weeks at 500 mg/kg/day evoked diuretic effects and prevented the formation of crystals of calcium oxalate in the kidney induced by glycolic acid (Khan et al., 2001).

Warning: The LD_{50} of intraperitoneal and oral administration for aqueous extract are 3.6 and 10.1 g/kg, respectively (Jouad et al., 2002). A 67-year-old man with diabetes and kidney insufficiency taking bottled boiled seeds suffered renal injuries (AL-Shoubaki et al., 2020).
Comments: (i) Khellin given to patients with mild to severe angina resulted in the complete cessation or a significant reduction in the frequency and intensity of angina pectoris (Hultgren et al., 1952). A number of pharmaceutical

DOI: 10.1201/9781003301455-15

preparations made with khella are available in Germany (Khalil et al., 2020), and such products could, under careful medical supervision, be beneficial against myocardial infarction and for controlling hypertension.

(ii) The purpose of this handbook is not to blindly reject all conventional medicine but to advocate the unique use of plants to delay the development of aging and related pathologies. The development of drugs since the mid-20th century has provided, for the first time in human history, a huge improvement in global health and delayed mortality. Consulting a medical doctor and complying with a prescription are absolute priorities if someone is unwell. Yet, acquiring knowledge about the many plants that may have potential effects to delay aging might allow a delay in the development of, for instance, hypertension, arthritis, and other age-related conditions. Over the years, I have witnessed in my visits to Asian villages numerous instances of longevity in deprived areas where medicines were not available. In Bangladesh, for instance, the intake of fruit juice from *Emblica officinalis* Gaertn. seems to delay the time for type-2 diabetics to be compelled to rely on insulin injections. In the Philippines, some elderly are able to work under harsh conditions by adding to their diet a mix of raw fish and moringa leaves, and those chewing "sireh" seem to keep perfect dentition over the age of 90. The Malays of Perak take seeds of *Scorodocarpus borneensis* (Baill.) Becc. for cancer and kidney failure. May more examples be provided. There is a window of opportunity for both conventional medicines and medicinal plants (as well as vegetables) to work together for prolonged survival.

REFERENCES

AL-Shoubaki, R., Akl, A., Sheikh, I. and Shaheen, F., 2020. Khella induced nephropathy: A case report and review of literature. *Urology & Nephrology Open Access Journal, 8*(3), pp. 62–64.

Anrep, G.V., Barsoum, G.S., Kenawy, M.R. and Misrahy, G., 1946. Ammi visnaga in the treatment of the anginal syndrome. *British Heart Journal, 8*(4), p. 171.

Duarte, J., Torres, A.I. and Zarzuelo, A., 2000. Cardiovascular effects of visnagin on rats. *Planta Medica, 66*(01), pp. 35–39.

Hultgren, H.N., Robertson, H.S. and Stevens, L.E., 1952. Clinical and experimental study of use of Khellin in treatment of angina pectoris. *Journal of the American Medical Association, 148*(6), pp. 465–469.

Jouad, H., Maghrani, M. and Eddouks, M., 2002. Hypoglycemic effect of aqueous extract of Ammi visnaga in normal and streptozotocin-induced diabetic rats. *Journal of Herbal Pharmacotherapy, 2*(4), pp. 19–29.

Khalil, N., Bishr, M., Desouky, S. and Salama, O., 2020. Ammi visnaga L., a potential medicinal plant: A review. *Molecules, 25*(2), p. 301.

Khan, Z.A., Assiri, A.M., Al-Afghani, H.M. and Maghrabi, T.M., 2001. Inhibition of oxalate nephrolithiasis with Ammi visnaga (AI-Khillah). *International Urology and Nephrology, 33*, pp. 605–608.

Rauwald, H.W., Brehm, O. and Odenthal, K.P., 1994. The involvement of a Ca2+ channel blocking mode of action in the pharmacology of Ammi visnaga fruits1. *Planta medica, 60*(02), pp. 101–105.

Ubeda, A. and Villar, A., 1989. Relaxant actions of Khellin on vascular smooth muscle. *Journal of Pharmacy and Pharmacology, 41*(4), pp. 236–241.

16 Sweet Almonds (*Amygdalus communis* L.)

Amygdalus communis L.

Etymology: From the Latin *amygdala* = almond and *communis* = common

Family: Rosaceae

Synonyms: *Amygdalus dulcis* Mill.; *Amygdalus sativa* Mill.; *Prunus communis* (L.) Arcang.; *Prunus dulcis* (Mill.) D.A. Webb

Common name: Sweet almonds; amandes douces (Fr.); süße mandel (Ger.); amêndoa doce (Port.); almendra dulce (Spa.); сладкий миндаль (Rus.)

Part used: Seed

Constituents: α-Tocopherol, magnesium, potassium, fixed oil (unsaturated fatty acids: oleic and linoleic acids) (Özcan et al., 2011).

Medical history: Almond trees are mentioned in the Book of Genesis as growing in the Holy Land as well as in the Book of Splendor (The Zohar). The plant was called *avellanae graece* in ancient Rome. Dioscorides recommends sweet almonds for internal pains, insomnia, ptysis, kidney diseases, lung diseases, and as a diuretic. In the 9th-century France, the Emperor Charlemagne ordered the trees to be introduced on imperial farms. In medieval France, sweet almonds were consumed abundantly by the wealthy. The German physicians of the 16th century, including Valerius Cordus (1590), recommended sweet almonds for cough and burning urination and defined them as dry ("*amygdala dulcia sicca*").

Blood pressure: Almonds given to rats on a high-fat diet at 3 g/kg for 4 weeks improved vascular reactivity by preserving endothelial NO synthase and promoting NO release (Jamshed & Gilani, 2014). Intake of 50g of sweet almonds/day for 4 weeks by healthy middle-aged volunteers improved flow-mediated dilatation and decreased blood pressure by about 6% (Choudhury et al., 2014).

Plasma lipids and glucose: Almonds given to rats on a high-fat diet at 3 g/kg for 4 weeks decreased plasma glucose by 61% and total cholesterol, LDL-cholesterol, and plasma triglycerides by 50%, 36%, and 33%, respectively (Jamshed & Gilani, 2014).

Healthy volunteer men with mild hyperlipidemia ingesting 60 g almond daily for 4 weeks experienced decreased LDL-cholesterol and cholesterol (Jalali-Khanabadi et al., 2010).

 DOI: 10.1201/9781003301455-16

Intake of 60 g per day of sweet almonds by patients with type-2 diabetes for 4 weeks resulted in decreases of total cholesterol and low-density lipoprotein cholesterol by 6 and 11.6%, respectively. Further, this regimen decreased plasma fasting insulin and fasting glucose by 4.1 and 0.8%, also respectively (Li et al., 2011).

Uric acid: Intake of 10 g of almonds per days (soaked overnight in water) by coronary heart disease patients for 12 weeks before breakfast decreased plasma uric acid by about 15% (Jamshed et al., 2015).

Comments: (i) Valerius Cordus (1515–1544), was a German physician and botanist, lecturer at the University of Wittenberg.

(ii) Sweet almonds contain up to 691 mg/100 g of potassium, 230 g/100 g of magnesium (Drogoudi et al., 2013), and α-tocopherol (Kodad et al., 2018).

(iii) Magnesium tends to attenuate blood pressure and is given in high doses parenterally to patients with profound cardiac arrhythmia. At the vascular smooth muscle level, magnesium attenuates calcium-dependent contractions; at the level of vascular endothelial cells, it stimulates the production of NO; and it inhibits catecholamine secretion from both the adrenal gland and peripheral adrenergic nerve terminals (Dominguez et al., 2020). In addition, intake of magnesium is inversely correlated with calcium deposition in arteries (Hruby et al., 2014). Magnesium also regulates collagen and elastin turnover in the vascular wall and matrix metalloproteinase activity and maintains the elasticity of the arteries by protecting elastic fibers from calcium deposition (Kostov & Halacheva, 2018). Intake of magnesium is also inversely correlated with insulin resistance and inflammation (Kim et al., 2010a). In streptozotocin-induced diabetic rats, intake of magnesium (1 mg/kg diet) for 4 weeks improved glucose tolerance from 74.4 to 52.4 mg/dL; improved cholesterol, HDL-cholesterol, and LDL-cholesterol; and increased the expression of insulin receptors and GLUT-4 in the skeletal muscles (Morakinyo et al., 2018).

(iv) Potassium ion intake is correlated with decreasing blood pressure (Whelton et al., 1997) via mechanism involving at least in part the enhanced expression of sodium in the urine but excessive intake of potassium is extremely dangerous. Caution must be taken with seeds (sweet almonds, pumpkin), vegetables (cucumbers, potatoes, radish), and medicinal plants rich in potassium must be absolutely avoided in patients with hyperkaliemia and those taking medicines that increase kaliemia such as beta-blockers. Note that hyperkaliemia is often without symptoms yet potentially lethal hence the need for complete blood check for electrolytes on regular basis (Smith et al., 1992)

(v) α-Tocopherol given to spontaneously hypertensive rats at 34 mg/kg diet reduced blood pressure from 209.56 to 128.8 mm Hg and increased the activity of NO synthetase in blood vessels and increased plasma nitrites (Newaz et al., 1999).

(vi) An increase in serum uric acid by 1 mg/dL contributes to 12% increase in risk of death by cardiovascular diseases (Kim et al., 2010b).

REFERENCES

Choudhury, K., Clark, J. and Griffiths, H.R., 2014. An almond-enriched diet increases plasma α-tocopherol and improves vascular function but does not affect oxidative stress markers or lipid levels. *Free Radical Research*, *48*(5), pp. 599–606.

Cordus, V. and Coudenberg, P., 1590. *Val. Cordi Dispensatorium siue Pharmacorum conficiendorum ratio: adietco Valerii Cordi nouo libello, aliisque paucis post praefationes annotatu.* ex Officina Plantiniana, apud Franciscum Raphelengium.

Dominguez, L.J., Veronese, N. and Barbagallo, M., 2020. Magnesium and hypertension in old age. *Nutrients*, *13*(1), p. 139.

Drogoudi, P.D., Pantelidis, G., Bacchetta, L., De Giorgio, D., Duval, H., Metzidakis, I. and Spera, D., 2013. Protein and mineral nutrient contents in kernels from 72 sweet almond cultivars and accessions grown in France, Greece and Italy. *International Journal of Food Sciences and Nutrition*, *64*(2), pp. 202–209.

Hruby, A., O'Donnell, C.J., Jacques, P.F., Meigs, J.B., Hoffmann, U. and McKeown, N.M., 2014. Magnesium intake is inversely associated with coronary artery calcification: The Framingham heart study. *JACC: Cardiovascular Imaging*, *7*(1), pp. 59–69.

Jalali-Khanabadi, B.A., Mozaffari-Khosravi, H. and Parsaeyan, N., 2010. Effects of almond dietary supplementation on coronary heart disease lipid risk factors and serum lipid oxidation parameters in men with mild hyperlipidemia. *The Journal of Alternative and Complementary Medicine*, *16*(12), pp. 1279–1283.

Jamshed, H. and Gilani, A.H., 2014. Almonds inhibit dyslipidemia and vascular dysfunction in rats through multiple pathways. *Journal of Nutrition*, *144*, pp. 1768–1774.

Jamshed, H., Gilani, A.U.H., Sultan, F.A.T., Amin, F., Arslan, J., Ghani, S. and Masroor, M., 2015. Almond supplementation reduces serum uric acid in coronary artery disease patients: A randomized controlled trial. *Nutrition Journal*, *15*, pp. 1–5.

Kim, D.J., Xun, P., Liu, K., Loria, C., Yokota, K., Jacobs Jr, D.R. and He, K., 2010a. Magnesium intake in relation to systemic inflammation, insulin resistance, and the incidence of diabetes. *Diabetes Care*, *33*(12), pp. 2604–2610.

Kim, S.Y., Guevara, J.P., Kim, K.M., Choi, H.K., Heitjan, D.F. and Albert, D.A., 2010b. Hyperuricemia and coronary heart disease: A systematic review and meta-analysis. *Arthritis Care & Research*, *62*, pp. 170–180.

Kodad, O., Socias I Company, R. and Alonso, J.M., 2018. Genotypic and environmental effects on tocopherol content in almond. *Antioxidants*, *7*(1), p. 6.

Kostov, K. and Halacheva, L., 2018. Role of magnesium deficiency in promoting atherosclerosis, endothelial dysfunction, and arterial stiffening as risk factors for hypertension. *International Journal of Molecular Sciences*, *19*(6), p. 1724.

Li, S.C., Liu, Y.H., Liu, J.F., Chang, W.H., Chen, C.M. and Chen, C.Y.O., 2011. Almond consumption improved glycemic control and lipid profiles in patients with type 2 diabetes mellitus. *Metabolism*, *60*(4), pp. 474–479.

Morakinyo, A.O., Samuel, T.A. and Adekunbi, D.A., 2018. Magnesium upregulates insulin receptor and glucose transporter-4 in streptozotocin-nicotinamide-induced type-2 diabetic rats. *Endocrine Regulations*, *52*(1), pp. 6–16

Newaz, M.A., Nawal, N.N.A., Rohaizan, C.H., Muslim, N. and Gapor, A., 1999. α-Tocopherol increased nitric oxide synthase activity in blood vessels of spontaneously hypertensive rats. *American Journal of Hypertension*, *12*(8), pp. 839–844.

Özcan, M.M., Ünver, A., Erkan, E. and Arslan, D., 2011. Characteristics of some almond kernel and oils. *Scientia Horticulturae*, *127*(3), pp. 330–333.

Smith, S.R., Klotman, P.E. and Svetkey, L.P., 1992. Potassium chloride lowers blood pressure and causes natriuresis in older patients with hypertension. *Journal of the American Society of Nephrology*, *2*(8), pp. 1302–1309.

Whelton, P.K., He, J., Cutler, J.A., Brancati, F.L., Appel, L.J., Follmann, D. and Klag, M.J., 1997. Effects of oral potassium on blood pressure: Meta-analysis of randomized controlled clinical trials. *Jama*, *277*(20), pp. 1624–1632.

17 Italian Alkanet (*Anchusa italica* Retz.)

Anchusa italica Retz.

Etymology: From the Greek *ankusa* = alkanet and the Latin *italica* = from Italy

Family: Boraginaceae

Synonym: *Anchusa azurea* Schur

Common names: Italian alkanet, Italian bugloss; buglosse d'Italie, buglosse azurée (Fr.); Italienische ochsenzunge (Ger.); lingua de- aca (Port.); воловик лазоревый (Rus.); lengua de buey (Spa.)

Part used: Flower

Constituents: Flavone glycosides (rutin), flavanone glycosides (hesperidin), flavones (quercetin), flavanones (naringenin) (Wang et al., 2020), hydroxycinnamic acid derivatives (rosmarinic acid) (Kuruüzüm-Uz et al., 2013).

Medical history: Galen asserts that Italian alkanet brings joy to the heart, and for Simeon Seth (12th-century Byzantium), it is diuretic. The Medical School of Salerno (10th century) says that Italian alkanet is good against depression, heart palpitations, and heart troubles. It was used for cooling in 19th-century France (Guibourt, 1836) as well as for cough and jaundice.

Medicinal uses: Diuretic (Iraq)

Heart: Ethanol extract of flowers given orally at 100 mg/kg to rats for 10 days intraperitoneally after transient bilateral carotid artery occlusion decreased serum MDA from about 220 to 110 nmol/g (normal: about 140 nmol/g) (Asgharzade et al., 2020). A flavonoid fraction given orally to mice at 50 mg/kg/day for 4 weeks evoked some levels of protection against coronary artery ligation as evidenced by improved cardiac function and decreased infarct size (Wang et al., 2020). Oleanane and ursane triterpenes in this plant protected neonatal rat myocytes against hypoxia/reoxygenation *in vitro* (Liu et al., 2020; Hu et al., 2020).

Warning: Produces hepatotoxic pyrrolizidine alkaloids (Röder, 2020).

Comments: (i) Oral route should be preferred for testing extracts or compounds as the intraperitoneal route is not therapeutically convenient and because it causes useless suffering in laboratory animals.

(ii) TNF α binding to the myocardial receptors increases sphingosine, which translates into contractile dysfunction (Meldrum, 1998).

(iii) Should pyrrolizidine alkaloids be absent from the flowers, they could be used to prevent myocardial infarction. More experiments are needed.

DOI: 10.1201/9781003301455-17

REFERENCES

Asgharzade, S., Sewell, R.D., Rabiei, Z., Forouzanfar, F., Sheikhshabani, S.K. and Rafieian-Kopaei, M., 2020. Hydroalcoholic extract of Anchusa Italica protects global cerebral ischemia-reperfusion injury via a nitrergic mechanism. *Basic and Clinical Neuroscience*, *11*(3), p. 323.

Hu, B.C., Liu, Y., Zheng, M.Z., Zhang, R.Y., Li, M.X., Bao, F.Y., Li, H. and Chen, L.X., 2020. Triterpenoids from Anchusa Italica and their protective effects on hypoxia/reoxygenation induced cardiomyocytes injury. *Bioorganic Chemistry*, *97*, p. 103714.

Kuruüzüm-Uz, A., Güvenalp, Z., Kazaz, C. and Demirezer, L.Ö., 2013. Phenolic compounds from the roots of Anchusa azurea var. azurea. *Turkish Journal of Pharmaceutical Sciences*, *10*(2), pp. 177–184.

Liu, Y., Hu, B., Wang, Y., Bao, F., Li, H. and Chen, L., 2020. Chemical constituents of Anchusa italica Retz. and their protective effects on cardiomyocytes injured by hypoxia/reoxygenation. *Phytochemistry Letters*, *38*, pp. 155–160.

Meldrum, D.R., 1998. Tumor necrosis factor in the heart. *American Journal of Physiology-Regulatory, Integrative and Comparative Physiology*, *274*(3), pp. R577–R595.

Röder, E., 2020. Medicinal plants containing pyrrolizidine alkaloids in the New Kreuterbuch by Leonhart Fuchs (1543). *Die Pharmazie-An International Journal of Pharmaceutical Sciences*, *75*(7), pp. 294–298.

Wang, S., Zhao, Y., Song, J., Wang, R., Gao, L., Zhang, L., Fang, L., Lu, Y. and Du, G., 2020. Total flavonoids from Anchusa italica Retz. Improve cardiac function and attenuate cardiac remodeling post myocardial infarction in mice. *Journal of Ethnopharmacology*, *257*, p. 112887.

18 Dill
(*Anethum graveolens* L.)

Anethum graveolens L.

Etymology: From the Greek *aneton* = anise and the Latin *gravis* = heavy and *olens* = smelling

Family: Apiaceae

Synonyms: *Anethum sowa* Roxb. ex Fleming; *Ferula marathrophylla* Walp.; *Peucedanum anethum* Baill.; *Peucedanum graveolens* (L.) Hiern; *Peucedanum sowa* (Roxb. ex Fleming) Kurz

Common names: Anethum, dill; aneth (French); dillkraut (Ger.); eneldo (Spa.); aneto (Port.); укрóп (Rus.)

Part used: Seed

Constituents: Essential oil (apiol, carvone) (Charles et al., 1995; Khalid, 2012; Rana, & Blazquez, 2014; Dimov et al., 2019).

Medical history: In the 18th-century Scotland, Alston (1770) advocated the use of dill against vomiting, to induce sleeping, and to calm flatulent colic, and defined it as hot to the third degree and dry to the second degree. Dill water (*aqua anethi*) was given with meals to infants to relieve flatulence and griping in 19th-century Europe and North America.

Medicinal uses: Hypertension, flatulence (Afghanistan); bronchitis, gastritis (Pakistan); high cholesterol (Iraq), chest discomfort, to invigorate the heart (India)

Plasma lipids and glucose: Ethanol extract of aerial parts given daily at 600 mg for 3 months to obese patients had no effect on blood pressure but decreased LDL-cholesterol from 122.1 to 116.9 mg/dL (Mansouri et al., 2012). In a subsequent study, taking tablets of dill 3 times a day for 2 months decreased plasma cholesterol and decreased triglycerides (Mirhosseini et al., 2014). Powder given to type-2 diabetic patients at 1 g, 3 times per day after meals for 8 weeks decreased insulin resistance, plasma triglycerides from 196.5 to 172.3 mg/dL., total cholesterol from 160.2 from 149.2 mg/dL, and LDL-cholesterol from 81 to 71.2 mg/dL, and increased HDL-cholesterol from 41.8 to 44.8 mg/dL (Haidari et al., 2020).

Warning: To be avoided at all cost in all stages of pregnancy (Talebi et al., 2020).

Comments: (i) It is clear that dill attenuates plasma lipids and glucose. WHO recommends not taking more than 3 g per day of dill seeds. Infusion of a teaspoon of dill in 150 mL of boiling water could be used.

DOI: 10.1201/9781003301455-18

(ii) According to Unani practitioners who use the Hippocratic system, dill being hot and dry might be welcomed by patients with cold temperament but not well tolerated by individuals with hot temperament.

REFERENCES

Alston, C., 1770. *Lectures on the Materia Medica: Containing the Natural History of Drugs, their Virtues and Doses: Also Directions for the Study of the Materia Medica; and an Appendix on the Method of Prescribing*. Edward and Charles Dilly.

Charles, D.J., Simon, J.E. and Widrlechner, M.P., 1995. Characterization of essential oil of dill (Anethum graveolens L.). *Journal of Essential Oil Research*, 7(1), pp. 11–20.

Dimov, M.D., Dobreva, K.Z. and Stoyanova, A.S., 2019. Chemical composition of the dill essential oils (Anethum graveolens L.) from Bulgaria. *Bulgarian Chemical Communications*, 51(1), pp. 214–216.

Guibourt, N.J.B.G., 1836. *Histoire abrégée des drogues simples*. Méquignon-Marvis Père et fils.

Haidari, F., Zakerkish, M., Borazjani, F., Ahmadi Angali, K. and Amoochi Foroushani, G., 2020. The effects of Anethum graveolens (dill) powder supplementation on clinical and metabolic status in patients with type 2 diabetes. *Trials*, 21(1), pp. 1–11.

Khalid, K.A., 2012 Influence of hydro-distillation time on the yield and quality of dill volatile constituents. *Medicinal and Aromatic Plant Science and Biotechnology*, 6(1), pp. 46–49.

Mansouri, M., Nayebi, N., Hasani-Ranjbar, S., Taheri, E. and Larijani, B., 2012. The effect of 12 weeks Anethum graveolens (dill) on metabolic markers in patients with metabolic syndrome: A randomized double blind controlled trial. *DARU Journal of Pharmaceutical Sciences*, 20(1), pp. 1–7.

Mirhosseini, M., Baradaran, A. and Rafieian-Kopaei, M., 2014. Anethum graveolens and hyperlipidemia: A randomized clinical trial. *Journal of Research in Medical Sciences: The Official Journal of Isfahan University of Medical Sciences*, 19(8), p. 758.

Rana, V.S. and Blazquez, M.A., 2014. Chemical composition of the essential oil of Anethum graveolens aerial parts. *Journal of Essential Oil Bearing Plants*, 17(6), pp. 1219–1223.

Talebi, F., Malchi, F., Abedi, P. and Jahanfar, S., 2020. Effect of dill (Anethum Graveolens Linn) seed on the duration of labor: A systematic review. *Complementary Therapies in Clinical Practice*, 41, p. 101251.

19 Garden Angelica (*Angelica archangelica* L.)

Angelica archangelica L.

Etymology: From the Greek *angelikos* = angel and the Latin *archangelus* = archangel

Family: Apiaceae

Synonyms: *Archangelica officinalis* (Moench) Hoffm.; *Archangelica officinalis* (Moench) Hoffm.

Common names: Garden angelica; angéliquee officinale (Fr.); echte engelwurz (Ger.); angélica archangelica (Spa.); angélica dos jardins (Port.); ду́дник лека́рственный (Rus.)

Part used: Root

Constituents: Furanocoumarins (osthol, bergapten, imperatonin) (Steck & Bailey, 1969; Härmälä et al., 1992).

Medical history: During the Middle Ages, it was believed that holding a piece of the root in the mouth could protect against plagues, and there was a legend about an angel who indicated the use of this plant. Fusch (1555) described as hot and dry to the third degree and as useful in cases of plague, for indigestion, and for blood clotting. It was used to prepare the "*baume du commandeur*" and sweets in 19th-century France and was also a carminative (Guibourt, 1836).

Medicinal uses: Indigestion and hypertension (Nepal); the Laplanders use the roots for cough, abdominal pain, indigestion, and hypertension.

Blood pressure: Osthol given as part of the diet (0.05%) of stroke-prone spontaneously hypertensive rats decreased systolic blood pressure and hepatic triglyceride after 3 weeks (Ogawa et al., 2007). Given orally to rats at 20 mg/kg/day for 4 weeks, it decreased blood pressure, heart weight, and myocardial MDA induced by renovascular hypertension (Zhou et al., 2012).

Plasma lipids and cholesterol: Given to rats on high-fat and -sugar diet at 40 mg/kg for 4 weeks, it evoked some protection of kidneys (García-Arroyo et al., 2021).

Uric acid: Osthol inhibited urate transporter 1 with an IC_{50} of 78.8 µM (Tashiro et al., 2018).

Bones and cartilages: Imperatonin given orally at 20 mg/kg/day for 12 weeks prevented osteoporosis in ovariectomized rats (Yan et al., 2020).

Brain: Imperatonin given orally at 790 mg/kg/day for 14 days to mice enhanced memory (Sigurdsson & Gudbjarnason, 2013).

DOI: 10.1201/9781003301455-19

Warning: Bergapten induces photo-dermatitis (Kelly, 1969).

Comments: (i) Urate transporter 1 reabsorbs uric acid from urine (Sun et al., 2021) in the brush border membrane of proximal tubular cells (Enomoto et al., 2002).

(ii) Preparation of garden archangelica evoked some levels of protection in patients with symptoms of dementia related to frontotemporal lobar degeneration (Kimura et al., 2011). Patients who suffered an ischemic or hemorrhage stroke are prone to cognitive disorders (Sensenbrenner et al., 2020), and those who had cognitive impairment and patients with Alzheimer's disease are prone to stroke (Chi et al., 2013).

(iii) The plant can be confused with the dreadfully poisonous water hemlock (*Cicuta virosa* L.) (Knutsen & Paszkowski, 1984), highlighting the need to teach botany (and materia medica) in schools of pharmacy.

REFERENCES

Chi, N.F., Chien, L.N., Ku, H.L., Hu, C.J. and Chiou, H.Y., 2013. Alzheimer disease and risk of stroke: A population-based cohort study. *Neurology*, *80*(8), pp. 705–711.

Enomoto, A., Kimura, H., Chairoungdua, A., Shigeta, Y., Jutabha, P., Cha, S.H., Hosoyamada, M., Takeda, M., Sekine, T., Igarashi, T., et al., 2002. Molecular identification of a renal urate anion exchanger that regulates blood urate levels. *Nature*, *417*, 447–452.

Fusch, L., 1555. *De Historia Stirpium Commetarii Insignes*. Lugduni Apud Ioan Tornaesium.

García-Arroyo, F.E., Gonzaga-Sánchez, G., Tapia, E., Muñoz-Jiménez, I., Manterola-Romero, L., Osorio-Alonso, H., Arellano-Buendía, A.S., Pedraza-Chaverri, J., Roncal-Jiménez, C.A., Lanaspa, M.A. and Johnson, R.J., 2021. Osthol ameliorates kidney damage and metabolic syndrome induced by a high-fat/high-sugar diet. *International Journal of Molecular Sciences*, *22*(5), p. 2431.

Guibourt, N.J.B.G., 1836. *Histoire abrégée des drogues simples*. Méquignon-Marvis Père et fils.

Härmälä, P., Vuorela, H., Hiltunen, R., Nyiredy, S., Sticher, O., Törnquist, K. and Kaltia, S., 1992. Strategy for the isolation and identification of coumarins with calcium antagonistic properties from the roots of Angelica archangelica. *Phytochemical Analysis*, *3*(1), pp. 42–48.

Kelly, A., 1969. Phyto-photo dermatitis. *The Ulster Medical Journal*, *38*(1), p. 51.

Kimura, T., Hayashida, H., Murata, M. and Takamatsu, J., 2011. Effect of ferulic acid and Angelica archangelica extract on behavioral and psychological symptoms of dementia in frontotemporal lobar degeneration and dementia with Lewy bodies. *Geriatrics & Gerontology International*, *11*(3), pp. 309–314.

Knutsen, O.H. and Paszkowski, P., 1984. New aspects in the treatment of water hemlock poisoning. *Journal of Toxicology: Clinical Toxicology*, *22*(2), pp. 157–166.

Ogawa, H., Sasai, N., Kamisako, T. and Baba, K., 2007. Effects of osthol on blood pressure and lipid metabolism in stroke-prone spontaneously hypertensive rats. *Journal of Ethnopharmacology*, *112*(1), pp. 26–31.

Sensenbrenner, B., Rouaud, O., Graule-Petot, A., Guillemin, S., Piver, A., Giroud, M., Béjot, Y. and Jacquin-Piques, A., 2020. High prevalence of social cognition disorders and mild cognitive impairment long term after stroke. *Alzheimer Disease & Associated Disorders*, *34*(1), pp. 72–78.

Sigurdsson, S. and Gudbjarnason, S., 2013. Effect of oral imperatorin on memory in mice. *Biochemical and Biophysical Research Communications*, *441*(2), pp. 318–320.

Steck, W. and Bailey, B.K., 1969. Leaf coumarins of Angelica archangelica. *Canadian Journal of Chemistry*, *47*(13), pp. 2425–2430.

Sun, H.L., Wu, Y.W., Bian, H.G., Yang, H., Wang, H., Meng, X.M. and Jin, J., 2021. Function of uric acid transporters and their inhibitors in hyperuricaemia. *Frontiers in Pharmacology*, *12*, p. 667753.

Tashiro, Y., Sakai, R., Hirose-Sugiura, T., Kato, Y., Matsuo, H., Takada, T., Suzuki, H. and Makino, T., 2018. Effects of osthol isolated from cnidium monnieri fruit on urate transporter 1. *Molecules*, *23*(11), p. 2837.

Yan, D.Y., Tang, J., Chen, L., Wang, B., Weng, S., Xie, Z., Wu, Z.Y., Shen, Z., Bai, B. and Yang, L., 2020. Imperatorin promotes osteogenesis and suppresses osteoclast by activating AKT/GSK3 β/β-catenin pathways. *Journal of Cellular and Molecular Medicine*, *24*(3), pp. 2330–2341.

Zhou, F., Zhong, W., Xue, J., Gu, Z.L. and Xie, M.L., 2012. Reduction of rat cardiac hypertrophy by osthol is related to regulation of cardiac oxidative stress and lipid metabolism. *Lipids*, *47*, pp. 987–994.

20 Chervil (*Anthriscus cerefolium* (L.) Hoffm.)

Anthriscus cerefolium (L.) Hoffm.

Etymology: From the Greek *anthriskon* = probably the Southern chervil *Sandix australis* L. and *khairéphullon* = chervil

Family: Apiaceae

Synonyms: *Anthriscus longirostris* Bertol.; *Scandix cerefolium* L.

Common names: Chervil, cerfeuil (Fr.); kerbel (Ger.); cerefolion (Port.) Perifollo (Spa.)

Part used: Leaf

Constituents: Essential oil (methyl chavicol = estragol) (Baser et al., 1998), flavone glycosides (Slimestad et al., 2022), lignans (deoxypodophyllotoxin) (2009).

Medical history: Chervilas was known as *gingidium* by Dioscorides, who asserts it is good for the stomach and the bladder, and as a diuretic. The Medical School of Salerno (10th century) calls it *cherefolium* and recommends it for external tumors, side pains, and the stomach.

Dry to the second degree (Fusch, 1555); known of Nicholas Culpeper in 17th-century England, fever, purifies the blood, kidney stones (Lémery, 1716); diuretic (Guibourt, 1836).

Medicinal uses: Diuretic, stomachic, deobstruent (India)

Bones and cartilages: Aqueous extract given orally protected rats against the destabilization of medial meniscus cartilage destruction and proteoglycan loss. *In vitro*, this extract protected primary chondrocytes against interleukin-1β (Lee et al., 2018).

Comments: (i) Nicholas Culpeper (1616–1654) was an English physician and astrologer and the author of a book titled *The English physician or an astrologo-physical discourse of the vulgar herbs of this nation* (1651)

(ii) Deoxypodophyllotoxin *in vitro* inhibited the expression of matrix metalloproteinase-9 and the migration of human aortic smooth muscle cells induced exposed to TNF α (Suh et al., 2009). The formation of wrinkles involves the degradation of collagen by matrix metalloproteinase-9 (Tang et al., 2019).

(iii) Deoxypodophyllotoxin given orally to mice fed a high-fat diet at 10 mg/kg/day for 4 weeks prevented hepatic accumulation of cholesterol (Kim et al., 2018).

DOI: 10.1201/9781003301455-20

REFERENCES

Baser, K.H.C., Ermin, N. and Demirçakmak, B., 1998. The essential oil of Anthriscus cerefolium (L.) Hoffm. (Chervil) growing wild in Turkey. *Journal of Essential Oil Research*, *10*(4), pp. 463–464.

Fusch, L., 1555. *De Historia Stirpium Commetarii Insignes*. Lugduni Apud Ioan Tornaesium.

Guibourt, N.J.B.G., 1836. *Histoire abrégée des drogues simples*. Méquignon-Marvis Père et fils.

Kim, K.Y., Park, K.I., Lee, S.G., Baek, S.Y., Lee, E.H., Kim, S.C., Kim, S.H., Park, S.G., Yu, S.N., Oh, T.W. and Kim, J.H., 2018. Deoxypodophyllotoxin in Anthriscus sylvestris alleviates fat accumulation in the liver via AMP-activated protein kinase, impeding SREBP-1c signal. *Chemico-Biological Interactions*, *294*, pp. 151–157.

Lee, S.A., Moon, S.M., Han, S.H., Hwang, E.J., Park, B.R., Kim, J.S., Kim, D.K. and Kim, C.S., 2018. Chondroprotective effects of aqueous extract of Anthriscus sylvestris leaves on osteoarthritis in vitro and in vivo through MAPKs and NF-κB signaling inhibition. *Biomedicine & Pharmacotherapy*, *103*, pp. 1202–1211.

Lémery, N., 1716. *Traité universel des drogues simples, mises en ordre alphabétique. Où l'on trouve leurs différens noms . . . et tout ce qu'il y a de particulier dans les animaux, dans les végétaux, et dans les minéraux*. Au dépend de la Companie.

Slimestad, R., Rathe, B.A., Aesoy, R., Diaz, A.E.C., Herfindal, L. and Fossen, T., 2022. A novel bicyclic lactone and other polyphenols from the commercially important vegetable Anthriscus cerefolium. *Scientific Reports*, *12*(1), p. 7805.

Suh, S.J., Kim, J.R., Jin, U.H., Choi, H.S., Chang, Y.C., Lee, Y.C., Kim, S.H., Lee, I.S., Moon, T.C., Chang, H.W. and Kim, C.H., 2009. Deoxypodophyllotoxin, flavolignan, from Anthriscus sylvestris Hoffm. inhibits migration and MMP-9 via MAPK pathways in TNF-α-induced HASMC. *Vascular Pharmacology*, *51*(1), pp. 13–20.

Tang, S.C., Tang, L.C., Liu, C.H., Liao, P.Y., Lai, J.C. and Yang, J.H., 2019. Glycolic acid attenuates UVB-induced aquaporin-3, matrix metalloproteinase-9 expression, and collagen degradation in keratinocytes and mouse skin. *Biochemical Journal*, *476*(10), pp. 1387–1400.

21 Celery (*Apium graveolens* L.)

Apium graveolens L.

Etymology: From the Latin *api* = bee, *gravis* = heavy, and *olens* = smelling

Family: Apiaceae

Synonyms: *Apium dulce* Mill; *Apium integrilobum* Hayata; *Celeri graveolens* (L.) Britton

Common names: Celery; céreli (French); sellerie (Ger.); salsão (Port.); сельдерей (Rus.); apio (Spa.)

Parts used: Leaf, seed

Constituent: Essential oil (3-butylphthalide) (Tsi & Tan, 1997).

Medical history: For the physicians of ancient Rome, celery was used as a diuretic and emmenagogue (Dioscorides). In his *History of plants*, published in 1879, the America naturalist Charles Pickering (1805–1878) states that celery was used for sea sickness in ancient Egypt. In 19th-century India, it was used as a tonic according to O'Shaughnessy (1842).

Medicinal uses: Diuretic (Turkey; Afghanistan); carminative, aromatic, and tonic (Iraq); rheumatoid arthritis, arthralgia, hypertension (Iran); hypertension, promote circulation, dropsy, indigestion (Myanmar)

Blood pressure: Six gram of seeds given to hypertensive patients decreased their systolic blood pressure from 171.35 to 94 mmHg to 154.3 and 89.6 mmHg, respectively (Gharooni & Sarkarati, 2000). Ethanol extract of seeds given at a dose of 1.3 g/day for 4 weeks to hypertensive patients reduced systolic blood pressure from 141.2 to 130 mmHg and diastolic blood pressure from 92.2 to 84.2 mmHg (Shayani Rad et al., 2022).

3-Butylphthalide is a vasorelaxant because it inhibits the entry of calcium into smooth muscle vascular cells (Tsi & Tan, 1997). Given orally to spontaneously hypertensive rats at 30 mg/kg/day for 20 weeks, 3-butylphthalide decreased blood pressure (Zhu et al., 2015).

Kidneys: Within 6 hours, a single 200 mg oral dose of ethanol extract of leaves given to rats increased urine volume from 0.8 to 1.9 mL and increased the urinary excretion of sodium ions from 55.8 to 61 mmol/L (Ali, 2018). Given orally to spontaneously hypertensive rats at 30 mg/kg/day for 20 weeks, 3-butylphthalide prevented glomerulosclerosis (Zhu et al., 2015).

Brain: 3-Butylphthalide given orally at 15 mg/kg/day for 18 weeks to a mouse transgenic model of Alzheimer's disease inhibited brain amyloid β plaque deposition as well as amyloid β levels and enhanced soluble amyloid precursor protein secretion (Tan et al., 2023).

 DOI: 10.1201/9781003301455-21

Warning: Anaphylactic shock after celery consumption can happen in allergic subjects (Pałgan et al., 2018).

Comments: (i) William Brooke O'Shaughnessy (1809–1889) was an Irish physician and professor of materia medica at the medical college of Calcutta. He wrote a book titled *Bengal dispensatory* (1842).

(ii) It is clear that celery has the ability to decrease blood pressure. Consuming celery regularly at normal dietary dose might assist with controlling hypertension and prevent premature cardiovascular aging.

REFERENCES

Ali, I., 2018. Comparative evaluation of diuretic activity of ethanolic extracts of (celery) Apium graveolens and (Parsley) Petroselinum crispum in male rats. *Indian Journal of Natural Sciences*, 9(50).

Gharooni, M. and Sarkarati, A.R., 2000. Application of Apium graveolens in treatment of hypertension. *Tehran University Medical Journal TUMS Publications*, 58(3), pp. 67–69.

O'Shaughnessy, W.B., 1842. *The Bengal Dispensatory and Companion to the Pharmacopœia... Chiefly Compiled ... by W. B. O'Shaughnessy, Etc. (1842)*. Calcutta Printed.

Pałgan, K., Żbikowska-Gotz, M. and Bartuzi, Z., 2018. Dangerous anaphylactic reaction to mustard. *Archives of Medical Science*, 14(2), pp. 477–479.

Shayani Rad, M., Moohebati, M. and Mohajeri, S.A., 2022. Effect of celery (Apium graveolens) seed extract on hypertension: A randomized, triple-blind, placebo-controlled, cross-over, clinical trial. *Phytotherapy Research*, 36(7), pp. 2889–2907.

Tan, T.Y.C., Lim, X.Y., Norahmad, N.A., Chanthira Kumar, H., Teh, B.P., Lai, N.M. and Syed Mohamed, A.F., 2023. Neurological applications of celery (Apium graveolens): A scoping review. *Molecules*, 28(15), p. 5824.

Tsi, D. and Tan, B.K.H., 1997. Cardiovascular pharmacology of 3-n-butylphthalide in spontaneously hypertensive rats. *Phytotherapy Research: An International Journal Devoted to Medical and Scientific Research on Plants and Plant Products*, 11(8), pp. 576–582.

Zhu, J., Zhang, Y. and Yang, C., 2015. Protective effect of 3-n-butylphthalide against hypertensive nephropathy in spontaneously hypertensive rats. *Molecular Medicine Reports*, 11(2), pp. 1448–1454.

22 Bearberry (*Arctostaphylos uva-ursi* (L.) Spreng.)

Arctostaphylos uva-ursi (L.) **Spreng.**

Etymology: From the Greek *arco* = bear, *staphyle* = grape and the Latin *uva* = grape and *ursi* = bears

Family: Ericaceae

Synonym: *Arbutus uva-ursi* L.

Common names: Bearberry; busserole, raisin d'ours (Fr.); bärentraubenblätter (Ger.); uva-ursina (Port.); толокнянка (Rus.); gayuba (Spain)

Part used: Leaf

Constituents: Phenolic glycosides (arbutin) (Quintus et al., 2005).

Medical history: In 17th-century France, bearberry was used to treat kidney diseases. The 18th-century Austrian physician Antonius De Haen Gravenhage in his book *"Rationis mediendi"* (1760) recommends bearberry for urinary stones. In 19th-century France it was used for urinary stones, diarrhea, gonorrhea, leucorrhea, urinary tract diseases, and enlargement of prostate and to tan skins (Guibourt, 1836);

Medicinal uses: Urinary tract infection, 1.5–4 g of leaves in 150 mL of boiling water 2 to 4 times daily (European Union); chest pain (Pakistan)

Kidneys: Aqueous extract displayed diuretic and natriuretic effects (Beaux et al., 1999).

Warning: Nausea, vomiting, and stomachache have been reported, as well as interaction with cytochrome p450 enzymes (Chauhan et al., 2007).

REFERENCES

Beaux, D., Fleurentin, J. and Mortier, F., 1999. Effect of extracts of Orthosiphon stamineus benth, Hieracium pilosella L., Sambucus nigra L. and Arctostaphylos uva-ursi (L.) spreng. in rats. *Phytotherapy Research*, 13(3), pp. 222–225.

Chauhan, B., Yu, C., Krantis, A., Scott, I., Arnason, J.T., Marles, R.J. and Foster, B.C., 2007. In vitro activity of uva-ursi against cytochrome P450 isoenzymes and P-glycoprotein. *Canadian Journal of Physiology and Pharmacology*, 85(11), pp. 1099–1107.

Guibourt, N.J.B.G., 1836. *Histoire abrégée des drogues simples*. Méquignon-Marvis Père et fils.

Quintus, J., Kovar, K.A., Link, P. and Hamacher, H., 2005. Urinary excretion of arbutin metabolites after oral administration of bearberry leaf extracts. *Planta Medica*, 71(02), pp. 147–152.

DOI: 10.1201/9781003301455-22

23 Burdock (*Arctium lappa* L.)

Arctium lappa L.

Etymology: From the Greek *arcktos* = bear and the Latin *lappa* = burdock

Family: Asteraceae

Synonyms: *Arctium majus* (Gaertn.) Bernh.*; Lappa major* Gaertn.

Common names: Burdock, great clot-bur; bardane à grosse tête, glouteron, herbe aux teigneux (Fr.); klettenwurzel (Ger.); bardana (Port.; Spa.); лопух (Rus.)

Parts used: Root, seed

Constituents: Lignans (arctigenin, arctiin), fructan-type polysaccharides (inulin) (Milani et al., 2011).

Medical history: Burdock was used for toothache, sciatica, diuretic, bronchitis, articular pains, and ulcers by Roman physicians, and Dioscorides says of it that it is good for wounds: *ictibus auxilliatur*. The Romans were comparing the leaves to a human face, which has the Latin name "*personata*". In 17th-century England, it was used to induce urination and to ease sciatica and for kidney stones (Parkinson, 1640). In 18th-century France, a beer of burdock was taken for aging. In 19th-century France, it was used for syphilis, skin diseases, and rheumatism (Guibourt, 1836).

Medicinal uses: 2–6 g of roots as an infusion 3 times daily to induce urination (European Union); blood pressure, joint pain including gout and rheumatism (Pakistan); pneumonia (China)

Blood pressure: Ethanol extract of seeds given orally at 200 mg/kg/day for 6 weeks to high-fat-diet poisoned rats decreased systolic blood pressure from about 145 to 120 mmHg (Lee et al., 2012). This regimen improved acetylcholine-induced relaxation of aortic rings and maintained smooth and flexible intimal endothelial layers via the reduction of fat-induced vascular inflammation (Lee et al., 2012). Arctigenin given orally at 50 mg/kg to spontaneously hypertensive rats once a day for a total of 8 weeks decreased systolic blood pressure and improved endothelial function via increased NO production by enhancing endothelial NO synthetase (Liu et al., 2015).

Three cups of a tea prepared by steeping 2 tea bags of roots in 150 mL of boiling water for 10 min taken daily 30 mins after meals for 6 weeks decreased plasma cholesterol, LDL-cholesterol, and blood pressure in patients with knee osteoarthritis (Maghsoumi-Norouzabad et al., 2019).

Plasma lipids and glucose: Ethanol extract of seeds given orally at 100 mg/kg/day for 6 weeks to high-fat diet poisoned rats decreased plasma triglycerides

DOI: 10.1201/9781003301455-23

from 67.2 to 33.8 mg/dL, LDL-cholesterol from 120.8 to 113.4 mg/dL, and increased HDL-cholesterol from 20.8 to 25 mg/dL (Lee et al., 2012). Three cups of a tea prepared by steeping 2 tea bags of roots in 150 mL of boiling water for 10 min taken daily 30 mins after meals for 6 weeks decreased plasma cholesterol and LDL-cholesterol (Maghsoumi-Norouzabad et al., 2019).

Aqueous extract of roots given orally at 250 mg/day for 8 weeks to rats on a high-fat diet decreased glycemia from 474 to 230 mg/dL without side effects (Bok et al., 2017).

Bones and cartilages: Aqueous extract of roots at the concentration of 100 µg/mL induced the proliferation of human bone marrow-derived mesenchymal stem cells (Wu et al., 2020). Infusion of roots (2 g/150 mL boiled water) given 3 times a day half an hour after meals for 42 days to patients with knee osteoarthritis caused decreased plasma interleukin-6, hr-CRP, and MDA, while plasma superoxide dismutase activity increased (Maghsoumi-Norouzabad et al., 2016).

Skin and hair: An extract of fruits used externally for 12 weeks increased procollagen synthesis and increased hyaluronan synthase-2 expression as well as hyaluronan levels and decreased the volume of facial wrinkles (Knott et al., 2008). Arctiin at 50 µM induced procollagen type-1 synthesis while inhibiting matrix metalloproteinase-1 by human dermal fibroblasts (Hwang et al., 2012).

Brain: Ethanol extract of roots given orally at 1.2 g/kg/day for 4 weeks protected rats against dementia, induced the intracerebrovascular injection of amyloid β proteins, and decreased cerebral MDA to normal levels (Kwon et al., 2016).

Arctigenin given orally at 60 mg/kg protected mice against scopolamine-induced memory deficits by 73% (Lee et al., 2011).

Warning: Some individuals are allergic to burdock with risk of anaphylactic shock (Sasaki et al., 2003).

Comments: (i) With aging and central obesity, the plasma levels of cytokines and inflammatory mediators increase, leading to cardiovascular diseases, type-2 diabetes, and chronic diseases (Ferrucci & Fabbri, 2018). Burdock is a gentle anti-inflammatory (Gao et al., 2018), and intake of roots of burdock as infusions could potentially delay "inflamma-aging", vascular aging, hyperlipidemia, and hyperglycemia.

(ii) With aging and ultraviolet B radiations, the concentration of ROS in the dermis increases leading to the activation of mitogen-activated protein kinase (MAPK) in fibroblasts and the subsequent activation of the transcription factor activator protein 1 (AP-1), the production of NF-kB, and downstream secretion of matrix metalloproteinase-1, -3, and -9 in the dermis and degradation of collagen. In parallel, activation of dermis macrophages by ultraviolet B and other pro-inflammatory stimuli produces matrix metalloproteinase-12, which catalyzes the degradation of elastin (Shin et al., 2019).

(iii) The degradation of elastin in carotids with aging by matrix metalloproteinase-2 results in progressive aortic stiffness (Diaz-Canestro et al., 2022).
(iv) John Parkinson (1567–1650) was an English apothecary, herbalist to the king, and author of a book titled *Theatrum botanicum* (1640).

REFERENCES

Bok, S.H., Cho, S.S., Bae, C.S., Park, D.H. and Park, K.M., 2017. Safety of 8-weeks oral administration of Arctium lappa L. *Laboratory Animal Research*, *33*, pp. 251–255.
Diaz-Canestro, C., Puspitasari, Y.M., Liberale, L., Guzik, T.J., Flammer, A.J., Bonetti, N.R., Wüst, P., Costantino, S., Paneni, F., Akhmedov, A. and Varga, Z., 2022. MMP-2 knockdown blunts age-dependent carotid stiffness by decreasing elastin degradation and augmenting eNOS activation. *Cardiovascular Research*, *118*(10), pp. 2385–2396.
Ferrucci, L. and Fabbri, E., 2018. Inflammageing: Chronic inflammation in ageing, cardiovascular disease, and frailty. *Nature Reviews Cardiology*, *15*(9), pp. 505–522.
Gao, Q., Yang, M. and Zuo, Z., 2018. Overview of the anti-inflammatory effects, pharmacokinetic properties and clinical efficacies of arctigenin and arctiin from Arctium lappa L. *Acta Pharmacologica Sinica*, *39*(5), pp. 787–801.
Guibourt, N.J.B.G., 1836. *Histoire abrégée des drogues simples.* Méquignon-Marvis Père et fils.
Hwang, J.Y., Park, T.S., Kim, D.H., Hwang, E.Y., Lee, J.N., Young Lee, J., Lee, G.T., Lee, K. and Son, J.H., 2012. Anti-wrinkle compounds isolated from the seeds of Arctium lappa L. 생명과학회지, *22*(8), pp. 1092–1098.
Knott, A., Reuschlein, K., Mielke, H., Wensorra, U., Mummert, C., Koop, U., Kausch, M., Kolbe, L., Peters, N., Stäb, F. and Wenck, H., 2008. Natural Arctium lappa fruit extract improves the clinical signs of aging skin. *Journal of Cosmetic Dermatology*, *7*(4), pp. 281–289.
Kwon, Y.K., Choi, S.J., Kim, C.R., Kim, J.K., Kim, Y.J., Choi, J.H., Song, S.W., Kim, C.J., Park, G.G., Park, C.S. and Shin, D.H., 2016. Antioxidant and cognitive-enhancing activities of Arctium lappa L. roots in Aβ1–42-induced mouse model. *Applied Biological Chemistry*, *59*(4), pp. 553–565.
Lee, I.A., Joh, E.H. and Kim, D.H., 2011. Arctigenin isolated from the seeds of Arctium lappa ameliorates memory deficits in mice. *Planta Medica*, *77*(13), pp. 1525–1527.
Lee, Y.J., Choi, D.H., Cho, G.H., Kim, J.S., Kang, D.G. and Lee, H.S., 2012. Arctium lappa ameliorates endothelial dysfunction in rats fed with high fat/cholesterol diets. *BMC Complementary and Alternative Medicine*, *12*(1), pp. 1–10.
Liu, Y., Wang, G., Yang, M., Chen, H., Yang, S. and Sun, C., 2015. Arctigenin reduces blood pressure by modulation of nitric oxide synthase and NADPH oxidase expression in spontaneously hypertensive rats. *Biochemical and Biophysical Research Communications*, *468*(4), pp. 837–842.
Maghsoumi-Norouzabad, L., Alipoor, B., Abed, R., Eftekhar Sadat, B., Mesgari-Abbasi, M. and Asghari Jafarabadi, M., 2016. Effects of Arctium lappa L. (Burdock) root tea on inflammatory status and oxidative stress in patients with knee osteoarthritis. *International Journal of Rheumatic Diseases*, *19*(3), pp. 255–261.
Maghsoumi-Norouzabad, L., Shishehbor, F., Abed, R., Javid, A.Z., Eftekhar-Sadat, B. and Alipour, B., 2019. Effect of Arctium lappa linne (Burdock) root tea consumption on lipid profile and blood pressure in patients with knee osteoarthritis. *Journal of Herbal Medicine*, *17*, p. 100266.

Milani, E., Koocheki, A. and Golimovahhed, Q.A., 2011. Extraction of inulin from Burdock root (Arctium lappa) using high intensity ultrasound. *International Journal of Food Science & Technology*, *46*(8), pp. 1699–1704.

Parkinson, J., 1640. *Theatrum Botanicum: The Theater of Plants: Or, An Herball of Large Extent: Containing Therein a More Ample and Exact History and Declaration of the Physicall Herbs and Plants . . . Distributed Into Sundry Classes Or Tribes, for the More Easie Knowledge of the Many Herbes of One Nature and Property*. Tho. Cotes.

Sasaki, Y., Kimura, Y., Tsunoda, T. and Tagami, H., 2003. Anaphylaxis due to burdock. *International Journal of Dermatology*, *42*(6), pp. 472–473.

Shin, J.W., Kwon, S.H., Choi, J.Y., Na, J.I., Huh, C.H., Choi, H.R. and Park, K.C., 2019. Molecular mechanisms of dermal aging and antiaging approaches. *International Journal of Molecular Sciences*, *20*(9), p. 2126.

Wu, K.C., Weng, H.K., Hsu, Y.S., Huang, P.J. and Wang, Y.K., 2020. Aqueous extract of Arctium lappa L. root (burdock) enhances chondrogenesis in human bone marrow-derived mesenchymal stem cells. *BMC Complementary Medicine and Therapies*, *20*(1), pp. 1–14.

24 Horseradish (*Armoracia rusticana* G. Gaertn., B. Mey. & Scherb.)

Armoracia rusticana G. Gaertn., B. Mey. & Scherb.

Etymology: From the Latin *armoracia* = horseradish and *rusticana* = rural

Family: Brassicaceae

Synonyms: *Cochlearia armoracia* L.; *Cochlearia rusticana* Lam.; *Rorippa rusticana* (G. Gaertn., B. Mey. & Scherb.) Godr.

Common names: Horsedish; cran de Bretagne, raifort sauvage (Fr.); meerrettich (Ger.); rábano (Port.); хрен (Rus.); rábano picante (Spa.)

Part used: Root

Constituents: Glucosinolates (sinigrin) (Ciska et al., 2017).

Medical history: Roman physicians, including Dioscorides, used horseradish as a diuretic. In 19th-century France, it was used as a strong tonic and antiscorbutic (Guibourt, 1836)

Plasma lipids and glucose: The addition of 1% of horseradish to a high-fat diet for 3 weeks prevented a rise in plasma cholesterol (Balasinska et al., 2005).

Kidneys: Ethanol extract of roots given to rats with partial nephrectomy at the dose of 400 mg/kg/day for 15 days decreased plasma creatinine, urea, and uric acid and increased urine excretion (Antony et al., 2010).

Warnings: Excessive intake of horseradish has the potential to induce fainting caused by vasodepressant effects (Rubin & Wu, 1988). Patients with thyroid disorders must avoid consuming glucosinolates.

REFERENCES

Antony, A.S., Jayasankar, K., Roy, P.D., Pankaj, N., Dhamodaran, P., Duraiswamy, B. and Elango, K., 2010. Renal protectant activity of Cochlearia armoracia in 5/6-nephrectomized rat model. *Research Journal of Pharmacology and Pharmacodynamics*, 2(4), pp. 300–302.

Balasinska, B., Nicolle, C., Gueux, E., Majewska, A., Demigne, C. and Mazur, A., 2005. Dietary horseradish reduces plasma cholesterol in mice. *Nutrition Research*, 25(10), pp. 937–945.

DOI: 10.1201/9781003301455-24

Ciska, E., Horbowicz, M., Rogowska, M., Kosson, R., Drabinska, N. and Honke, J., 2017. Evaluation of seasonal variations in the glucosinolate content in leaves and roots of four European horseradish (Armoracia rusticana) landraces. *Polish Journal of Food and Nutrition Sciences*, *67*(4).

Guibourt, N.J.B.G., 1836. *Histoire abrégée des drogues simples*. Méquignon-Marvis Père et fils.

Rubin, H.R. and Wu, A.W., 1988. The bitter herbs of Seder: More on horseradish horrors. *JAMA*, *259*(13), pp. 1943–1943.

25 Southernwood (*Artemisia abrotanum* L.)

Artemisia abrotanum L.

Etymology: From the Greek *artemisia* because the plant might have been sacred to the Greek deity *Artemis* and *a-* = without and *broton* = mortal

Family: Asteraceae

Common names: Southernwood, soothing wort; aurone des jardins, aurone citronelle, garderobe (Fr.); eberautte (Ger.); abrótono (Port.); abrótano (Spa.)

Part used: Leaf

Constituent: Essential oil (Piperitone) (Saunoriūtė et al., 2020).

Medical history: Southernwood was considered as a dry and hot medicine, diuretic, and useful for inflamed eyes by Dioscorides.

Medicinal uses: Jaundice (Bosnia); sedative, anthelmintic, fever orexigenic, antipyretic, diuretic (Turkey); anthelminthic (Iran)

Gallbladder: The plant produces choleretic principles (Nieschulz & Schmersahl, 1968).

REFERENCES

Nieschulz, O. and Schmersahl, P., 1968. On choleretic agents from Artemisia abrotanum L. *Arzneimittel-Forschung*, *18*(10), pp. 1330–1336.

Saunoriūtė, S., Ragažinskienė, O., Ivanauskas, L. and Marksa, M., 2020. Essential oil composition of Artemisia abrotanum L. during different vegetation stages in Lithuania. *Chemija*, *31*(1).

DOI: 10.1201/9781003301455-25

26 Wormwood (*Artemisia absinthium* L.)

Artemisia absinthium L.

Etymology: From the Greek *artemisia* because the plant might have been sacred to the Greek deity *Artemis* and *apsinthion* = wormwood

Family: Asteraceae

Synonyms: *Absinthium officinale* Brot.; *Absinthium vulgare* Lam.

Common names: Wormwood; armoise amère, absinthe amere, grande absinthe, aluyne (Fr.); wermuth (Ger.); absinto (Port.); полыньгорькая (Rus.); absintio (Spa.)

Part used: Leaf

Constituents: Essential oil (thujone) (Nguyen et al., 2019), sesquiterpene lactones (Turak et al., 2014).

Medical history: Wormwood was used by Roman physicians as a diuretic, carminative, as a remedy for jaundice, toothache, dropsy, and headache (Dioscorides). The Medical School of Salerno (10th century) advocates wormwood for seasickness, for the stomach and nerves, and for ringing of the ear. In 19th-century France, it was used as an anthelminthic and a carminative (Guibourt, 1836), but it was responsible for a vicious form of intoxication, or "absinthism" (including fetal microcephaly), in the working class of France until its ban in 1915.

Medicinal uses: Appetite stimulant, carminative at 1–1.5 g in 150 mL of boiling water for 2 weeks maximum (European Union); diabetes, hyperlipidemia, hypertension (Afghanistan); sedative, anthelmintic, for fever, diuretic, carminative, hypertension, diabetes (Pakistan)

Heart: Methanol extract exhibited protective effects against supraventricular tachyarrhythmia in isolated heart of rats (Khori & Nayebpour, 2007).

Plasma lipids and glucose: Two capsules containing 500 mg of wormwood powder given twice a day to type-2 diabetic patients for 30 days reduced fasting glucose from 211 to 191 mg/dL (Li et al., 2015).

Gallbladder: Aqueous extract given to rats increased the secretion of bile (Sharma & Shukla, 2012).

Immune system: Patients suffering from Crohn's disease given 750 mg of powder 3 times a day for 6 weeks had a decrease in plasma TNF α from 24.5 to 8 pg/mL (Krebs et al., 2010).

Warnings: Essential oil is neurotoxic because of thujone (Lachenmeier et al., 2006), and in rats, a dose of 25 mg/day for 13 weeks induced seizures (Hudson et al., 2018). Liquors of absinth taken in excess induce

     DOI: 10.1201/9781003301455-26

Wenckebach-type atrioventricular block (Benezet-Mazuecos & de la Fuente, 2006). Drug interaction with warfarin (Açıkgöz & Açıkgöz, 2013).

REFERENCES

Açıkgöz, S.K. and Açıkgöz, E., 2013. Gastrointestinal bleeding secondary to interaction of Artemisia absinthium with warfarin. *Drug Metabolism and Drug Interactions*, *28*(3), pp. 187–189.

Benezet-Mazuecos, J. and de la Fuente, A., 2006. Electrocardiographic findings after acute absinthe intoxication. *International Journal of Cardiology*, *113*(2), pp. E48–E50.

Guibourt, N.J.B.G., 1836. *Histoire abrégée des drogues simples*. Méquignon-Marvis Père et fils.

Hudson, A., Lopez, E., Almalki, A.J., Roe, A.L. and Calderón, A.I., 2018. A review of the toxicity of compounds found in herbal dietary supplements. *Planta Medica*, *84*(09/10), pp. 613–626.

Khori, V. and Nayebpour, M., 2007. Effect of Artemisia absinthium on electrophysiological properties of isolated heart of rats. *Physiology and Pharmacology*, *10*(4), pp. 303–311.

Krebs, S., Omer, T.N. and Omer, B., 2010. Wormwood (Artemisia absinthium) suppresses tumour necrosis factor alpha and accelerates healing in patients with Crohn's disease–a controlled clinical trial. *Phytomedicine*, *17*(5), pp. 305–309.

Lachenmeier, D.W., Emmert, J., Kuballa, T. and Sartor, G., 2006. Thujone—cause of absinthism? *Forensic Science International*, *158*(1), pp. 1–8.

Li, Y., Zheng, M.I.N., Zhai, X., Huang, Y., Khalid, A., Malik, A., Shah, P., Karim, S., Azhar, S. and Hou, X., 2015. Effect of Gymnema sylvestre, Citrullus colocynthis and Artemisia absinthium on blood glucose and lipid profile in diabetic human. *Acta Pol Pharm*, *72*(5), pp. 981–985.

Nguyen, H.T., Radácsi, P., Rajhárt, P. and Németh, É., 2019. Variability of thujone content in essential oil due to plant development and organs from Artemisia absinthium L. and Salvia officinalis L. *Journal of Applied Botany and Food Quality*, *92*, pp. 100–105.

Sharma, M. and Shukla, S., 2012. Reversal of carbon tetrachloride-induced hepatic injury by aqueous extract of Artemisia absinthium in rats. *Journal of Environmental Pathology, Toxicology and Oncology*, *31*(4).

Turak, A., Shi, S.P., Jiang, Y. and Tu, P.F., 2014. Dimeric guaianolides from Artemisia absinthium. *Phytochemistry*, *105*, pp. 109–114.

27 Tarragon (*Artemisia dracunculus* L.)

Artemisia dracunculus L.

Etymology: From Greek *artemisia* because the plant might have been sacred to the Greek deity *Artemis* and Latin *draco* = snake

Family: Asteraceae

Synonym: *Oligosporus dracunculus* (L.) Poljakov

Common names: Tarragon; estragon (Fr.; Ger.); serpentine, targon (Fr.); estragão (Port.); эстрагон (Rus.); estragón (Spa.)

Part used: Leaf

Constituents: Coumarins (herniarin, capillarin), alkylamine alkaloids (pellitorine) (Saadali et al., 2001), essential oil (methyl chavicol) (Eisenman et al., 2013), flavanones (Sayyah et al., 2004; Bhutia et al., 2008).

Medical history: Matthioli (1572) defines tarragon as hot, and it was used as a salad during his time. In 19th-century France, it was used as a tonic and antiseptic (Guibourt, 1836).

Medicinal uses: Appetite stimulant, menstrual disorders, constipation (Turkey); stomachache (India); carminative (Myanmar)

Blood pressure: Tarragon given orally at 1 g before breakfast and dinner for 90 days to patients with impaired glucose tolerance induced a decrease in systolic blood pressure from 120 to 113 mmHg (Mendez-del Villar et al., 2016).

Plasma lipid and glucose: Ethanol extract given orally to streptozotocin-induced diabetic rats at 250 mg/kg for 28 days decreased fasting blood glucose from 215.3 to 149 mg/dL and LDL from 113.3 to 74.2 mg/dL (Samyal et al., 2011). Alcoholic extract stimulated insulin release from β cells (Aggarwal et al., 2015) and increased uptake of glucose by primary skeletal muscle cells in subjects with type 2 diabetes mellitus (Wang et al., 2008). Tarragon given orally at 1 g before breakfast and dinner for 90 days to patients with impaired glucose tolerance induced a decrease in insulin secretion (Mendez-del Villar et al., 2016).

Platelets: A coumarin-enriched extract evoked anticoagulant effects (Duric et al., 2015).

Brain: Aqueous extract given orally to mice decreased symptoms of multiple sclerosis together with decreasing interleukin-17 and -23 and increased plasmatic antioxidant capacity (Safari et al., 2021).

Warnings: Maximum daily dose 10 g (Obolskiy et al., 2011).

 DOI: 10.1201/9781003301455-27

Comments: (i) Tarragon could be added to salads, as is done in France, or taken occasionally as a light tea to control glycemia and blood pressure.

(ii) Hydroalcoholic extract of tarragon increased plasma magnesium in mice (Hosseini et al., 2016). With aging (above the age of 50), there is a decrease in mitochondrial function (measured by mitochondrial or mtDNA) that is involved, at least in part, in insulin secretion (Nile et al., 2014). Magnesium ions are essential for mitochondrial function (Killilea & Maier, 2008), and intake of magnesium might be able to preserve pancreatic insulin production in elderly persons (Barbagallo et al., 1997). Food supplements and other herbal products produced under strict pharmaceutical control must be taken according to the posology recommended by the manufacturer and it is highly recommended to consult a medical doctor before taking any herbal or even food supplements. As for self-medication and especially making teas of any plants self-collected in gardens forests, fields, etc., must be completely avoided in order to prevent dreadful poisonings and the possibility of fatalities. One must avoid at all cost to play the sorcerer's apprentice with plants. Medicinal plant identification, prescription, dosage, mode of administration is an art that should not fall under the responsibility of quacks and charlatans but well-trained pharmacists.

REFERENCES

Aggarwal, S., Shailendra, G., Ribnicky, D.M., Burk, D., Karki, N. and Wang, M.Q., 2015. An extract of Artemisia dracunculus L. stimulates insulin secretion from β cells, activates AMPK and suppresses inflammation. *Journal of Ethnopharmacology*, *170*, pp. 98–105.

Barbagallo, M., Resnick, L.M., Dominguez, L.J. and Licata, G., 1997. Diabetes mellitus, hypertension and ageing: The ionic hypothesis of ageing and cardiovascular-metabolic diseases. *Diabetes & Metabolism*, *23*(4), pp. 281–294.

Bhutia, T.D. and Valant-Vetschera, K.M., 2008. Chemodiversity of Artemisia dracunculus L. from Kyrgyzstan: Isocoumarins, coumarins, and flavonoids from aerial parts. *Natural Product Communications*, *3*(8), p. 1934578X0800300811.

Duric, K., Kovac-Besovic, E.E., Niksic, H., Muratovic, S. and Sofic, E., 2015. Anticoagulant activity of some Artemisia dracunculus leaf extracts. *Bosnian Journal of Basic Medical Sciences*, *15*(2), p. 9.

Eisenman, S.W., Juliani, H.R., Struwe, L. and Simon, J.E., 2013. Essential oil diversity in North American wild tarragon (Artemisia dracunculus L.) with comparisons to French and Kyrgyz tarragon. *Industrial Crops and Products*, *49*, pp. 220–232.

Guibourt, N.J.B.G., 1836. *Histoire abrégée des drogues simples*. Méquignon-Marvis Père et fils.

Hosseini, S.E., Mehrabani, D. and Mirshekari, S., 2016. The effect of tarragon hydroalcoholic extract on the rate of blood cations and infants sexual ratio in adult mice. *Medical Journal of Tabriz University of Medical Sciences*, *38*(5), pp. 22–27.

Killilea, D.W. and Maier, J.A., 2008. A connection between magnesium deficiency and aging: New insights from cellular studies. *Magnesium Research: Official Organ of the International Society for the Development of Research on Magnesium*, *21*(2), p. 77.

Matthioli, P.A., 1572. *Commentaires sur les Six Livres de Pedacius Dioscorides Anazarbeen de la matière medicinale*. A l'Escue de Milan.

Mendez-del Villar, M., Puebla-Perez, A.M., Sanchez-Pena, M.J., Gonzalez-Ortiz, L.J., Martinez-Abundis, E. and Gonzalez-Ortiz, M., 2016. Effect of Artemisia dracunculus administration on glycemic control, insulin sensitivity, and insulin secretion in patients with impaired glucose tolerance. *Journal of Medicinal Food*, *19*(5), pp. 481–485.

Nile, D.L., Brown, A.E., Kumaheri, M.A., Blair, H.R., Heggie, A., Miwa, S., Cree, L.M., Payne, B., Chinnery, P.F., Brown, L. and Gunn, D.A., 2014. Age-related mitochondrial DNA depletion and the impact on pancreatic Beta cell function. *PLoS One*, *9*(12), p. e115433.

Obolskiy, D., Pischel, I., Feistel, B., Glotov, N. and Heinrich, M., 2011. Artemisia dracunculus L. (tarragon): A critical review of its traditional use, chemical composition, pharmacology, and safety. *Journal of Agricultural and Food Chemistry*, *59*(21), pp. 11367–11384.

Saadali, B., Boriky, D., Blaghen, M., Vanhaelen, M. and Talbi, M., 2001. Alkamides from Artemisia dracunculus. *Phytochemistry*, *58*(7), pp. 1083–1086.

Safari, H., Anani Sarab, G. and Naseri, M., 2021. Artemisia dracunculus L. modulates the immune system in a multiple sclerosis mouse model. *Nutritional Neuroscience*, *24*(11), pp. 843–849.

Samyal, M.L., Kumar, H., Khokra, S.L., Parashar, B., Sahu, R.K. and Ahmed, Z., 2011. Evaluation of antidiabetic and antihyperlipidemic effects of Artemisia dracunculus extracts in streptozotocin-induced-diabetic rats. *Pharmacologyonline*, *2*, pp. 1230–1237.

Sayyah, M., Nadjafnia, L. and Kamalinejad, M., 2004. Anticonvulsant activity and chemical composition of Artemisia dracunculus L. essential oil. *Journal of Ethnopharmacology*, *94*(2–3), pp. 283–287.

Wang, Z.Q., Ribnicky, D., Zhang, X.H., Raskin, I., Yu, Y. and Cefalu, W.T., 2008. Bioactives of Artemisia dracunculus L enhance cellular insulin signaling in primary human skeletal muscle culture. *Metabolism*, *57*, pp. S58–S64.

28 Mugwort (*Artemisia vulgaris* L.)

Artemisia vulgaris L.

Etymology: From the Greek *artemisia* because the plant might have been sacred to the Greek deity *Artemis* and the Latin *vulgaris* = common

Family: Asteraceae

Synonym: *Artemisia opulenta* Pamp.

Common names: Mugwort; armoise vulgaire, herbe de Saint-Jean (Fr.); beifusswurzel (Gr.); artemísia (Port.); полынь (Rus.); artemisia (Spa.)

Part used: Aerial part

Constituents: Essential oil (borneol, 1,8-cineole, camphor, thujone) (Jerkovic et al., 2003), sesquiterpene lactones (Chen et al., 2022), flavanones (eriodyctiol), flavone glycosides (vitexin) (Lee et al., 1998; Malik et al., 2019).

Medical history: Mugwort was used to induce menses, as a remedy for inflammation and kidney stones by Romans physicians. In 16th-century Germany, Fusch (1555) considers mugwort as hot and dry and recommends its use to expels worms from the intestines. In 19th-century France, it was used as carminative, for intestinal worms, jaundice, liver diseases, and dropsy (Guibourt, 1836).

Medicinal uses: Carminative (Iraq); cuts, bruises, nose bleeding, antiseptic (India); helminth (Nepal); menstrual pain (Vietnam); cough, fever, sore throat, colds, phlegm (the Philippines)

Plasma lipids and cholesterol: Methanol extract given orally at 100 mg/kg/day for 4 weeks to rats poisoned with a high-fat diet reduced plasma triglycerides from 93 to 59.3 mg/dL, total cholesterol from 124 to 72.2 mg/dL, and LDL-cholesterol from 89 to 39 mg/dL and increased HDL-cholesterol from 21.5 to 28.1 mg/dL (El-Tantawy, 2015).

Essential oil given at 50 mg/kg/day for 30 days to rats on a high-fat diet decreased plasma cholesterol and triglycerides (Khan, 2015).

Immune system: Oral administration of ethanol extract at 400 mg/kg/day for 7 days to rats caused an increase in phagocytic activity and increased leukocytes count (neutrophils) (Marbun & Suwarso, 2018).

Skin and hair: Aqueous extract induced collagen synthesis *in vitro* (Park et al., 2019).

Warning: The plant contains thujone (Hudson et al., 2018).

Comment: Flavonoids in this herb have estrogenic effects (Lee et al., 1998). In post-menopausal women, lack of estradiol translates into vascular dysfunction (Wingrov & Stevenson, 1997).

DOI: 10.1201/9781003301455-28

REFERENCES

Chen, X.Y., Liu, T., Hu, Y.Z., Qiao, T.T., Wu, X.J., Sun, P.H., Qian, C.W., Ren, Z., Zheng, J.X. and Wang, Y.F., 2022. Sesquiterpene lactones from Artemisia vulgaris L. as potential NO inhibitors in LPS-induced RAW264.7 macrophage cells. *Frontiers in Chemistry*, *10*, p. 948714.

El-Tantawy, W.H., 2015. Biochemical effects, hypolipidemic and anti-inflammatory activities of Artemisia vulgaris extract in hypercholesterolemic rats. *Journal of Clinical Biochemistry and Nutrition*, *57*(1), pp. 33–38.

Fusch, L., 1555. *De Historia Stirpium Commetarii Insignes*. Lugduni Apud Ioan Tornaesium.

Guibourt, N.J.B.G., 1836. *Histoire abrégée des drogues simples*. Méquignon-Marvis Père et fils.

Hudson, A., Lopez, E., Almalki, A.J., Roe, A.L. and Calderón, A.I., 2018. A review of the toxicity of compounds found in herbal dietary supplements. *Planta Medica*, *84*(09/10), pp. 613–626.

Jerkovic, I., Mastelic, J., Milos, M., Juteau, F., Masotti, V. and Viano, J., 2003. Chemical variability of Artemisia vulgaris L. essential oils originated from the Mediterranean area of France and Croatia. *Flavour and Fragrance Journal*, *18*(5), pp. 436–440.

Khan, K.A., 2015. A preclinical antihyperlipidemic evaluation of Artemisia vulgaris root in diet induced hyperlipidemic animal model. *International Journal of Pharmacology Research*, *5*, pp. 110–114.

Lee, S.J., Chung, H.Y., Maier, C.G.A., Wood, A.R., Dixon, R.A. and Mabry, T.J., 1998. Estrogenic flavonoids from Artemisia vulgaris L. *Journal of Agricultural and Food Chemistry*, *46*(8), pp. 3325–3329.

Malik, S., de Mesquita, L.S.S., Silva, C.R., de Mesquita, J.W.C., de Sá Rocha, E., Bose, J., Abiri, R., de Maria Silva Figueiredo, P. and Costa-Júnior, L.M., 2019. Chemical profile and biological activities of essential oil from Artemisia vulgaris L. cultivated in Brazil. *Pharmaceuticals*, *12*(2), p. 49.

Marbun, R. and Suwarso, E., 2018. Immunomodulatory effects of ethanol extract of Artemisia vulgaris L. in male rats. *Asian Journal of Pharmaceutical and Clinical Research*, *11* (Special Issue 1).

Park, Y.S., Nam, G.H., Jo, K.J., Kawk, H.W., Yoo, J.G., Jang, J.D., Kang, S.M., Kim, S.Y. and Kim, Y.M., 2019. Adequacy of the anti-aging and anti-wrinkle effects of the Artemisia vulgaris fermented solvent fraction. *KSBB Journal*, *34*(3), pp. 199–206.

Wingrove, C.S. and Stevenson, J.C., 1997. 17 beta-Oestradiol inhibits stimulated endothelin release in human vascular endothelial cells. *European Journal of Endocrinology*, *137*(2), pp. 205–208.

29 Asparagus (*Asparagus officinalis* L.)

Asparagus officinalis L.

Etymology: From the Greek *sparasso* = to tear and the Latin *officinalis* = of medicinal value

Synonym: *Asparagus polyphyllus* Steven

Common names: Asparagus; asperges (Fr.); spargel (Ger.); espargos (Port.); спаржа (Rus.); espárrago (Spa.)

Constituents: Steroidal saponins (protodioscin) (Shao et al., 1997), fibers (Agudelo Cadavid et al., 2015).

Medical history: Roman physicians, including Dioscorides, used asparagus as a diuretic, for nephritis and inflammation. Pliny the Elder advocates asparagus as an aphrodisiac. Simeon Seth (12th-century Byzantium) considered asparagus as useful against heart palpitation: "*Cordis palpitationibus succurunt*". In 16th-century Italy, it was used as a diuretic, for jaundice, painful kidneys, and sciatica but had the reputation of making men and women sterile (Matthioli, 1572). Guibourt (1836) recommended asparagus to facilitate digestion. In 19th-century North America, it was used as a diuretic, emmenagogue, and aphrodisiac, to dissolve kidney stones (Pereira, 1843) as well as blood purifier (Shoemaker, 1895).

Medicinal uses: Toothache (Iran); diuretic, laxative, cardiotonic, for rheumatism (India); anemia, liver, gallbladder, urinary tract disorders (Myanmar); toothache, cardiac dropsy (the Philippines)

Blood pressure: Hypertensive rats given a diet comprising 5% asparagus for 10 weeks had a decrease in systolic blood pressure from 192 to 159 mmHg and decreased renal angiotensin-converting enzyme (ACE) (Sanae et al., 2013).

Plasma lipids and glucose: In streptozotocin-induced diabetic rats, a methanolic extract of seeds (2500 mg/kg per day) given for 28 days increased fasting serum insulin levels and decreased fasting glycemia from about 8 to 5.5 mmol/L as efficiently as glibenclamide and protected pancreatic β-cells (Hafizur et al., 2012).

Kidneys: Hypertensive rats given a diet comprising 5% asparagus for 10 weeks had a decrease in plasma creatinine clearance (Sanae & Yasuo, 2013).

Immune system: Butanol extract at 100 mg/kg in an animal model of multiple Sclerosis evoked beneficial effects (Aliomrani et al., 2022).

Brain: An extract given orally to mice models of Alzheimer's at 1 g/kg/day for 1 month prevented cognitive impairment and at the cerebral level increased

DOI: 10.1201/9781003301455-29

heat shock protein (Hsp 27) and decreased amyloid β proteins, tau proteins, and caspsase-3 (Peng et al., 2021).

Warning: Some individuals develop allergic reaction to asparagus (Rademaker & Yung, 2000).

Comments: (i) Heat shock protein (Hsp 27) blocks apoptosis and decreases cellular ROS by increasing glutathione. Increased concentrations of ROS in cells results in mitochondrial dysfunction, and as direct consequence generate more ROS, and all of this in exponential manner until "combustion" of cells (Venugopal et al., 2019).

(ii) The pathophysiology of Alzheimer's disease involves the cleavage of Tau proteins by caspase 3 to form neurofibrillary tangles that with amyloid β proteins form senile plaques (Jana et al., 2013).

(iii) John Veitch Shoemaker (1852–1910) was a professor of materia medica at the medico-chirurgical college of Philadelphia and the author of *A Practical treatrise on materia medica and therapeutics* (1895).

(iv) With aging (and especially with diabetes), the renal blood microperfusion flow decreases. As a consequence, the arterioles of Bowman's capsules constrict, resulting in increased glomerular filtration that increases glomerular pressure and risks glomerular injuries or glomerulosclerosis, translating into proteinuria (Kanasaki et al., 2012).

(v) Tubular atrophy and tubulointerstitial fibrosis are also common manifestations of aging (Kanasaki et al., 2012). Kidney fibrosis results, at least in part, from increased concentrations of ROS in renal tissues (Kim et al., 2009). Upon ROS stimulation, pericytes detach from the vasculature and differentiate into myofibroblasts, contributing to capillary rarefaction, inflammation, and kidney fibrosis (Campanholle et al., 2013).

(vi) In aging diabetic patients, extracellular matrix proteins such as fibronectin and collagen accumulate in mesangial cells. A direct consequence of this is a thickening of glomerular basement membranes, kidney fibrosis and diabetic nephropathy (Kanwar et al., 2011). Therefore, antioxidant and anti-inflammatory medicinal plants, like asparagus, could potentially hamper the progression of kidney fibrosis. Addition of asparagus in the diet of rats for 18 days decreased renal lipid peroxidation (Yan & Zhenhua, 1999).

(vii) ACE catalyzes the conversion of angiotensin I into angiotensin II, which increases blood pressure by triggering vasoconstriction as well as sodium and water retention via aldosterone secretion (Gintoni et al., 2021).

REFERENCES

Agudelo Cadavid, E.L., Restrepo Molina, D.A. and Cartagena Valenzuela, J.R., 2015. Chemical, physicochemical and functional characteristics of dietary fiber obtained from asparagus byproducts (Asparagus officinalis L.). *Revista Facultad Nacional de Agronomía Medellín*, 68(1), pp. 7533–7544.

Aliomrani, M., Rezaei, M., Dinani, M.S. and Mesripour, A., 2022. Effects of Asparagus officinalis on immune system mediated EAE model of multiple sclerosis. *Toxicology Research*, *11*(6), pp. 931–939.

Campanholle, G., Ligresti, G., Gharib, S.A. and Duffield, J.S., 2013. Cellular mechanisms of tissue fibrosis. 3. Novel mechanisms of kidney fibrosis. *American Journal of Physiology-Cell Physiology*, *304*(7), pp. C591–C603.

Gintoni, I., Adamopoulou, M. and Yapijakis, C., 2021. The angiotensin-converting enzyme insertion/deletion polymorphism as a common risk factor for major pregnancy complications. *In Vivo*, *35*(1), pp. 95–103.

Guibourt, N.J.B.G., 1836. *Histoire abrégée des drogues simples*. Méquignon-Marvis Père et fils.

Hafizur, R.M., Kabir, N. and Chishti, S., 2012. Asparagus officinalis extract controls blood glucose by improving insulin secretion and β-cell function in streptozotocin-induced type 2 diabetic rats. *British Journal of Nutrition*, *108*(9), pp. 1586–1595.

Jana, K., Banerjee, B. and Parida, P.K., 2013. Caspases: A potential therapeutic targets in the treatment of Alzheimer's disease. *Translational Medicine*, 2. https://doi.org/10.4172/2161-1025.S2-006.

Kanasaki, K., Kitada, M. and Koya, D., 2012. Pathophysiology of the aging kidney and therapeutic interventions. *Hypertension Research*, *35*(12), pp. 1121–1128.

Kanwar, Y., Sun, L., Xie, P., Liu, F. and Chen, S., 2011. A glimpse of various pathogenetic mechanisms of diabetic nephropathy. *Annual Review of Pathology*, *6*, pp. 395–423.

Kim, J., Seok, Y.M., Jung, K.J. and Park, K.M., 2009. Reactive oxygen species/oxidative stress contributes to progression of kidney fibrosis following transient ischemic injury in mice. *American Journal of Physiology-Renal Physiology*, *297*(2), pp. F461–F470.

Matthioli, P.A., 1572. *Commentaires sur les Six Livres de Pedacius Dioscorides Anazarbeen de la matière medicinale*. A l'Escue de Milan.

Peng, Z., Bedi, S., Mann, V., Sundaresan, A., Homma, K., Gaskey, G., Kowada, M., Umar, S., Kulkarni, A.D., Eltzschig, H.K. and Doursout, M.F., 2021. Neuroprotective effects of Asparagus officinalis stem extract in transgenic mice overexpressing amyloid precursor protein. *Journal of Immunology Research*, *2021*, pp. 1–10.

Pereira, J., 1843. *The Elements of Materia Medica and Therapeutics*. Lea and Blanchard.

Rademaker, M. and Yung, A., 2000. Contact dermatitis to Asparagus officinalis. *Australasian Journal of Dermatology*, *41*(4), pp. 262–263.

Sanae, M. and Yasuo, A., 2013. Green asparagus (Asparagus officinalis) prevented hypertension by an inhibitory effect on angiotensin-converting enzyme activity in the kidney of spontaneously hypertensive rats. *Journal of Agricultural and Food Chemistry*, *61*(23), pp. 5520–5525.

Shao, Y.U., Poobrasert, O., Kennelly, E.J., Chin, C.K., Ho, C.T., Huang, M.T., Garrison, S.A. and Cordell, G.A., 1997. Steroidal saponins from Asparagus officinalis and their cytotoxic activity. *Planta Medica*, *63*(03), pp. 258–262.

Shoemaker, J.V., 1895. *A Practical Treatise on Materia Medica and Therapeutics: With Especial Reference to the Clinical Application of Drugs*. F.A. Davis Company.

Venugopal, A., Sundaramoorthy, K. and Vellingiri, B., 2019. Therapeutic potential of Hsp27 in neurological diseases. *Egyptian Journal of Medical Human Genetics*, *20*, pp. 1–8.

Yan, H. and Zhenhua, T., 1999. The effect of Asparagus officinalis L to MDA cotnent and SOD activity of the rats. *Journal of Guiyang Medical College*, *24*(2), pp. 122–124.

30 Beetroot (*Beta vulgaris* L.)

Beta vulgaris L.

Etymology: From the Latin *beta* = beet and *vulgaris* = common
Family: Betulaceae
Synonym: *Beta orientalis* L.
Common names: Beetroot; betterave (Fr.); rote beete (Ger.); beterraba (Port.) свёкла (Rus.); raíz de remolacha (Spa.)
Part used: Root
Constituents: Nitrates, betacyanins (Wruss et al., 2015), phenolic acids (Ben Haj Koubaier et al., 2014), saponins (Mikołajczyk-Bator et al., 2016).
Medicinal uses: According to Simeon Seth (12th-century Byzantium) beetroot is hot and dry to the third degree. Matthioli (1572) says it is good for the liver and spleen.
Medicinal uses: Cough, inflammation (India)

Blood pressure: Intake of 500 g of beetroot juice by normotensive subjects decreased systolic and diastolic blood pressure (Hobbs et al., 2012). Beetroot juice given to elderly subjects for 28 days reduced blood pressure by about 7 mm Hg (Jajja et al., 2014). Elderly subjects taking beetroot juice daily (corresponding to 6.1 mmol of inorganic nitrate) for 7 days had improved aerobic endurance and lower blood pressure (Eggebeen et al., 2016).
Plasma lipids and glucose: Beetroot juice given for 15 days to physically active volunteers evoked an increase of HDL-cholesterol from 42.9 to 50.2 mg/dl and a decrease of LDL-cholesterol from 129.7 to 119.5 mg/dl (Singh et al., 2015).
Uric acid: Hyperuricemic rats given beetroot powder 3.1 g/kg/bw for 14 days showed a decrease in plasmatic MDA (Wulandari et al., 2021).

Warnings: Nitrates are carcinogenic (Zamani et al., 2021), and 700 mg of nitrates are poisonous for a 70 kg adult and much less for a child (Maynard & Barker, 1972). None of the plants listed here are for pediatric use.
Comment: It is clear that eating boiled beetroots regularly could contribute to controlling hypertension and plasma lipids.

 DOI: 10.1201/9781003301455-30

REFERENCES

Ben Haj Koubaier, H., Snoussi, A., Essaidi, I., Chaabouni, M.M., Thonart, P. and Bouzouita, N., 2014. Betalain and phenolic compositions, antioxidant activity of Tunisian red beet (Beta vulgaris L. conditiva) roots and stems extracts. *International Journal of Food Properties*, *17*(9), pp. 1934–1945.

Eggebeen, J., Kim-Shapiro, D.B., Haykowsky, M., Morgan, T.M., Basu, S., Brubaker, P., Rejeski, J. and Kitzman, D.W., 2016. One week of daily dosing with beetroot juice improves submaximal endurance and blood pressure in older patients with heart failure and preserved ejection fraction. *JACC: Heart Failure*, *4*(6), pp. 428–437.

Hobbs, D.A., Kaffa, N., George, T.W., Methven, L. and Lovegrove, J.A., 2012. Blood pressure-lowering effects of beetroot juice and novel beetroot-enriched bread products in normotensive male subjects. *British Journal of Nutrition*, *108*(11), pp. 2066–2074.

Jajja, A., Sutyarjoko, A., Lara, J., Rennie, K., Brandt, K., Qadir, O. and Siervo, M., 2014. Beetroot supplementation lowers daily systolic blood pressure in older, overweight subjects. *Nutrition Research*, *34*(10), pp. 868–875.

Maynard, D.N. and Barker, A.V., 1972. Nitrate content of vegetable crops. *HortScience*, *7*(3), pp. 224–226.

Matthioli, P.A., 1572. *Commentaires sur les Six Livres de Pedacius Dioscorides Anazarbeen de la matière medicinale*. A l'Escue de Milan.

Mikołajczyk-Bator, K., Błaszczyk, A., Czyżniejewski, M. and Kachlicki, P., 2016. Characterisation and identification of triterpene saponins in the roots of red beets (Beta vulgaris L.) using two HPLC–MS systems. *Food Chemistry*, *192*, pp. 979–990.

Singh, A., Verma, S., Singh, V., Nanjappa, C., Roopa, N., Raju, P.S. and Singh, S.N., 2015. Beetroot juice supplementation increases high density lipoprotein-cholesterol and reduces oxidative stress in physically active individuals. *Journal of Pharmacy and Nutrition Sciences*, *5*(3), pp. 179–185.

Wruss, J., Waldenberger, G., Huemer, S., Uygun, P., Lanzerstorfer, P., Müller, U., Höglinger, O. and Weghuber, J., 2015. Compositional characteristics of commercial beetroot products and beetroot juice prepared from seven beetroot varieties grown in Upper Austria. *Journal of Food Composition and Analysis*, *42*, pp. 46–55.

Wulandari, A., Dirgahayu, P. and Wiboworini, B., 2021. Beetroot powder (Beta vulgaris L.) decrease oxidative stress by reducing of malondialdehyde (MDA) levels in hyperuricemia. *Journal of International Conference Proceedings (JICP)*, *4*(1), July, pp. 290–299.

Zamani, H., De Joode, M.E.J.R., Hossein, I.J., Henckens, N.F.T., Guggeis, M.A., Berends, J.E., de Kok, T.M.C.M. and van Breda, S.G.J., 2021. The benefits and risks of beetroot juice consumption: A systematic review. *Critical Reviews in Food Science and Nutrition*, *61*(5), pp. 788–804.

31 Borage (*Borago officinalis* L.)

Borago officinalis L.

Etymology: From the Latin *cor* = heart, *ago* = I bring, and *officinalis* = of medicinal value

Family: Boraginaceae

Common names: Borage; bourrache officinale (Fr.); gurkenkraut (Ger.); borragem (Port.); бораго (Rus.); borraja (Spa.)

Parts used: Leaf, seed

Constituents: Fixed oil in seeds (unsaturated fatty acids: γ-linolenic acid) (Tanwar et al., 2021), pyrrolizidine alkaloids (leave and seeds) (Dodson & Stermitz, 1986).

Medical history: According to Galen, borage brings joy to the heart. The Medical School of Salerno (10th century) asserts that borage is good against depression, stomachache, heart palpitations, and heart troubles. Simeon Seth (12th-century Byzantium) describes it as a diuretic. The plant was recommended against the black plague by the Faculty of Medicine of Paris in the year 1348. According to Fusch (1555), borage is classified as humid. In 18th-century Scotland, it was used as diuretic, for fever, and depression (Alston, 1770); In 19th-century France, borage had the reputation to be good for the lungs, and used for inducing urination (Moquin-Tandon, 1861) and to refresh the body (Guibourt, 1836).

Medicinal uses: Hypertension (Iran)

Blood pressure: Hypertensive and normotensive rats given a diet comprising 11% borage oil for 5 weeks had a decrease in systolic blood pressure via, at least in part, interference with the renin-angiotensin-aldosterone system (Engler et al., 1992; Engler & Engler, 1998).

Heart: Ethanol extract of leaves decreased potassium ions-induced contractions of isolated rabbit jejunum preparations via the inhibition of calcium ions channels blockade (Gilani et al., 2007).

Plasma lipids and glucose: Rat fed with the fixed oil of seeds were protected against obesity and experienced an increase in HDL-cholesterol (Navarro-Herrera et al., 2018).

Skin and hair: Mice exposed to ultraviolet B irradiation and orally given a borage extract experienced an increase in skin collagen. This extract also attenuated wrinkle formation, epidermal thickness, and skin dehydration (Seo et al., 2018).

DOI: 10.1201/9781003301455-31

Warnings: Pyrrolizidine alkaloids are liver poisons (Dodson & Stermitz, 1986; Chojkier, M., 2003). A maximum of 5 g/day of oil of seeds is tolerated (Schirmer & Phinney, 2007), but the safest dose is 3 g. Not to be taken by patients on anticoagulants such as warfarin (Ulbricht et al., 2008).

Comments: (i) Alfred Moquin-Tandon (1804–1863) was a professor of medical natural history in the Faculty of Medicine of Paris and author of a book titled *Element de Botanique Médicale* (1861).

(ii) Oil of borage is available as a food supplement in pharmacies and could be taken to control blood pressure.

REFERENCES

Alston, C., 1770. *Lectures on the Materia Medica: Containing the Natural History of Drugs, their Virtues and Doses: Also Directions for the Study of the Materia Medica; and an Appendix on the Method of Prescribing.* Edward and Charles Dilly.

Chojkier, M., 2003. Hepatic sinusoidal-obstruction syndrome: Toxicity of pyrrolizidine alkaloids. *Journal of Hepatology*, *39*(3), pp. 437–446.

Dodson, C.D. and Stermitz, F.R., 1986. Pyrrolizidine alkaloids from borage (Borago officinalis) seeds and flowers. *Journal of Natural Products*, *49*(4), pp. 727–728.

Engler, M.M. and Engler, M.B., 1998. Dietary borage oil alters plasma, hepatic and vascular tissue fatty acid composition in spontaneously hypertensive rats. *Prostaglandins, Leukotrienes and Essential Fatty Acids*, *59*(1), pp. 11–15.

Engler, M.M., Engler, M.B. and Paul, S.M., 1992. Effects of dietary borage oil rich in gamma-linolenic acid on blood pressure and vascular reactivity. *Nutrition Research*, *12*(4–5), pp. 519–528.

Fusch, L., 1555. *De Historia Stirpium Commetarii Insignes.* Lugduni Apud Ioan Tornaesium.

Gilani, A.H., Bashir, S. and Khan, A.U., 2007. Pharmacological basis for the use of Borago officinalis in gastrointestinal, respiratory and cardiovascular disorders. *Journal of Ethnopharmacology*, *114*(3), pp. 393–399.

Guibourt, N.J.B.G., 1836. *Histoire abrégée des drogues simples.* Méquignon-Marvis Père et fils.

Moquin-Tandon, A., 1861. *Element de Botanique Médicale.* J.Baillière et Fils.

Navarro-Herrera, D., Aranaz, P., Eder-Azanza, L., Zabala, M., Romo-Hualde, A., Hurtado, C., Calavia, D., López-Yoldi, M., Martínez, J.A., González-Navarro, C.J. and Vizmanos, J.L., 2018. Borago officinalis seed oil (BSO), a natural source of omega-6 fatty acids, attenuates fat accumulation by activating peroxisomal beta-oxidation both in C. elegans and in diet-induced obese rats. *Food & Function*, *9*(8), pp. 4340–4351.

Schirmer, M.A. and Phinney, S.D., 2007. γ-linolenate reduces weight regain in formerly obese humans. *The Journal of Nutrition*, *137*(6), pp. 1430–1435.

Seo, S.A., Park, B., Hwang, E., Park, S.Y. and Yi, T.H., 2018. Borago officinalis L. attenuates UVB-induced skin photodamage via regulation of AP-1 and Nrf2/ARE pathway in normal human dermal fibroblasts and promotion of collagen synthesis in hairless mice. *Experimental Gerontology*, *107*, pp. 178–186.

Tanwar, B., Goyal, A., Kumar, V., Rasane, P. and Sihag, M.K., 2021. Borage (Borago officinalis) seed. *Oilseeds: Health Attributes and Food Applications*, pp. 351–371.

Ulbricht, C., Chao, W., Costa, D., Rusie-Seamon, E., Weissner, W. and Woods, J., 2008. Clinical evidence of herb-drug interactions: A systematic review by the natural standard research collaboration. *Current Drug Metabolism*, *9*(10), pp. 1063–1120.

32 White Mustard (*Brassica alba* L.)

Brassica alba L.

Etymology: From the Celtic *bresic* = cabbage and the Latin *alba* = white
Family: Brassicaceae
Synonyms: *Brassica hirta* Moench.; *Eruca alba* (L.) Noulet; *Sinapis alba* L.; *Raphanus albus* (L.) Crantz
Common names: White mustard, yellow mustard; moutarde blanche (Fr.); weißer senf (Ger.); mostarda branca (Port.) горчица белая (Rus.); mostaza blanca (Spa.)
Part used: Seed
Constituents: Glucosinolates (sinigrin and sinalbin) (Popova & Morra, 2014), fixed oil (unsaturated fatty acids: erucic acid, oleic acid, linoleic acid) (Antova et al., 2017), phenolic acid including sinapoyl choline (Bopp & Lüdicke, 1980).

Medical history: White mustard was a popular food in the Middle Ages. In 19th-century North America, it was used as a diuretic, laxative, stomachic, and for preserving the health of elderly persons (Pereira, 1843)

Blood pressure: Sinigrin given orally at 40 mg/kg/day for 28 days to hypertensive rats decreased systolic blood pressure from about 180 to 130 mmHg and diastolic blood pressure from about 125 to 110 mmHg (Cong et al., 2021).
Kidneys: Sinigrin given orally at 40 mg/kg/day for 28 days to hypertensive rats decreased urinary proteins, thickness of glomerular basement membranes, and the renal expression of fibronectin and collagen (Cong et al., 2021). Ethanol extract of seeds given orally at 200 mg/kg/day protected rats against gentamycin-induced kidney necrosis (Suvarna et al., 2019).

Warnings: Some individuals will develop anaphylactic shock after mustard consumption (Pałgan et al., 2018). Glucosinolates must not be ingested by patients suffering from thyroid problems. Erucic acid causes myocardial lesions (Galanty et al., 2023)

REFERENCES

Antova, G.A., Angelova-Romova, M.I., Petkova, Z.Y., Teneva, O.T. and Marcheva, M.P., 2017. Lipid composition of mustard seed oils (Sinapis alba L.). *Bulgarian Chemical Communications*, **49**, pp. 55–60.

DOI: 10.1201/9781003301455-32

Bopp, M. and Lüdicke, W., 1980. Synthesis of sinapine during seed development of Sinapis alba. *Zeitschrift für Naturforschung C*, *35*(7–8), pp. 539–543.

Cong, C., Yuan, X., Hu, Y., Chen, W., Wang, Y. and Tao, L., 2021. Sinigrin attenuates angiotensin II-induced kidney injury by inactivating nuclear factor-κB and extracellular signal-regulated kinase signaling in vivo and in vitro. *International Journal of Molecular Medicine*, *48*(2), pp. 1–10.

Galanty, A., Grudzińska, M., Paździora, W. and Paśko, P., 2023. Erucic acid—both sides of the story: A concise review on its beneficial and toxic properties. *Molecules*, *28*(4), p. 1924.

Pałgan, K., Żbikowska-Gotz, M. and Bartuzi, Z., 2018. Dangerous anaphylactic reaction to mustard. *Archives of Medical Science*, *14*(2), pp. 477–479.

Pereira, J., 1843. *The Elements of Materia Medica and Therapeutics*. Lea and Blanchard.

Popova, I.E. and Morra, M.J., 2014. Simultaneous quantification of sinigrin, sinalbin, and anionic glucosinolate hydrolysis products in Brassica juncea and Sinapis alba seed extracts using ion chromatography. *Journal of Agricultural and Food Chemistry*, *62*(44), pp. 10687–10693.

Suvarna, J.N., Parida, A., Poojar, B., Daggupati, S., Manju, V. and Rao, R.R., 2019. Nephroprotective activity of Sinapis alba in gentamicin induced murine model of renotoxicity. *International Journal of Research in Pharmaceutical Sciences*, *10*(4), pp. 3761–3767.

33 Indian Mustard (*Brassica juncea* (L.) Czern)

Brassica juncea (L.) Czern

Etymology: From the Celtic *bresic* = cabbage and the Latin *juncea* = sedge-like

Family: Brassicaceae

Synonyms: *Sinapis juncea* L.; *Brassica japonica* (Thunb.) Siebold ex Miq.

Common names: Indian mustard; moutarde Indienne (Fr.); Indischen senf (Ger.) mostarda indiana (Port.); индийская горчица (Rus.); mostaza india (Spa.)

Parts used: Leaf, seed

Constituents: Fixed oil (unsaturated fatty acids: oleic acid, erucic acid) (Sutariya et al., 2016), glucosinolates (sinigrin, sinalbin) (Popova & Morra, 2014).

Medical history: The seeds were used as carminative and counter-irritant in 19th-century India (Watt, 1889).

Medicinal uses: Asthma, phlegm, skin infection (India); phlegm, thorax congestion (Korea)

Blood pressure: Sinigrin given orally at 40 mg/kg/day for 28 days to rats decreased systolic blood pressure from about 180 to 130 mmHg and diastolic blood pressure from about 125 to 110 mmHg (Cong et al., 2021).

Blood lipids and glucose: The intake of seeds (10% of diet) by rats poisoned with a coconut-oil-enriched diet for 90 days prevented an increase in plasma cholesterol (Khan et al., 1996). Intake of seeds (10% of diet) by rats poisoned with a high-fructose diet for 30 days mitigated increases in fasting plasma glucose, insulin, and cholesterol (Yadav et al., 2004).

Ethylacetate extract of seeds given orally to streptozotocin-induced diabetic rats at 200 mg/kg/day for 10 days reduced plasma glucose from 598 to 486 mg/dL (normal: 112 mg/dL) as well as plasmatic, real, and mitochondrial concentration of ROS (Yokozawa et al., 2003).

Kidneys: Sinigrin given orally at 40 mg/kg/day for 28 days to rats decreased urinary proteins, thinning of the glomerular basement membrane, and decreased the renal expression of fibronectin and collagen (Cong et al., 2021).

Bones and cartilages: An extract given to rats inhibited Complete Freund's Adjuvant (CFA)-induced arthritis (Lakshmanan et al., 2022)

Warnings: Glucosinolates must not be ingested by patients with thyroid dysfunction. Erucic acid causes myocardial lesions (Galanty et al., 2023).

Comment: With aging tend to appear pathologies of the thyroid (Levy, 1991). Oil of seeds given at 2 mL/kg for 8 weeks to rats with hypothyroid increased

DOI: 10.1201/9781003301455-33

plasma thyroxine from 1.7 to 4.4 ng/dL and triiodothyronine from 1.3 to 5.4 ng/mL. Oil of seeds given at 2 mL/kg to hyperthyroidic rats for 8 weeks decreased thyroxine from 11.5 to 7.2 ng/dL and triiodothyronine from 13.5 to 7.6 ng/mL (Udovcic et al., 2017).

REFERENCES

Cong, C., Yuan, X., Hu, Y., Chen, W., Wang, Y. and Tao, L., 2021. Sinigrin attenuates angiotensin II-induced kidney injury by inactivating nuclear factor-κB and extracellular signal-regulated kinase signaling in vivo and in vitro. *International Journal of Molecular Medicine*, 48(2), pp. 1–10.

Eissa, E., Ibrahim, H.S. and Rabeh, N., 2021. The impact of mustard oil (Brassica Juncea) ingestion on thyroid gland hormones in experimental rats. *Middle East Journal of Therapeutic Nutrition and Complementary Medicine*, 1(1), pp. 1–10.

Galanty, A., Grudzińska, M., Paździora, W. and Paśko, P., 2023. Erucic acid—both sides of the story: A concise review on its beneficial and toxic properties. *Molecules*, 28(4), p. 1924.

Khan, B.A., Abraham, A. and Leelamma, S., 1996. Biochemical response in rats to the addition of curry leaf (Murraya koenigii) and mustard seeds (Brassica juncea) to the diet. *Plant Foods for Human Nutrition*, 49, pp. 295–299.

Lakshmanan, D.K., Murugesan, S., Rajendran, S., Ravichandran, G., Elangovan, A., Raju, K., Prathiviraj, R., Pandiyan, R. and Thilagar, S., 2022. Brassica juncea (L.) Czern. leaves alleviate adjuvant-induced rheumatoid arthritis in rats via modulating the finest disease targets-IL2RA, IL18 and VEGFA. *Journal of Biomolecular Structure and Dynamics*, 40(18), pp. 8155–8168.

Levy, E.C., 1991. Thyroid disease in the elderly. *Medical Clinics of North America*, 75(1), pp. 151–167.

Popova, I.E. and Morra, M.J., 2014. Simultaneous quantification of sinigrin, sinalbin, and anionic glucosinolate hydrolysis products in Brassica juncea and Sinapis alba seed extracts using ion chromatography. *Journal of Agricultural and Food Chemistry*, 62(44), pp. 10687–10693.

Stabouli, S., Papakatsika, S. and Kotsis, V., 2010. Hypothyroidism and hypertension. *Expert Review of Cardiovascular Therapy*, 8(11), pp. 1559–1565.

Sutariya, D.A., Patel, K.M., Bhadauria, H.S., Vaghela, P.O., Prajapati, D.V. and Parmar, S.K., 2016. Genetic diversity for quality traits in Indian mustard [Brassica juncea (L.)]. *Journal of Oilseed Brassica*, 1(1), pp. 44–47.

Udovcic, M., Pena, R.H., Patham, B., Tabatabai, L. and Kansara, A., 2017. Hypothyroidism and the heart. *Methodist DeBakey Cardiovascular Journal*, 13(2), p. 55.

Watt, G., 1889. *A Dictionary of the Economic Products of India*. Printed by the Superintendentof Government Printing, India.

Yadav, S.P., Vats, V., Ammini, A.C. and Grover, J.K., 2004. Brassica juncea (Rai) significantly prevented the development of insulin resistance in rats fed fructose-enriched diet. *Journal of Ethnopharmacology*, 93(1), pp. 113–116.

Yokozawa, T., Kim, H.Y., Cho, E.J., Yamabe, N. and Choi, J.S., 2003. Protective effects of mustard leaf (Brassica juncea) against diabetic oxidative stress. *Journal of Nutritional Science and Vitaminology*, 49(2), pp. 87–93.

34 Rapeseed (*Brassica napus* L.)

Brassica napus L.

Etymology: From the Celtic *bresic* = cabbage and the Latin *napus* = turnip

Family: Brassicaceae

Synonyms: *Brassica napobrassica* (L.) Mill.; *Brassica rugosa* (Roxb.) L.H. Bailey

Common names: Canola, rapeseed; colza (Fr.; Port.; Spa.); raps (Ger.); апс (Rus.)

Part used: Seed

Constituents: Fixed oil (unsaturated fatty acids: oleic acid, linoleic acid, linolenic acid, erucic acid) (Dupont et al., 1989).

Medical history: The Medical School of Salerno (10th century) recommends rapeseed for urination: "*provocat urinam*". In 16th-century Italy, it was used for jaundice, hydropisie (Matthioli, 1572). Canola was used a s a diuretic in 19th-century Europe.

Medicinal uses: Aphrodisiac, tonic (Pakistan)

Blood lipids and glucose: In type-2 diabetic women over 50, ingesting 30 g of oil daily in their food for 8 weeks decreased systolic blood pressure from 128.5 to 127.1 mmHg and diastolic blood pressure from 79.5 to 78.6 mmHg (Atefi et al., 2018).

Plasma lipids and glucose: In type-2 diabetic women over 50, ingesting 30 g of oil daily in their food for 8 weeks decreased plasma cholesterol from 163.2 to 155.5 mg/dL and plasma triglycerides from 148.2 to 131.9 mg/dL (Atefi et al., 2018). Hypercholesterolemic and/or hypertriglyceridemic subjects taking 30 mL/day of oil for 4 months for cooking had a decrease of LDL-cholesterol from 173 to 160 mg/dL (Bierenbaum et al., 1991).

Warnings: Rats fed a chow containing 600 mg of rapeseed oil (containing erucic acid) for 8 weeks showed elevated creatinine (Hasan et al., 2018) and when on a diet containing 20% of oil developed alopecia and hemorrhagic tails (Hulan et al., 1976). Fried oil induces vascular dysfunction via inflammatory insults (Jaarin et al., 2016). Erucic acid causes myocardial lesions (Galanty et al., 2023)

 DOI: 10.1201/9781003301455-34

REFERENCES

Atefi, M., Pishdad, G.R. and Faghih, S., 2018. Canola oil and olive oil impact on lipid profile and blood pressure in women with type 2 diabetes: A randomized, controlled trial. *Progress in Nutrition*, *20*(Suppl 1), pp. 102–109.

Bierenbaum, M.L., Reichstein, R.P., Watkins, T.R., Maginnis, W.P. and Geller, M., 1991. Effects of canola oil on serum lipids in humans. *Journal of the American College of Nutrition*, *10*(3), pp. 228–233.

Dupont, J., White, P.J., Johnston, K.M., Heggtveit, H.A., McDonald, B.E., Grundy, S.M. and Bonanome, A., 1989. Food safety and health effects of canola oil. *Journal of the American College of Nutrition*, *8*(5), pp. 360–375.

Galanty, A., Grudzińska, M., Paździora, W. and Paśko, P., 2023. Erucic acid—both sides of the story: A concise review on its beneficial and toxic properties. *Molecules*, *28*(4), p. 1924.

Hasan, K.M.M., Tamanna, N. and Haque, M.A., 2018. Biochemical and histopathological profiling of Wistar rat treated with *Brassica napus* as a supplementary feed. *Food Science and Human Wellness*, *7*(1), pp. 77–82.

Hulan, H.W., Hunsaker, W.G., Kramer, J.K.G. and Mahadevan, S., 1976. The development of dermal lesions and alopecia in male rats fed rapeseed oil. *Canadian Journal of Physiology and Pharmacology*, *54*(1), pp. 1–6.

Jaarin, K., Masbah, N. and Nordin, S.H., 2016. Heated cooking oils and its effect on blood pressure and possible mechanism: A review. *International Journal of Clinical & Experimental Medicine*, *9*(2).

Matthioli, P.A., 1572. *Commentaires sur les Six Livres de Pedacius Dioscorides Anazarbeen de la matière medicinale*. A l'Escue de Milan.

35 Black Mustard (*Brassica nigra* (L.) W.D.J. Koch)

Brassica nigra (L.) W.D.J. Koch

Etymology: From the Celtic *bresic* = cabbage and the Latin *nigra* = black

Family: Brassicaeae

Synonyms: *Sinapis nigra* L.; *Sisymbrium nigrum* (L.) Prant

Common names: Black mustard; moutarde noire, sénevé (Fr.); schwarze senf (Ger.); mostarda preta (Port.); горчица черная (Rus.); mostaza negra (Spa.)

Part used: Seed

Constituents: Fixed oil (unsaturated fatty acids: oleic acid, linoleic acid, erucic acid) (Kaur et al., 2022), glucosinolates (Sodhi et al., 2002), phenolic acids (sinapic acid) (Fang et al., 2008).

Medical history: Black mustard was known to Hippocrates. The Medical School of Salerno (10th century) recommended black mustard as an antidote for poisons and for flu: *"purgatque caput"*. In 16th-century Germany, it was considered a hot and dry remedy to the fourth degree (Fusch, 1555). In 19th-century India, black mustard was given as a counter-irritant and digestive (Pharmacopoeia of India, 1868). In Europe, it was given for dropsy (Moquin-Tandon, 1861). In North America, alck mustard was used as carminative, emetic, diuretic, and rubefacient (Pereira, 1843).

Medicinal uses: Emetic, treats narcotic poisoning (Iran, Iraq); stomachache, indigestion, muscular pain (Pakistan); muscular pain, digestive, carminative (Bangladesh)

Plasma lipids and glucose: Aqueous extract of seeds given orally to streptozotocin-induced diabetic rats at 200 mg/kg/day for a month brought down fasting glycemia from 254 to 127 mg/dL (stronger activity than Glibenclamide at 200 mg/kg/day) as well as plasma cholesterol from 231 to 181 mg/dL, LDL-cholesterol from 157 to 100 mg/dL, and triglycerides from 141 to 98 mg/dL (Anand et al., 2007).

Blood pressure: Sinapic acid prevented hypertension and cardiovascular remodeling in rats (Silambarasan et_al., 2014). Sinapic acid given orally at 25 mg/kg/day for 28 days to diabetic rats decreased plasma LDL-cholesterol and cardiac ROS concentration (Cherng et_al., 2013; Zych et_al., 2019).

Bile secretion: Intake of seeds powder (250 mg in daily chow) for 4 weeks increased the bile flow in rats from 0.5 to 0.7 m/hr with secretion of

DOI: 10.1201/9781003301455-35

cholesterol from 0.2 to 0.4 µmol/hr and increased total bile acid from 6.3 to 9.4 µmol/hr (Sambaiah & Srinivasan, 1991).

Kidneys: Methanol extract of leaves given to rats at 400 mg/kg/day for 21 days evoked hepatoprotection and nephroprotection against the effect of a single intraperitoneal injection of D-galactosamine (Rajamurugan et al., 2012).

Brain: Rats given fixed oil at 925 mg/kg/day for 19 days were protected against amyloid β proteins-induced-dementia (Nazari et al., 2020).

Warning: Glucosinolates must not be ingested by patients with thyroid dysfunction.

REFERENCES

Anand, P., Murali, K.Y., Tandon, V., Chandra, R. and Murthy, P.S., 2007. Preliminary studies on antihyperglycemic effect of aqueous extract of Brassica nigra (L.) Koch in streptozotocin induced diabetic rats. *IJEB 45*(8) [August 2007]

Cherng, Y.G., Tsai, C.C., Chung, H.H., Lai, Y.W., Kuo, S.C. and Cheng, J.T., 2013. Antihyperglycemic action of sinapic acid in diabetic rats. *Journal of Agricultural and Food Chemistry*, *61*(49), pp. 12053–12059.

Fang, Z., Hu, Y., Liu, D., Chen, J. and Ye, X., 2008. Changes of phenolic acids and antioxidant activities during potherb mustard (Brassica juncea, Coss.) pickling. *Food Chemistry*, *108*(3), pp. 811–817.

Fusch, L., 1555. *De Historia Stirpium Commetarii Insignes*. Lugduni Apud Ioan Tornaesium.

Kaur, G., Kaur, R. and Kaur, S., 2022. Studies on physiochemical properties of oil extracted from Brassica nigra and Brassica rapa toria. *Materials Today: Proceedings*, *48*, pp. 1645–1651.

Moquin-Tandon, A., 1861. *Element de Botanique Médicale*. J.Baillière et Fils.

Nazari, E., Khanavi, M., Amani, L., Sharifzadeh, M., Vazirian, M., Saeedi, M., Sanati, M. and Lamardi, S.N.S., 2020. Beneficial effect of Brassica nigra fixed oil on the changes in memory caused by β-amyloid in an animal model. *Pharmaceutical Sciences*, *26*(3), pp. 261–269.

Pereira, J., 1843. *The Elements of Materia Medica and Therapeutics*. Lea and Blanchard.

Rajamurugan, R., Suyavaran, A., Selvaganabathy, N., Ramamurthy, C.H., Reddy, G.P., Sujatha, V. and Thirunavukkarasu, C., 2012. Brassica nigra plays a remedy role in hepatic and renal damage. *Pharmaceutical Biology*, *50*(12), pp. 1488–1497.

Sambaiah, K. and Srinivasan, K., 1991. Secretion and composition of bile in rats fed diets containing spices. *Journal of Food Science and Technology*, *28*(1), pp. 35–38.

Silambarasan, T., Manivannan, J., Krishna Priya, M., Suganya, N., Chatterjee, S. and Raja, B., 2014. Sinapic acid prevents hypertension and cardiovascular remodeling in pharmacological model of nitric oxide inhibited rats. *PloS One*, *9*(12), p. e115682.

Sodhi, Y.S., Mukhopadhyay, A., Arumugam, N., Verma, J.K., Gupta, V., Pental, D. and Pradhan, A.K., 2002. Genetic analysis of total glucosinolate in crosses involving a high glucosinolate Indian variety and a low glucosinolate line of Brassica juncea. *Plant Breeding*, *121*(6), pp. 508–511.

Zych, M., Wojnar, W., Borymski, S., Szałabska, K., Bramora, P. and Kaczmarczyk-Sedlak, I., 2019. Effect of rosmarinic acid and sinapic acid on oxidative stress parameters in the cardiac tissue and serum of type 2 diabetic female rats. *Antioxidants*, *8*(12), p. 579.

36 Cabbage (*Brassica oleracea* L.)

Brassica oleracea L.

Etymology: From the Celtic *bresic* = cabbage and the Latin *oleraceus* = a pot herb

Family: Brassicaeae

Synonyms: *Brassica alboglabra* L.H. Bailey; *Brassica maritima* Tardent; *Crucifera brassica* E.H.L. Krause; *Napus oleracea* (L.) K.F. Schimp. & Spenn.

Common names: Broccoli, cabbage; choux (Fr.); kohl (Ger.); repolho (Port.); кап ý ста (Rus.); repollo (Spa.)

Part used: Aerial part

Constituents: Glucosinolates (sinigrin, glucoraphanin) (Moyes et al., 2000), flavone glycosides, hydroxycinnamic acid derivatives (Velasco et al., 2011).

Medical history: Cabbage was held in high esteem by Dioscorides as a diuretic, a cure for arthritis, and a cure for all, while Galen advised its use for malignant ulcers. It was a gentle laxative for the Medical School of Salerno (10th century) and used for inflammation by Simeon Seth (12th-century Byzantium). In 16th-century Germany, it was considered a dry and hot medicine to the first degree (Fusch, 1555)

Medicinal uses: Laxative (Turkey); hypertension, high cholesterol (Afghanistan); diuretic, laxative (Myanmar); liver troubles, gout, cancer (Korea)

Plasma lipids and glucose: Methanol extract given to alloxan-induced diabetic rats at 500 mg/kg/day for 30 days decreased glycemia from 246.3 to 139.7 mg/mL, cholesterol from 240.3 to 146 mg/dL, and triglycerides from 198.7 to 143 mg/dL (Assad et al., 2014).

Ethanol extract of leaves given to diabetic rodent at 1 g/kg/day for 8 weeks prevented weight loss and decreased glycemia (Kataya & Hamza, 2008). Aqueous extract given to hamsters on a hypercholesterolemic diet reduced plasma cholesterol (Rodríguez-Cantú et al., 2011).

Gallbladder: Bile acids bind to cabbage fibers *in vitro* (Kahlon et al., 2007).

Kidneys: Ethanol extract of leaves given to diabetic rodent at 1 g/kg/day for 8 weeks normalized kidney weight and decreased serum creatinine (Kataya & Hamza, 2008).

Bones and cartilages: Ethanol extract given to rats at 2 g/kg/day for 14 days attenuated CFA-induced arthritis with decreases in periarticular MDA concentration (Prabowo, 2019).

Brain: Chloroform extract given to mice with amyloid β proteins-induced dementia preserved cognitive function (Park et al., 2016).

DOI: 10.1201/9781003301455-36

REFERENCES

Assad, T., Khan, R.A. and Feroz, Z., 2014. Evaluation of hypoglycemic and hypolipidemic activity of methanol extract of Brassica oleracea. *Chinese Journal of Natural Medicines*, *12*(9), pp. 648–653.

Fusch, L., 1555. *De Historia Stirpium Commetarii Insignes*. Lugduni Apud Ioan Tornaesium.

Kahlon, T.S., Chapman, M.H. and Smith, G.E., 2007. In vitro binding of bile acids by spinach, kale, brussels sprouts, broccoli, mustard greens, green bell pepper, cabbage and collards. *Food Chemistry*, *100*(4), pp. 1531–1536.

Kataya, H.A. and Hamza, A.A., 2008. Red cabbage (Brassica oleracea) ameliorates diabetic nephropathy in rats. *Evidence-Based Complementary and Alternative Medicine*, *5*(3), pp. 281–287.

Moyes, C.L., Collin, H.A., Britton, G. and Raybould, A.F., 2000. Glucosinolates and differential herbivory in wild populations of Brassica oleracea. *Journal of Chemical Ecology*, *26*, pp. 2625–2641.

Park, S.K., Ha, J.S., Kim, J.M., Kang, J.Y., Lee, D.S., Guo, T.J., Lee, U., Kim, D.O. and Heo, H.J., 2016. Antiamnesic effect of broccoli (Brassica oleracea var. italica) leaves on amyloid beta (Aβ) 1–42-induced learning and memory impairment. *Journal of Agricultural and Food Chemistry*, *64*(17), pp. 3353–3361.

Prabowo, S., 2019. Broccoli extract (Brassica oleracea) decrease periarticular malondialdehyde level and disease activity score in rats (Rattus norvegicus) with adjuvant arthritis. In *IOP Conference Series: Earth and Environmental Science* (Vol. 217, No. 1, p. 012046). IOP Publishing.

Rodríguez-Cantú, L.N., Gutiérrez-Uribe, J.A., Arriola-Vucovich, J., Díaz-De La Garza, R.I., Fahey, J.W. and Serna-Saldivar, S.O., 2011. Broccoli (Brassica oleracea var. Italica) sprouts and extracts rich in glucosinolates and isothiocyanates affect cholesterol metabolism and genes involved in lipid homeostasis in hamsters. *Journal of Agricultural and Food Chemistry*, *59*(4), pp. 1095–1103.

Velasco, P., Francisco, M., Moreno, D.A., Ferreres, F., García-Viguera, C. and Cartea, M.E., 2011. Phytochemical fingerprinting of vegetable Brassica oleracea and Brassica napus by simultaneous identification of glucosinolates and phenolics. *Phytochemical Analysis*, *22*(2), pp. 144–152.

37 Turnip (*Brassica rapa* L.)

Brassica rapa L.

Etymology: From the Celtic *bresic* = cabbage and the Latin *rapa* = turnip

Family: Brassicaeae

Synonym: *Brassica campestris* L.

Common names: Turnip, bird's rape; rabioule, grosse rave (Fr.); rübe (Ger.); nabo (Port.; Spa.); репа (Rus.)

Parts used: Leaf, root, seed

Constituents: Fixed oil (unsaturated fatty acids: oleic acid, linoleic acid, erucic acid) (Kaur et al., 2022); glucosinolates (Lee et al., 2013).

Medical history: Aphrodisiac, diuretic (Dioscorides); According to Galen, turnip induces blood clothing if eaten in excess (Galen). Simeon Seth (12th-century Byzantium) advocates the use of turnip to induce urination. Hot to the second degree and humid to the first (Fusch, 1555).

Medicinal uses: Eczema, blood purification (Pakistan)

Blood lipids and glucose: Ethanol extract of roots given to obese mice at 260 mg/100 g of chow reduced fasting plasma glucose, insulin, and cholesterol as well as hepatic lipids (Jung et al., 2008). Aqueous extract of leaves given orally to alloxan-induced diabetic rats at 200 mg/kg/day for 28 days decreased fasting glycemia from 515 to 63.2 mg/dL (normal: 4.9 mg/dL), cholesterol from 112.2 to 62.1 mg/dL (normal value: 81.6 mg/dL), triglycerides from 114 to 109 mg/dL (normal: 73.3 mg/dL), and LDL-cholesterol from 55.1 to 17.7 mg/dL (normal: 30.6 mg/dL) (Fard et al., 2015). Aqueous extract was hypolipidemic in rats (Birjand, 2015). Aqueous extract of roots given to alloxan-induced diabetic rats at 200 mg/kg/day for 8 weeks decreased fasting plasma glucose (Hassanzadeh-Taheri et al., 2016)

In obese patients, intake of ethanol extract at 2 g/day for 10 weeks increased HDL-cholesterol and total cholesterol (Jeon et al., 2013).

Kidneys: Aqueous extract of roots given to alloxan-induced diabetic rats at 200 mg/kg/day for 8 weeks decreased urine volume and protected kidneys, as evidenced by fewer glomerular lesions and thickened Bowman's capsules (Hassanzadeh-Taheri et al., 2016).

Bones and cartilages: In methotrexate-induced osteoporosis in rat, oil of seeds given orally at 400 mg/kg/day for 28 days lengthened femur shaft cortical bones and femur head trabecular bones (El-Makawy et al., 2020).

DOI: 10.1201/9781003301455-37

REFERENCES

Birjand, I., 2015. Hypolipidemic activity of aqueous extract of turnip (Brassica rapa) root in hyperlipidemic rats. *Ofogh-E-Danesh*, *21*, pp. 45–51.

El-Makawy, A.I., Ibrahim, F.M., Mabrouk, D.M., Abdel-Aziem, S.H., Sharaf, H.A. and Ramadan, M.F., 2020. Efficiency of turnip bioactive lipids in treating osteoporosis through activation of Osterix and suppression of Cathepsin K and TNF-α signaling in rats. *Environmental Science and Pollution Research*, *27*(17), pp. 20950–20961.

Fard, M.H., Naseh, G., Lotfi, N., Hosseini, S.M. and Hosseini, M., 2015. Effects of aqueous extract of turnip leaf (Brassica rapa) in alloxan-induced diabetic rats. *Avicenna Journal of Phytomedicine*, *5*(2), p. 148.

Fusch, L., 1555. *De Historia Stirpium Commetarii Insignes*. Lugduni Apud Ioan Tornaesium.

Hassanzadeh-Taheri, M., Hosseini, M., Hassanpour-Fard, M., Ghiravani, Z., Vazifeshenas-Darmiyan, K., Yousefi, S. and Ezi, S., 2016. Effect of turnip leaf and root extracts on renal function in diabetic rats. *Oriental Pharmacy and Experimental Medicine*, *16*, pp. 279–286.

Jeon, S.M., Kim, J.E., Shin, S.K., Kwon, E.Y., Jung, U.J., Baek, N.I., Lee, K.T., Jeong, T.S., Chung, H.G. and Choi, M.S., 2013. Randomized double-blind placebo-controlled trial of powdered Brassica rapa ethanol extract on alteration of body composition and plasma lipid and adipocytokine profiles in overweight subjects. *Journal of Medicinal Food*, *16*(2), pp. 133–138.

Jung, U.J., Baek, N.I., Chung, H.G., Bang, M.H., Jeong, T.S., Lee, K.T., Kang, Y.J., Lee, M.K., Kim, H.J., Yeo, J. and Choi, M.S., 2008. Effects of the ethanol extract of the roots of Brassica rapa on glucose and lipid metabolism in C57BL/KsJ-db/db mice. *Clinical Nutrition*, *27*(1), pp. 158–167.

Kaur, G., Kaur, R. and Kaur, S., 2022. Studies on physiochemical properties of oil extracted from Brassica nigra and Brassica rapa toria. *Materials Today: Proceedings*, *48*, pp. 1645–1651.

Lee, J.G., Bonnema, G., Zhang, N., Kwak, J.H., de Vos, R.C. and Beekwilder, J., 2013. Evaluation of glucosinolate variation in a collection of turnip (Brassica rapa) germplasm by the analysis of intact and desulfo glucosinolates. *Journal of Agricultural and Food Chemistry*, *61*(16), pp. 3984–3993.

38 Tea (*Camellia sinensis* (L.) Kuntze)

Camellia sinensis (L.) **Kuntze**

Etymology: After the Czech botanist Georg Joseph Kamel (1661–1706), who was the first European to study the plants of the Philippines and the author of *Historia stirpium insula Luzonis et reliquarum Philippinarum*, and from the Latin *sinensis* = from China

Family: Theaeae

Synonym: *Thea sinensis* L.

Common names: Tea; thé (Fr.); tee (Ger.); chá (Port.); чай (Rus.); té (Spa.)

Part used: Dried leaf after fermentation (black tea) or leafs after heating (green tea)

Constituents: Tannins (theaflavins in black tea) (Nishimura et al., 2007), phenolic acids (gallic acid), flavanes (epicatechin, epicatechin gallate, epigallocatechin, epigallocatechin gallate in green tea) (Chen et al., 2001), purine alkaloids (caffeine) (Zhang et al., 2022).

DOI: 10.1201/9781003301455-38

Medical history: The Dutch brough tea to Europe during the 17th century. The surgeon Nicolaes Tulp in his book *Amstelredamensis observationes medicae* (1672) writes that tea is good against sleepiness, invigorates the body, cures headache and flu, and removes kidney stones. The French physician Jonquet called tea "*herbe divine*" in 1657 while the apothecary Philippe Sylvestre Dufour (1688) writes in his *Traitez nouveaux et curieux du café, du thé et du chocolat* that tea can provoke agitation. A description of the plant is found in a book titled *Amoenitatum exoticarum* (1712) written by the German naturalist Engelbert Kaempfer. In 19th-century France, tea used was a mild stimulant and a diuretic and was used to help digestion and stimulate blood circulation (Moquin-Tandon, 1861).

Medicinal uses: Globally, tea is used as a drink with mild stimulant properties. Diarrhea (India); liver diseases, diabetes (Vietnam); aphrodisiac for men (Indonesia). Hakeems of Pakistan consider tea to be a "cold medicine" to be given to people with hot temperament but do not recommend it for people with cold temperament. I personally met Hakeems and other Asian traditional practitioners, and their medical skills should not be derided or viewed with contempt as they are improving the lives of many in destitute areas where specialists will not open their clinics. In developing countries, the use of medicinal plants to treat illnesses often saves lives.

Blood pressure: A number of studies have demonstrated that black tea, particularly with milk added, does not by itself reduce blood pressure to an appreciable degree. However, if taken regularly over a long period, tea does mildly reduce blood pressure (Bingham et al., 1997). The effects of tea without milk include a mild and transitory rise in blood pressure (caffeine) and a decrease of arterial stiffness. Another benefit of tea without milk is to mitigate the vasoconstrictions occurring after fatty meals.

For instance, 13 patients with mild systolic hypertension (130 to 150 mmHg) taking 200 mL of tea/day for 7 days (7.6 g of tea leaves in 400 mL of boiled water) experienced mild transient increases in systolic and diastolic blood pressure and a small decrease in heart rate (Hodgson et al., 1999). Hypertensive patients ingesting a polyphenol extract of tea (150 mg) twice a day for 8 days were protected against arterial stiffness and subsequent increased blood pressure induced by intake of fatty meals (Grassi et al., 2015).

Chronic daily intake of tea is much more beneficial. Volunteers taking 3 cups/day of black tea without milk for 4 weeks did not experience decrease in blood pressure but at night time had reduced blood pressure variation as well as a mild decrease in heart rate (Hodgson et al., 2013, 2013a). Volunteers with a systolic blood pressure of about 115 to 150 mmHg who drank tea for a period of 6 months at 3 cups/day had decreases in systolic and diastolic blood pressure of between 2 and 3 mm Hg (Hodgson et al., 2012).

The effects of green tea and black tea are about the same. Twenty-eight obese patients taking 1 capsule of green tea extract (containing 208 mg of epigallocatechin gallate/day for 3 months) had blood pressure decreases of about 4 mmHg (Bogdanski et al., 2012).

Plasma lipids and glucose: Rats fed green tea polyphenols (200 mg/kg body weight for 18 weeks) were protected against high-fat diet-induced vascular oxidative injuries (Xiao et al., 2022).

Twenty-eight obese patients taking 1 capsule of green tea extract (containing 208 mg of epigallocatechin gallate) per day for 3 months had mild decreases in LDL- and HDL-cholesterol and triglycerides (Bogdanski et al., 2012). Subjects taking 5 servings/day of black tea for 3 weeks had reductions in total and LDL-cholesterol of 3.8% and 7.5%, respectively, while plasma triglycerides where not affected (Davies et al., 2003). Further, black tea polyphenols improve insulin sensitivity altered by ROS and saturated fatty acids (Tong et al., 2018), and this is of value for diabetic patients, who tend to have high blood pressure.

Kidneys: Diuresis is a means of lowering blood pressure. Rats given black tea extract at 2.4 g/kg had an increase in urine flow from 2.2 to 3.8 mL/100 g/6 h with a peak at around 1 h and enhanced urinary secretion of sodium ions. This effect was superior to that of furosemide (13 mg/kg) (Abeywickrama et al., 2010). The consumption of tea tends to decrease the formation of kidney stones (Wang et al., 2021). Caffeine is a mild diuretic (Rieg et al., 2005).

Bones and cartilages: Aqueous extract given orally to ovariectomized mice daily for 28 days decreased plasma levels of alkaline phosphatase and increased the calcification of bones (Das et al., 2004). Ethanol extract given orally at 400 mg/kg/day to rodents with collagen-induced arthritis alleviated joint deformity, pannus formation, and neutrophils infiltration (Tanwar et al., 2017).

Skin and hair: Aqueous extract (2%) applied to the skin of mice exposed to ultraviolet B irradiation normalized total wrinkle area and decreased mast cells count in the dermis (Lee et al., 2014).

Brain: Rats given green tea for 8 weeks prior to intrahippocampal injection of amyloid β proteins were protected against cognitive dysfunction and oxidative stress and damage in the hippocampus (Schimidt et al., 2017).

Warnings: Caffeine in tea increases blood pressure, and polyphenols decrease iron absorption (Jain et al., 2013). Caffeine should be avoided in patients with panic attacks, high blood pressure, heart condition, anemia, and thalassemia. Black tea consumption increased the development of colon cancer in a population of 7,833 subjects, although it decreased prostate cancer incidence (Heilbrun et al., 1987). Posadzki et al. (2013) recommend using tea cautiously in patients taking analgesics, antilipemics, antiseizures, antivirals, and/or β-adrenoceptor blockers.

Comments: (i) The beneficial effect of tea on blood pressure seems to be the maintenance of vascular homeostasis and elasticity, which are impaired with the chronic high consumption of saturated fats (palm oil, animal fats) and cholesterol (Jaeger, 2001). With aging, ROS decrease the ability of endothelial cells to release NO (Bachschmid et al., 2013). Intake of fatty food increases arterial stiffness and contributes to post-prandial hypertension (Wilkinson & Cockcroft, 2007), and this is mitigated by intake of black tea without milk.

(ii) Consumption of saturated fats is detrimental to the arterial lining, and *in vitro*, palmitic acid (4 mM/L) (a major fatty acid in palm oil) is cytotoxic for vascular endothelial cells (Constantinides & Kiser, 1981).

(iii) Blood vessels are lined with a monolayer of vascular endothelium that maintains normal platelet aggregation and produces NO from L-arginine via endothelial nitric acid synthase. NO relaxes adjacent vascular smooth muscle cells, resulting in vasodilation and subsequent decreased blood pressure (Seals et al., 2011). Aortas, with time, tend to become still owed to continuous deposition of calcium ions and cholesterol, and this phenomenon is very much increased in smokers (Seals et al., 2011). Subjects ingesting black tea without milk (5 grams brewed for 3 min with 500 mL of boiled water) showed increased arterial elasticity (Lorenz et al., 2007).

(iv) With aging, endothelial NO synthetase expression and the production of NO decrease while superoxide anions production increases, and this oxidative imbalance results in decreased vasorelaxation and increasing blood pressure with age. Is the degeneration of mitochondria involved at the root of this? Free radicals interfere with the vasodilating effects of NO (Brandes et al., 2005).

(v) The longer tea leaves steep in boiling water, the more polyphenols there are and presumably the best the effects on the vascular system. However, tea must be enjoyed, and extensive maceration yields a bitter drink. In the end, it is clear that drinking black tea without sugar and milk is beneficial for the aging cardiovascular system.

REFERENCES

Abeywickrama, K.R.W., Ratnasooriya, W.D. and Amarakoon, A.M.T., 2010. Oral diuretic activity of hot water infusion of Sri Lankan black tea (Camellia sinensis L.) in rats. *Pharmacognosy Magazine*, 6(24), p. 271.

Bachschmid, M.M., Schildknecht, S., Matsui, R., Zee, R., Haeussler, D., A. Cohen, R., Pimental, D. and Loo, B.V.D., 2013. Vascular aging: Chronic oxidative stress and impairment of redox signaling—consequences for vascular homeostasis and disease. *Annals of Medicine*, 45(1), pp. 17–36.

Bingham, S.A., Vorster, H., Jerling, J.C., Magee, E., Mulligan, A., Runswick, S.A. and Cummings, J.H., 1997. Effect of black tea drinking on blood lipids, blood pressure and aspects of bowel habit. *British Journal of Nutrition*, 78(1), pp. 41–55.

Bogdanski, P., Suliburska, J., Szulinska, M., Stepien, M., Pupek-Musialik, D. and Jablecka, A., 2012. Green tea extract reduces blood pressure, inflammatory biomarkers, and oxidative stress and improves parameters associated with insulin resistance in obese, hypertensive patients. *Nutrition Research*, 32(6), pp. 421–427.

Brandes, R.P., Fleming, I. and Busse, R., 2005. Endothelial aging. *Cardiovascular Research*, 66(2), pp. 286–294.

Chen, Z.Y., Zhu, Q.Y., Tsang, D. and Huang, Y., 2001. Degradation of green tea catechins in tea drinks. *Journal of Agricultural and Food Chemistry*, 49(1), pp. 477–482.

Constantinides, P. and Kiser, M., 1981. Arterial effects of palmitic, linoleic and acetoacetic acid. *Atherosclerosis*, *38*(3–4), pp. 309–319.

Das, A.S., Mukherjee, M. and Mitra, C., 2004. Evidence for a prospective anti-osteoporosis effect of black tea (Camellia Sinensis) extract in a bilaterally ovariectomized rat model. *Asia Pacific Journal of Clinical Nutrition*, *13*(2), pp. 210–216.

Davies, M.J., Judd, J.T., Baer, D.J., Clevidence, B.A., Paul, D.R., Edwards, A.J., Wiseman, S.A., Muesing, R.A. and Chen, S.C., 2003. Black tea consumption reduces total and LDL cholesterol in mildly hypercholesterolemic adults. *The Journal of Nutrition*, *133*(10), pp. 3298S–3302S.

Grassi, D., Draijer, R., Desideri, G., Mulder, T. and Ferri, C., 2015. Black tea lowers blood pressure and wave reflections in fasted and postprandial conditions in hypertensive patients: A randomised study. *Nutrients*, *7*(2), pp. 1037–1051.

Heilbrun, L.K., Nomura, A. and Stemmermann, G.N., 1987. Black tea consumption and cancer risk: A prospective study. *British Journal of Cancer*, *54*(4), pp. 677–683.

Hodgson, J.M., Croft, K.D., Woodman, R.J., Puddey, I.B., Fuchs, D., Draijer, R., Lukoshkova, E. and Head, G.A., 2013. Black tea lowers the rate of blood pressure variation: A randomized controlled trial. *The American of Clinical Nutrition*, *97*(5), pp. 943–950.

Hodgson, J.M., Puddey, I.B., Burke, V., Beilin, L.J. and Jordan, N., 1999. Effects on blood pressure of drinking green and black tea. *Journal of Hypertension*, *17*(4), pp. 457–463.

Hodgson, J.M., Puddey, I.B., Woodman, R.J., Mulder, T.P., Fuchs, D., Scott, K. and Croft, K.D., 2012. Effects of black tea on blood pressure: A randomized controlled trial. *Archives of Internal Medicine*, *172*(2), pp. 186–188.

Hodgson, J.M., Woodman, R.J., Puddey, I.B., Mulder, T., Fuchs, D. and Croft, K.D., 2013a. Short-term effects of polyphenol-rich black tea on blood pressure in men and women. *Food & Function*, *4*(1), pp. 111–115.

Jaeger, B.R., 2001. Evidence for maximal treatment of atherosclerosis: Drastic reduction of cholesterol and fibrinogen restores vascular homeostasis. *Therapeutic Apheresis*, *5*(3), pp. 207–211.

Jain, A., Manghani, C., Kohli, S., Nigam, D. and Rani, V., 2013. Tea and human health: The dark shadows. *Toxicology Letters*, *220*(1), pp. 82–87.

Lee, K.O., Kim, S.N. and Kim, Y.C., 2014. Anti-wrinkle effects of water extracts of teas in hairless mouse. *Toxicological Research*, *30*, pp. 283–289.

Lorenz, M., Jochmann, N., von Krosigk, A., Martus, P., Baumann, G., Stangl, K. and Stangl, V., 2007. Addition of milk prevents vascular protective effects of tea. *European Heart Journal*, *28*(2), pp. 219–223.

Moquin-Tandon, A., 1861. *Element de Botanique Médicale*. J.Baillière et Fils.

Nishimura, M., Ishiyama, K., Watanabe, A., Kawano, S., Miyase, T. and Sano, M., 2007. Determination of theaflavins including methylated theaflavins in black tea leaves by solid-phase extraction and HPLC analysis. *Journal of Agricultural and Food Chemistry*, *55*(18), pp. 7252–7257.

Rieg, T., Steigele, H., Schnermann, J., Richter, K., Osswald, H. and Vallon, V., 2005. Requirement of intact adenosine A1 receptors for the diuretic and natriuretic action of the methylxanthines theophylline and caffeine. *Journal of Pharmacology and Experimental Therapeutics*, *313*(1), pp. 403–409.

Schimidt, H.L., Garcia, A., Martins, A., Mello-Carpes, P.B. and Carpes, F.P., 2017. Green tea supplementation produces better neuroprotective effects than red and black tea in Alzheimer-like rat model. *Food Research International*, *100*, pp. 442–448.

Seals, D.R., Jablonski, K.L. and Donato, A.J., 2011. Aging and vascular endothelial function in humans. *Clinical Science*, *120*(9), pp. 357–375.

Tanwar, A., Chawla, R., Ansari, M.M., Thakur, P., Chakotiya, A.S., Goel, R., Ojha, H., Asif, M., Basu, M., Arora, R. and Khan, H.A., 2017. In vivo anti-arthritic efficacy of Camellia sinensis (L.) in collagen induced arthritis model. *Biomedicine & Pharmacotherapy*, *87*, pp. 92–101.

Tong, T., Ren, N., Soomi, P., Wu, J., Guo, N., Kang, H., Kim, E., Wu, Y., He, P., Tu, Y. and Li, B., 2018. Theaflavins improve insulin sensitivity through regulating mitochondrial biosynthesis in palmitic acid-induced HepG2 cells. *Molecules*, *23*(12), p. 3382.

Wang, H., Fan, J., Yu, C., Guo, Y., Pei, P., Yang, L., Chen, Y., Du, H., Meng, F., Chen, J. and Chen, Z., 2021. Consumption of tea, alcohol, and fruits and risk of kidney stones: A prospective cohort study in 0.5 million Chinese adults. *Nutrients*, *13*(4), p. 1119.

Wilkinson, I. and Cockcroft, J.R., 2007. Cholesterol, lipids and arterial stiffness. *Atherosclerosis, Large Arteries and Cardiovascular Risk*, *44*, pp. 261–277.

Xiao, X.T., He, S.Q., Wu, N.N., Lin, X.C., Zhao, J. and Tian, C., 2022. Green tea polyphenols prevent early vascular aging induced by high-fat diet via promoting autophagy in young adult rats. *Current Medical Science*, pp. 1–10.

Zhang, S., Jin, J., Chen, J., Ercisli, S. and Chen, L., 2022. Purine alkaloids in tea plants: Component, biosynthetic mechanism and genetic variation. *Beverage Plant Research*, *2*(1), pp. 1–9.

39 Shepherd's Purse (*Capsella bursa-pastoris* (L.) Medik)

Capsella bursa-pastoris (L.) Medik

Etymology: From the Latin *capsa* = a box and *bursa pastoris* = shepherd's purse

Family: Brassicaceae

Synonyms: *Bursa bursa-pastoris* (L.) Britton; *Lepidium bursa-pastoris* (L.) Willd.; *Thlaspi bursa-pastoris* L.

Common names: Lady's purse, mother's heart, shepherd's purse; bourse à pasteur (Fr.); gänsekresse (Ger.); bolsa do pastor (Port.); пастýшья сумка (Rus.); bolsa de pastor (Spa.)

Part used: Leaf

Constituents: Flavone glycosides, choline, acetylcholine (Miyazama et al., 1979; Al-Snafi, 2015).

Medical history: Shepherd's purse was used to warm the bile and break internal abscesses by Dioscorides, as choleretic by Galen, and to stop bleeding and sciatica by Paul of Aegina. In the 16th-century Germany, Jacobus Theodorus Tabernaemontanus (1590) recommended shepherd's purse for bleeding. Diuretic, for diarrhea and menorrhagia in 19th-century Europe and North America.

Medicinal uses: Heavy menstrual bleeding 1–5 g in 150 mL of boiling water, 2–4 times daily (European Union); kidney stones, diuretic, diabetes (Turkey); inflamed joints (Pakistan); cardiostimulant, anti-inflammatory, tonic, diuretic (India); hemostasis (Korea)

Blood lipids and glucose: Ethanol extract given at 1% of Western diet to mice for 12 weeks caused a reduction of plasma cholesterol from 195.7 to 145.5 mg/dL and LDL-cholesterol from 48 to 24 mg/dL (Hwang et al., 2021). Given to diabetic rats, ethanol extract reduced plasma cholesterol (Sook, 1992). Ethanol extract given to mice on a high-fat diet afforded some protection against fat accumulation in the liver (Choi et al., 2017).

Warning: Extracts of the plant promote blood coagulation (Vermathen & Glasl, 1993).

Comments: (i) Paul of Aegina (625–690) was a Greek surgeon and author of a medical text titled *Epitomae medicae libri septem* that was used as a reference in European and the Middle East during the Middle Ages.

DOI: 10.1201/9781003301455-39

(ii) Jacobus Theodorus Tabernaemontanus (1525–1590) was a German physician and herbalist and the author of a book titled *Eicones plantarum seu stirpium* (1590).

(iii) Blood stasis, vessel walls injuries, enhanced fibrin formation and delayed fibrinolysis contributes to a higher incidence of thromboembolic disorders in the aged (Hager et al., 1989).

REFERENCES

Al-Snafi, A.E., 2015. The chemical constituents and pharmacological effects of Capsella bursa-pastoris-A review. *International Journal of Pharmacology and Toxicology*, 5(2), pp. 76–81.

Choi, H.K., Shin, E.J., Park, S.J., Hur, H.J., Park, J.H., Chung, M.Y., Kim, M.S. and Hwang, J.T., 2017. Ethanol extract of Capsella bursa-pastoris improves hepatic steatosis through inhibition of histone acetyltransferase activity. *Journal of Medicinal Food*, 20(3), pp. 251–257.

Hager, K., Setzer, J., Vogl, T., Voit, J. and Platt, D., 1989. Blood coagulation factors in the elderly. *Archives of Gerontology and Geriatrics*, 9(3), pp. 277–282.

Hwang, J.T., Choi, E., Choi, H.K., Park, J.H. and Chung, M.Y., 2021. The cholesterol-lowering effect of Capsella Bursa-Pastoris is mediated via SREBP2 and HNF-1α-regulated PCSK9 inhibition in obese mice and HepG2 cells. *Foods*, 10(2), p. 408.

Miyazama M., Uetake, A. and Kamoka, H., 1979. The constituents of the essential oils from Capsella bursa-pastoris Menk. *Yakugaku Zasshi*, 10(99), pp. 1041–1043.

Sook, J.L., 1992. Hypoglycemic effects of korean wild vegetables. *Journal of Nutrition and Health*, 25(6), pp. 511–517.

Theodorus, Jacobus, and Basse, Nikolaus. *Eicones plantarum, seu stirpium, arborum nempe, fructicum, herbarum, fructuum lignorum*. N.p., Basse, 1590.

Vermathen, M. and Glasl, H., 1993. Effect of the herb extract of Capsella bursa-pastoris on blood coagulation. *Planta Medica*, 59(S 1), pp. A670–A670.

40 Chili Pepper (*Capsicum annuum* L.)

Capsicum annuum L.

Etymology: From the Latin *capsa* = a box and *annuum* = yearly

Family: Solanaceae

Synonyms: *Capsicum conoides* Mill.; *Capsicum frutescens* L.; *Capsicum longum* A. DC.

Common names: Capsicum; bell pepper, cayenne pepper, chili pepper, garden pepper, green pepper, paprika pepper; corail des jardins, piment, poivre d'Inde (Fr.); paprika (Ger.); pimentão (Port.); Болгарский перец (Rus.); pimiento (Spa.)

Part used: Fruit

Constituents: Amide alkaloids (capsaicin, dihydrocapsaicin) (Leete & Louden, 1968).

Medical history: The plant was known to ancient Roman physicians. Pliny the Elder calls chili pepper *siliquatrum* and *piperitis* and recommends it for gums and teeth problems. Dioscorides says it helps those suffering from angina pectoris and was used as an analgesic. Carminative, sore, local irritant in 19th-century Europe.

Medicinal uses: Appetite stimulant, diuretic, stimulant (Turkey); carminative (Iraq and Iran); putrid sore throat, stimulant, rubefacient (India); rubefacient (Myanmar)

Plasma lipids and glucose: Mixing chili pepper juice in the drink of mice on a high-fat diet decreased cholesterol and triglycerides but had no effects on glycemia (Kim & Park, 2015).

Warnings: A 59-year-old man with high thyroid-stimulating hormone was admitted to ICU in hypertensive crisis with acute myocardial infarction 24 hours after ingesting enormous amounts of chili pepper (Patanè et al., 2009). In rats, intravenous injection of capsaicin at 10 µg/kg caused short periods of hypotension followed by hypertension, probably mediated by endothelin (Dutta & Deshpande, 2010). In dogs, intravenous injection of capsaicin causes an increase in blood pressure (Toda et al., 1972).

Comment: Chili pepper does not decrease blood pressure (Kim & Park, 2015). Capsules containing 500 mg of capsaicin given to hypertensive patients on medication daily for 4 weeks did not cause a decrease in blood pressure (Salvador et al., 2015).

DOI: 10.1201/9781003301455-40

REFERENCES

Dutta, A. and Deshpande, S.B., 2010. Mechanisms underlying the hypertensive response induced by capsaicin. *International Journal of Cardiology*, *145*(2), pp. 358–359.

Kim, N.H. and Park, S.H., 2015. Evaluation of green pepper (Capsicum annuum L.) juice on the weight gain and changes in lipid profile in C57BL/6 mice fed a high-fat diet. *Journal of the Science of Food and Agriculture*, *95*(1), pp. 79–87.

Leete, E. and Louden, M.C., 1968. Biosynthesis of capsaicin and dihydrocapsaicin in Capsicum frutescens. *Journal of the American Chemical Society*, *90*(24), pp. 6837–6841.

Patanè, S., Marte, F., Di Bella, G., Cerrito, M. and Coglitore, S., 2009. Capsaicin, arterial hypertensive crisis and acute myocardial infarction associated with high levels of thyroid stimulating hormone. *International Journal of Cardiology*, *134*(1), pp. 130–132.

Salvador, F.A.P., Salvaña, C.A., Sedano, A.S., Yap, F.J.C., Yap, L.A.G., Yap, M.D.R.B., Yee, A.M.G., Yu, C.A.G. and Zamesa III, P.N.H., 2015 A randomized double-blind placebo-controlled trial on the effects of Capsicum annum as an adjunct for reducing blood pressure among hypertensive patients. *UER*, *4*(2), pp. 110–116.

Toda, N., Usui, H., Nishino, N. and Fujiwara, M., 1972. Cardiovascular effects of capsaicin in dogs and rabbits. *Journal of Pharmacology and Experimental Therapeutics*, *181*(3), pp. 512–521.

41 Papaya
(*Carica papaya* L.)

Carica papaya **L.**

Etymology: From the Latin *carica* = fig and the Amerindian word *papaya*

Family: Caricaceae

Synonym: *Papaya carica* Gaertn.

Common names: Melon tree, papaya; papaye (Fr.); papaya (Ger.; Spa.); mamão (Port.); пап á йя (Rus.)

Part used: Fruit

Constituents: Isothiocyanates (benzyl isothiocyanate in the seeds) (Burdick, 1971; Nakamura et al., 2007), pectin (fruit flesh) (Westerlund et al., 1991), proteolytic enzyme (papain in young fruits) (Drenth et al., 1971)

Medical history: The Spanish physician Francisco Hernandez de Toledo (1514–1596) explored Mexico from 1570 to 1577 and describes papaya to Europeans for the first time on pages 99 and 870 of his monumental book entitled *Nova plantarum, animalium, et mineralium Mexicanorum* (1651) under the name *Papaya orientalis* or *Pepo arborescent*. He stated that natives used papaya for intestinal worms, impetigo, and stomachache. The French naturalist Charles de Rochefort (1605–1683), who spent a decade of his life in the Caribbean region, wrote a book titled *Histoire naturelle et mo rale des Antilles de l'Amerique* (1658), where one can read that natives used papaya to help digestion. In 19th-century North America, latex was used for dyspepsia and warts and to dissolve the false membrane of diphtheria (Shoemaker, 1895).

Medicinal uses: Digestion, ringworms (Bangladesh); digestion, kidneys, heart, gallstone, blood, spleen, liver (Myanmar). In the Philippines, the immature fruits are eaten as famine food and boiled in chicken soup to increase milk in lactating mothers.

Blood pressure: Daily intake of 200 g of papaya fruits by elderly people with hypertension for 7 days caused a mild decrease in blood pressure (Wahdi et al., 2020).

In rats fed a high-fat diet, papaya juice given for 12 weeks at 0.5 mL/100 g of body weight decreased blood pressure from 134.8 to 129.2 mg/dL (Od-Ek et al., 2020).

Plasma lipids and glucose: Aqueous extract of seeds given orally to rats at 400 mg/kg/day for 30 days decreased fasting blood glucose from about 90 to 40 mg/dL as well as plasma cholesterol from 100 to 71 mg/dL and triglycerides from 100.5 to 68.3 mg/dL (100 mg/kg/day) (Adeneye & Olagunju, 2009).

DOI: 10.1201/9781003301455-41

In rats fed a high-fat diet, papaya juice given for 12 weeks at 0.5 mL/100 g of body weight decreased plasma insulin from 5.5 to 3.5 ng/dL as well as plasma cholesterol and triglycerides (Od-Ek et al., 2020).

Kidneys: Aqueous extract of unripe papaya seeds given to streptozotocin-induced diabetic rats at 200 mg/kg/day for 21 days mildly decreased glycemia and plasma urea from 42.7 to 32.3 mg/dL, creatinine from 2.4 to 2.2 mg/dL, and blood urea nitrogen from 60.7 to 38.7 mg/dL (Nnaemeka et al., 2023).

Warnings: The latex is dangerous in pregnancy (Gopalakrishnan & Rajasekharasetty, 1978). It literally dissolves proteins and causes severe esophageal injuries when used to treat food impaction (Weiner et al., 1978). The glycemic index of papaya is 59, and diabetics should avoid eating too much of it (Foster-Powell et al., 2002) because they need to consume foods with glycemic indices <55 (Atkinson et al., 2008).

Comments: (i) It is clear that the fresh papaya juice has beneficial effects on blood pressure and plasma lipids. It must however be avoided with patients with hyperkaliemia.

(ii) Pectin in the gut interacts with cholesterol and bile acids to decrease serum and hepatic cholesterol. Consumption of 15 g/day of pectin increased fecal bile acids by 35%. Intake of pectin after carbohydrate meals decreased glycemia and plasma insulin in type-2 diabetic patients (Endress, 1991).

(iii) Papaya is beneficial for fatty livers. The juice given at 1 mL/100 g for 12 weeks to rats fed a high-fat diet decreased body weight from 536 to 471.3 g and hepatic cholesterol from 152.6 to 92.3 mg/dL. This treatment also prevented hepatic steatosis, lobular inflammation, and decreased hepatic triglycerides, MDA, TNF α, and interleukin-6 (Deenin et al., 2021).

REFERENCES

Adeneye, A.A. and Olagunju, J.A., 2009. Preliminary hypoglycemic and hypolipidemic activities of the aqueous seed extract of Carica papaya Linn in Wistar rats. *Biology and Medicine*, *1*(1), pp. 1–10.

Atkinson, F.S., Foster-Powell, K. and Brand-Miller, J.C., 2008. International tables of glycemic index and glycemic load values. *Diabetes Care*, *31*(12), pp. 2281–2283.

Burdick, E.M., 1971. Carpaine: An alkaloid of Carica papaya: Its chemistry and pharmacology. *Economic Botany*, pp. 363–365.

Deenin, W., Malakul, W., Boonsong, T., Phoungpetchara, I. and Tunsophon, S., 2021. Papaya improves non-alcoholic fatty liver disease in obese rats by attenuating oxidative stress, inflammation and lipogenic gene expression. *World Journal of Hepatology*, *13*(3), p. 315.

Drenth, J., Jansonius, J.N., Koekoek, R. and Wolthers, B.G., 1971. The structure of papain. *Advances in Protein Chemistry*, *25*, pp. 79–115.

Endress, H.U., 1991. Nonfood uses of pectin. *The Chemistry and Technology of Pectin*, pp. 251–268.

Foster-Powell, K., Holt, S.H. and Brand-Miller, J.C., 2002. International table of glycemic index and glycemic load values: 2002. *The American Journal of Clinical Nutrition*, *76*(1), pp. 5–56.

Gopalakrishnan, M. and Rajasekharasetty, M.R., 1978. Effect of papaya (Carica papaya linn) on pregnancy and estrous cycle in albino rats of Wistar strain. *Indian Journal of Physiology and Pharmacology*, 22(1), pp. 66–70.

Nakamura, Y., Yoshimoto, M., Murata, Y., Shimoishi, Y., Asai, Y., Park, E.Y., Sato, K. and Nakamura, Y., 2007. Papaya seed represents a rich source of biologically active isothiocyanate. *Journal of Agricultural and Food Chemistry*, 55(11), pp. 4407–4413.

Nnaemeka, U.M., Ukamaka, I.A., God'swealth, U.S. and Resame, G.R., 2023. Comparative study of aqueous, methanol and petroleum ether extracts of unripe Carica papaya seed on liver and kidney function in streptozotocin-induced diabetic rats. *GSC Biological and Pharmaceutical Sciences*, 22(1), pp. 038–047.

Od-Ek, P., Deenin, W., Malakul, W., Phoungpetchara, I. and Tunsophon, S., 2020. Anti-obesity effect of Carica papaya in high-fat diet fed rats. *Biomedical Reports*, 13(4), pp. 1–1.

Wahdi, A., Astuti, P., Puspitosari, D.R., Maisaroh, S. and Pratiwi, T.F., 2020. The Effectiveness of giving papaya fruit (Carica papaya) toward blood pressure on elderly hypertension patients. In *IOP Conference Series: Earth and Environmental Science* (Vol. 519, No. 1, p. 012007). IOP Publishing, June.

Weiner, B., Curtis, R., Lovejoy, D. and Rheinlander, H.F., 1978. Three case reports. *Drug Intelligence & Clinical Pharmacy*, 12(8), pp. 458–460.

Westerlund, E., Åman, P., Andersson, R., Andersson, R.E. and Rahman, S.M.M., 1991. Chemical characterization of water-soluble pectin in papaya fruit. *Carbohydrate Polymers*, 15(1), pp. 67–78.

42 Caraway (*Carum carvi* L.)

Carum carvi L.

Etymology: From the Arab *karwiya* and the Latin *carvi* = caraway

Family: Apiaceae

Synonyms: *Bunium carvi* (L.) M. Bieb.; *Carum gracile* Lindl; *Foeniculum carvi* (L.) Link

Common names: Caraway; carvi (Fr.); kümmel (Ger.); alcaravia (Port.); тмин (Rus.); alcaravea (Spa.)

Part used: Seed

Constituent: Essential oil (carvone, limonene) (Baysal & Starmans, 1999).

Medical history: Known as *karos* by Dioscorides for being diuretic and carminative and for perfuming the mouth. It is the *careum* of Pliny the Elder. Galen designates caraway as hot and dry to the third degree and recommends it to induce urination and for flatulence. The plant was known to the 13th-century Arab physician Ibn Baytar. In his 18th-century lectures, Alston (1770) recommends the use of caraway for vertigo. In 19th-century Europe, it was used as a carminative (Guibourt, 1836).

Medicinal uses: Relief of bloating and flatulence (0.5–2 g of seeds in 150 mL of boiling water as an herbal infusion, 1–3 times daily) (European Union); kidney stones (Turkey); carminative (Afghanistan); internal wounds, dysentery (Pakistan); inflammation (China).

Plasma cholesterol and glucose: Aqueous extract of seeds given orally at 60 mg/kg/day for 8 weeks to rats poisoned with cholesterol-enriched diet decreased plasma cholesterol from 152.8 to 91.8 mg/dL, triglycerides from 166.5 to 110 mg/dL, LDL-cholesterol from 103.6 to 23.7 mg/dL and these effects where as effective as Simvastatin at 1 mg/kg (Saghir et al., 2012).

Obese women taking daily 30 mL of an aqueous extract made from mixing 1 g of essential oil of seeds in 1 L of water for 12 weeks had no effect on plasma glucose and blood pressure but decreased plasma cholesterol from 209.3 to 199 mg/dL (Kazemipoor et al., 2014).

Aqueous extract of seeds given orally at 20 mg/kg/day for 15 days decreased plasma cholesterol from about 2.8 to 2.4 mmol/L in normal rats and from about 4 to 3 mmol/L in diabetic rats. In diabetic rats, this regimen attenuated weight loss (Lemhadri et al., 2006).

DOI: 10.1201/9781003301455-42

Gallbladder: The seeds contain components that promote the flow of bile from the gallbladder to the intestines by inhibiting spasms in the bile ducts and sphincter (Spiridonov, 2012).

Kidneys: In streptozotocin-induced diabetic rats, oral administration of essential oil at 10 mg/kg/day for 21 days decreased glycemia from 162.5 to 61.7 mg/dL and protected kidneys against glomerular degeneration (Abou El-Soud et al., 2014). Aqueous extract of seeds given to rats orally at 100 mg/kg/day for 8 days increased the secretion of urine and the urinary excretion of sodium ions (Lahlou et al., 2007).

Brain: Aqueous extract at the concentration of 100 μg/mL protected murine microglial cells (BV-2) against the lipopolysaccharide-induced activation of NF-kB and downstream secretion of interleukin-6 and TNF α (Kopalli & Koppula, 2015).

Warnings: Patients with gallstones should not take caraway (EU herbal monographs), and nor should those with thyroid troubles. Intake of caraway at 40 mg/kg/day for 2 weeks caused an increase in thyroid stimulating hormone in a hypothyroidic patient taking levothyroxine (Naghibi et al., 2015). The maximal daily amount of caraway seeds is 15 g (Ministry of Health, Canada).

Comments: (i) Addition of caraway in diet at normal dietary amounts and intake of caraway infusion could potentially be a mean of controlling cholesterol.

(ii) Microglial cells are macrophages in the brain that on activation secrete pro-inflammatory cytokines and activate a respiratory burst that destroys dopaminergic neurons, leading to Parkinson's disease (Qian & Flood, 2008).

REFERENCES

Abou El-Soud, N.H., El-Lithy, N.A., El-Saeed, G., Wahby, M.S., Khalil, M.Y., Morsy, F. and Shaffie, N., 2014. Renoprotective effects of caraway (Carum carvi L.) essential oil in streptozotocin induced diabetic rats. *Journal of Applied Pharmaceutical Science*, *4*(2), pp. 027–033.

Alston, C., 1770. *Lectures on the Materia Medica: Containing the Natural History of Drugs, their Virtues and Doses: Also Directions for the Study of the Materia Medica; and an Appendix on the Method of Prescribing*. Edward and Charles Dilly.

Baysal, T. and Starmans, D.A.J., 1999. Supercritical carbon dioxide extraction of carvone and limonene from caraway seed. *The Journal of Supercritical Fluids*, *14*(3), pp. 225–234.

Guibourt, N.J.B.G., 1836. *Histoire abrégée des drogues simples*. Méquignon-Marvis Père et fils.

Kazemipoor, M., Radzi, C.W.J.B.W.M., Hajifaraji, M. and Cordell, G.A., 2014. Preliminary safety evaluation and biochemical efficacy of a Carum carvi extract: Results from a randomized, triple-blind, and placebo-controlled clinical trial. *Phytotherapy Research*, *28*(10), pp. 1456–1460.

Kopalli, S.R. and Koppula, S., 2015. Carum carvi Linn (Umbelliferae) attenuates lipopolysaccharide-induced neuroinflammatory responses via regulation of NF-κB signaling in BV-2 Microglia. *Tropical Journal of Pharmaceutical Research*, *14*(6), pp. 1041–1047.

Lahlou, S., Tahraoui, A., Israili, Z. and Lyoussi, B., 2007. Diuretic activity of the aqueous extracts of Carum carvi and Tanacetum vulgare in normal rats. *Journal of Ethnopharmacology*, *110*(3), pp. 458–463.

Lemhadri, A., Hajji, L., Michel, J.B. and Eddouks, M., 2006. Cholesterol and triglycerides lowering activities of caraway fruits in normal and streptozotocin diabetic rats. *Journal of Ethnopharmacology*, *106*(3), pp. 321–326.

Naghibi, S.M., Ramezani, M., Ayati, N. and Zakavi, S.R., 2015. Carum induced hypothyroidism: An interesting observation and an experiment. *DARU Journal of Pharmaceutical Sciences*, *23*, pp. 1–4.

Qian, L. and Flood, P.M., 2008. Microglial cells and Parkinson's disease. *Immunologic Research*, *41*, pp. 155–164.

Saghir, M.R., Sadiq, S., Nayak, S. and Tahir, M.U., 2012. Hypolipidemic effect of aqueous extract of Carum carvi (black Zeera) seeds in diet induced hyperlipidemic rats. *Pakistan Journal of Pharmaceutical Sciences*, *25*(2).

Spiridonov, N.A., 2012. Mechanisms of action of herbal cholagogues. *Medicinal and Aromatic Plants*, *1*, p. 107.

43 Indian Pennywort (*Centella asiatica* (L.) Urb.)

Centella asiatica (L.) Urb.

Etymology: From the local Indian name *gotu kola* = Indian pennywort and the Latin *asiatica* = from Asia

Family: Apiaceae

Synonym: *Hydrocotyle asiatica* L.

Common names: Indian pennywort; hydrocotyle d'Asie (Fr.); Asiatisches was sernabelkraut (Ger.); centela (Port.); centella (Spa.)

Part used: Leaf Constituents: Ursane-type triterpenes (madecassic acid, asiatic acid) and saponins (madecassoside) (Inamdar et al., 1996).

Medical history: In ancient India, Ayurvedic physicians used the plant for fever, jaundice, leprosy, and gonorrhea according to Dymock (1884). Udoy Chand Dutt (1877) informs us that Indian pennywort was called in Sanskrit *mandulkaparni* and used as a tonic and for diseases of the blood as well as being parts of formulations for diabetes and urinary tract disorders. In 19th-century Europe, it was used for leprosy and syphilis.

Medicinal uses: This is one of the most used medicinal plants in Asia. In India, uses include those mentioned earlier plus anxiolytic, high blood pressure, and to promote production of milk. The fresh plant is a *jamu* mixed with rice by elderly Malays and Indonesian and consumed daily to maintain good health. Urinary disorders, stimulate digestion (Nepal). Diabetes, urinary troubles, induce sleep (Bangladesh); syphilitic ulcers, dysentery, headache (Vietnam)

Plasma cholesterol and glucose: Aqueous extract given at 50 mg/kg/day to alloxan-induced diabetic rats for 10 days decreased plasma glucose from 247.7 to 187.2 mg/dL and cholesterol from 237.2 to 116.1 mg/dL and attenuated plasmatic urea (Emran et al., 2015).

Blood pressure: Intake of a tea of Indian pennywort by hypertensive elderly patients caused a decrease in diastolic and systolic blood pressure (Astutik et al., 2021).

Ethanol extract of leaves given orally to rats at 500 mg/kg/day for 8 weeks attenuated hypertension induced by L-NAME with a decrease in systolic blood pressure from 148.5 to 131.7 mmHg, and these effects were comparable with those of captopril. This regimen decreased cardiac ACE

DOI: 10.1201/9781003301455-43

activity and cardiac and aortic lipid oxidation content, increased circulating NO, and normalized serum levels of brain natriuretic peptide (Bunaim et al., 2021).

Heart: Extract given orally at 100 mg/kg/day for 21 days protected rats against subcutaneous injection of isoproterenol on the 20th day, brought plasma lactate dehydrogenase and creatinine kinase close to normal, and increased heart glutathione as well as myocardial cytoarchitecture (Kumar et al., 2015).

Kidneys: A single oral dose of aqueous extract given to rats at 500 mg/kg increased urine excretion from 3.2 to 9.4 mL within 18 hours, and this effect was comparable with that of furosemide at 10 mg/kg (Al Huda & Debnath, 2017). Asiaticoside at 32 mg/kg mitigated adriamycin-induced nephropathy in rats (Wang et al., 2013).

Bones and cartilages: Asiatic acid given orally for 16 weeks at 50 mg/kg/day attenuated the deterioration of bone microstructure in rats (Chen et al., 2022). Madecassoside given orally at 40 mg/kg/day for 20 days protected mice against infiltration of inflammatory cells, synovial hyperplasia, and joint destruction (Liu et al., 2008).

Skin and hair: Ethanol extract at the concentration of 50 μg/mL induced the secretion of vascular endothelial cell growth factor (VEGF) by human follicle dermal papilla cells (Saansoomchai et al., 2018).

Brain: Extract given orally at 2.5 g/kg/day for 8 months to mice model of Alzheimer's disease decreased hippocampal levels of amyloid β proteins by 67% (Dhanasekaran et al., 2009). Aqueous extract given orally to rats at the daily dose of 300 mg/kg evoked some protection against 1-methyl-4-phenyl-1,2,3,6—tetrahydropyridine-induced Parkinsonism (Haleagrahara & Ponnusamy, 2010).

Warning: Tablets containing the plant evoked liver poisoning (Jorge & Jorge, 2005).

Comments: (i) William Dymock (1834–1892) was an English physician who spent most of his life in India. He wrote a book titled *The vegetable materia medica of Western India* (1884).

(ii) Udoy Chand Dutt (1834–1884) was an Indian physician and civil medical officer who wrote a book titled *The materia medica of the Hindus* (1877).

(iii) A number of over-the-counter products containing triterpene saponin fractions of Indian pennywort are commercially available in Europe for the treatment of chronic veinous insufficiency and varicose veins. Such products might be beneficial for the aging cardiovascular system.

(iv) VEGF induces the proliferation of human follicle dermal papilla cells through extracellular signal-regulated kinase activation (Li et al., 2012).

REFERENCES

Al Huda, E. and Debnath, J., 2017. Evaluation of diuretic activity of aqueous extract of leaves of Centella asiatica. *World Journal of Pharmaceutical Research*, 6(10), pp. 494–500.

Astutik, F.E.F., Zuhroh, D.F. and Ramadhan, M.R.L., 2021. The effect of gotu kola (Centella asiatica L.) tea on blood pressure of hypertension. *Enfermeria Clinica, 31*, pp. S195–S198.

Bunaim, M.K., Kamisah, Y., Mohd Mustazil, M.N., Fadhlullah Zuhair, J.S., Juliana, A.H. and Muhammad, N., 2021. Centella asiatica (L.) Urb. prevents hypertension and protects the heart in chronic nitric oxide deficiency rat model. *Frontiers in Pharmacology*, p. 3311.

Chen, X., Han, D., Liu, T., Huang, C., Hu, Z., Tan, X. and Wu, S., 2022. Asiatic acid improves high-fat-diet-induced osteoporosis in mice via regulating SIRT1/FOXO1 signaling and inhibiting oxidative stress. *Histology and Histopathology, 37*(8), pp. 769–777.

Dhanasekaran, M., Holcomb, L.A., Hitt, A.R., Tharakan, B., Porter, J.W., Young, K.A. and Manyam, B.V., 2009. Centella asiatica extract selectively decreases amyloid β levels in hippocampus of alzheimer's disease animal model. *Phytotherapy Research: An International Journal Devoted to Pharmacological and Toxicological Evaluation of Natural Product Derivatives, 23*(1), pp. 14–19.

Dutt, U.C., 1877. *The Materia Medica of the Hindus*. Thacker, Spink, & Co.

Dymock, W., 1884. *The Vegetable Materia Medica of Western India*. Education Society Press.

Emran, T.B., Dutta, M., Uddin, M.M.N., Nath, A.K. and Uddin, M.Z., 2015. Antidiabetic potential of the leaf extract of Centella asiatica in alloxaninduced diabetic rats. *Jahangirnagar University Journal of Biological Sciences, 4*(1), pp. 51–59.

Haleagrahara, N. and Ponnusamy, K., 2010. Neuroprotective effect of Centella asiatica extract (CAE) on experimentally induced parkinsonism in aged rats. *The Journal of Toxicological Sciences, 35*(1), pp. 41–47.

Inamdar, P.K., Yeole, R.D., Ghogare, A.B. and De Souza, N.J., 1996. Determination of biologically active constituents in Centella asiatica. *Journal of Chromatography A, 742*(1–2), pp. 127–130.

Jorge, O.A. and Jorge, A.D., 2005. Hepatotoxicity associated with the ingestion of Centella asiatica. *Revista Espanola de Enfermedades Digestivas, 97*(2), pp. 115–124.

Kumar, V., Babu, V., Nagarajan, K., Machawal, L. and Bajaj, U., 2015. Protective effects of centella asiatica against isoproterenol-induced myocardial infarction in rats: Biochemical, mitochondrial and histological findings. *The Journal of Phytopharmacology, 4*(2), pp. 80–86.

Li, W., Man, X.Y., Li, C.M., Chen, J.Q., Zhou, J., Cai, S.Q., Lu, Z.F. and Zheng, M., 2012. VEGF induces proliferation of human hair follicle dermal papilla cells through VEGFR-2-mediated activation of ERK. *Experimental Cell Research, 318*(14), pp. 1633–1640.

Liu, M., Dai, Y., Yao, X., Li, Y., Luo, Y., Xia, Y. and Gong, Z., 2008. Anti-rheumatoid arthritic effect of madecassoside on type II collagen-induced arthritis in mice. *International Immunopharmacology, 8*(11), pp. 1561–1566.

Saansoomchai, P., Limmongkon, A., Surangkul, D., Chewonarin, T. and Srikummool, M., 2018. Enhanced VEGF expression in hair follicle dermal papilla cells by Centella asiatica Linn. *Chiang Mai University Journal of Natural Sciences, 17*(1), pp. 25–37.

Wang, Z., Liu, J. and Sun, W., 2013. Effects of asiaticoside on levels of podocyte cytoskeletal proteins and renal slit diaphragm proteins in adriamycin-induced rat nephropathy. *Life Sciences, 93*(8), pp. 352–358.

44 Chicory (*Cichorium intybus* L.)

Cichorium intybus L.

Etymology: From the Arabic *shikorieh* and the Latin *intubus* = chicory

Family: Asteraceae

Common names: Chicory; chicorée (Fr.); wegwartenwurzel (Ger.); chicória (Port) цикорий (Rus.); achicoria (Spa.)

Parts used: Root, leaf, seed

Constituents: Sesquiterpene lactones (Kisiel & Zielińska, 2001), fructan-type polysaccharides (inulin) (Nishimura et al., 2015).

Medical history: The plant was known to ancient Egyptians. Dioscorides calls chicory *seris* and recommends it as cooling, astringent, and beneficial for the stomach and for weak heart. Pliny the Elder asserts that it is good for the kidneys and the bladder. For Simeon Seth (12th-century Byzantium), it is good for the stomach and for dispelling inflammation of the liver as well as to "calm the blood" (?), induce sleepiness, extinguish sexual desire, and treat bilious fevers. According to Fusch (1555), chicory is cold and dry to the second degree. It was a tonic, stomachic, and aperitive in 19th-century France (Moquin-Tandon, 1861). According to Dymock (1884), hakeems used the plant for bilious complains in 19th-century India. It was used as a depurative and laxative in 19th-century France, and its use as a substitute for coffee started during the Napoleonic era.

Medicinal uses: Carminative (2–4 g of roots in 250 mL of boiling water (European Union)

Blood lipids and glucose: Ethanol extract of the whole plant given to streptozotocin-induced diabetic rats at 125 mg/kg decreased glycemia after 30 mins on an oral glucose tolerance test. The extract given for 14 days decreased glycemia by 20%, triglycerides by 91%, and total cholesterol by 16% (Pushparaj et al., 2007).

A diet containing 10% chicory given to alloxan-induced diabetic rats for 50 days decreased glycemia from 305.6 to 173.6 mg/dL, plasma cholesterol from 170.3 to 118.8 mg/dL, and LDL-cholesterol from 112 to 60.5 mg/dL, as well as plasma triglycerides (Draz et al., 2010).

Gallbladder: Rats fed an aqueous extract of chicory at 5% of diet for 4 weeks had increased fecal lipids, cholesterol, and bile acid, likely attributable to inulin (Kim & Shin, 1998).

DOI: 10.1201/9781003301455-44

Kidneys: In rats, aqueous extract given orally at 16.7 g/kg for 40 days decreased serum uric acid via increased intestinal excretion of uric acid (Wang et al., 2017).

Heart: Aqueous extract of seeds given orally at 500 mg/kg/day for 3 weeks to streptozotocin-induced diabetic rats prevented myocardial degeneration, decreased lipid peroxides, interleukin-6, TNF α, and SOD, and increased CAT level in heart (Sharma et al., 2019).

Bones and cartilages: Aqueous extract of leaves given orally at 400 mg/kg/day for 8 weeks to rats with dexamethasone-induced osteoporosis increased the calcification and osteogenesis in bones (Hozayen et al., 2016). A sesquiterpene fraction from the roots given orally at 200 mg/kg/day for 2 months to mice with collagen-induced arthritis evoked beneficial effects (Ripoll et al., 2007).

Skin and hair: Extract given topically to human volunteers exhibited protective and restructuring effects on the skin (Maia Campos et al., 2017).

Comments: (i) Endives are a variety of chicory: *Cichorium intybus* var. *endivia* (L.) C.B. Clarke (synonym: *Cichorium endivia* L.). *Cichorium intybus* var. *endivia* L. (endives) has demonstrated beneficial effects in streptozotocin-induced diabetic rats (Kamel et al., 2011).

(ii) It is clear that chicory is beneficial for aging cardiovascular system (as well as fatty liver). A number of over-the-counter products are available in Europe. Chicory can be taken as an infusion, and roasted chicory can be added to coffee. Intake of endives could be beneficial for controlling glycemia.

(iii) Chicory is safe at normal dietary dose (Perović et al., 2021).

REFERENCES

Draz, S.N., Abo-Zid, M.M., Ally, A.F. and El-Debas, A.A., 2010. Hypoglycemic and hypolipidemic effect of chicory (Cichorium intybus L.) herb in diabetic rats. *Minufiya Journal of Agricultural Research*, 35(4), pp. 1201–1208.

Dymock, W., 1884. *The Vegetable Materia Medica of Western India*. Education Society Press.

Fusch, L., 1555. *De Historia Stirpium Commetarii Insignes*. Lugduni Apud Ioan Tornaesium.

Hozayen, W.G., El-Desouky, M.A., Soliman, H.A., Ahmed, R.R. and Khaliefa, A.K., 2016. Antiosteoporotic effect of Petroselinum crispum, Ocimum basilicum and Cichorium intybus L. in glucocorticoid-induced osteoporosis in rats. *BMC Complementary and Alternative Medicine*, 16, pp. 1–11.

Kamel, Z.H., Daw, I. and Marzouk, M., 2011. Effect of Cichorium endivia leaves on some biochemical parameters in streptozotocin-induced diabetic rats. *Australian Journal of Basic and Applied Sciences*, 5, pp. 387–396.

Kim, M. and Shin, H.K., 1998. The water-soluble extract of chicory influences serum and liver lipid concentrations, cecal short-chain fatty acid concentrations and fecal lipid excretion in rats. *The Journal of Nutrition*, 128(10), pp. 1731–1736.

Kisiel, W. and Zielińska, K., 2001. Guaianolides from Cichorium intybus and structure revision of Cichorium sesquiterpene lactones. *Phytochemistry*, 57(4), pp. 523–527.

Maia Campos, P.M.B.G., Mercurio, D.G., Melo, M.O. and Closs-Gonthier, B., 2017. Cichorium intybus root extract: A "vitamin D-like" active ingredient to improve skin barrier function. *Journal of Dermatological Treatment*, 28(1), pp. 78–81.

Moquin-Tandon, A., 1861. *Element de Botanique Médicale*. J.Baillière et Fils.

Nishimura, M., Ohkawara, T., Kanayama, T., Kitagawa, K., Nishimura, H. and Nishihira, J., 2015. Effects of the extract from roasted chicory (Cichorium intybus L.) root containing inulin-type fructans on blood glucose, lipid metabolism, and fecal properties. *Journal of Traditional and Complementary Medicine*, 5(3), pp. 161–167.

Perović, J., Šaponjac, V.T., Kojić, J., Krulj, J., Moreno, D.A., García-Viguera, C., Bodroža-Solarov, M. and Ilić, N., 2021. Chicory (*Cichorium intybus* L.) as a food ingredient–nutritional composition, bioactivity, safety, and health claims: A review. *Food Chemistry*, 336, p. 127676.

Pushparaj, P.N., Low, H.K., Manikandan, J., Tan, B.K.H. and Tan, C.H., 2007. Anti-diabetic effects of Cichorium intybus in streptozotocin-induced diabetic rats. *Journal of Ethnopharmacology*, 111(2), pp. 430–434.

Ripoll, C., Schmidt, B.M., Ilic, N., Poulev, A., Dey, M., Kurmukov, A.G. and Raskin, I., 2007. Anti-inflammatory effects of a sesquiterpene lactone extract from chicory (Cichorium intybus L.) roots. *Natural Product Communications*, 2(7), p. 1934578X0700200702.

Sharma, M., Afaque, A., Dwivedi, S., Jairajpuri, Z.S., Shamsi, Y., Khan, M.F., Khan, M.I. and Ahmed, D., 2019. Cichorium intybus attenuates streptozotocin induced diabetic cardiomyopathy via inhibition of oxidative stress and inflammatory response in rats. *Interdisciplinary Toxicology*, 12(3), p. 111.

Wang, Y., Lin, Z., Zhang, B., Nie, A. and Bian, M., 2017. Cichorium intybus L. promotes intestinal uric acid excretion by modulating ABCG2 in experimental hyperuricemia. *Nutrition & Metabolism*, 14(1), pp. 1–11.

45 True Cinnamon (*Cinnamomum zeylanicum* Bl.)

Cinnamomum zeylanicum Bl.

Etymology: From the Singhalese *cacyn nama* = sweet wood and the Latin *zeylanicum* = from Sri Lanka

Family: Lauraceae

Synonym: *Cinnamomum verum* J. Presl

Common names: Ceylon cinnamon, true cinnamon; canelle de Ceylan, canellier (Fr.); zimtrinde (Ger.); canela verdadeira (Port.); корица настоящая (Rus.); verdadera canella (Spa.)

Part used: Bark (after slight fermentation)

Constituents: Essential oil (cinnamaldehyde, cinnamy acetate, eugenol) (Angmor et al., 1979), tannins (oligomeric anthocyanidins: epicatechin-($2\beta \rightarrow O \rightarrow 7,4\beta \rightarrow 8$)-epicatechin-($4\beta \rightarrow 6$)-epicatechin-($2\beta \rightarrow O \rightarrow 7,4\beta \rightarrow 8$)-catechin, parameritannin A1, cinnamtannin B1) (Nam et al., 2020).

Medical history: Dioscorides writes that true cinnamon is good for digestion, edema, kidney diseases, and as a diuretic. Avicenna defines it as hot to the second degree and dry to the third degree and recommends it for the heart. The 17th-century French physician Moyse Charas (1681) recommends taking true cinnamon in cases of weakness and to stimulate the stomach and the brain. Alston (1770) states in his lectures that the natives of Sri Lanka used the oil for inflammation and pains and asserts that it is good for depression and for all cold and phlegmatic tempers. Lémery (1716) advocates it for the heart and for hysterics and informs us that Sri Lankan natives used it as a remedy for the stomach. In 19th-century North America true cinnamon was used for diarrhoea, fatigue, indigestion, and spasms (Pereira, 1843).

Medicinal uses: Loss of appetite, carminative, flatulence (European Union); carminative, tonic to the liver, inflammation, abdominal pain (Pakistan, India)

Blood pressure: Aqueous extract given to healthy adults at 85 mg/day for 30 days and then 250 mg/day for 30 days followed by 500 mg daily for 30 days reduced systolic blood pressure from 124 to 117.3 mmHg and diastolic blood pressure from 76 to 72 mmHg without effect on heartbeat (Ranasinghe et al., 2017). Intake of 1.5 g of true cinnamon powder daily for 20 days by hypertensive women caused a decrease in systolic blood pressure from 131.2 to 118.7 mmHg and diastolic blood pressure from 100 to 83.7 mmHg (Farrukh, 2019).

DOI: 10.1201/9781003301455-45

Plasma lipids and glucose: Aqueous extract given at 250 mg/kg/day for 15 days to streptozotocin-induced diabetic rats decreased cholesterol, triglycerides, LDL-cholesterol, and very low-density lipoprotein (VLDL)-cholesterol by 12.5, 23.86, 14.96, and 20%, respectively (Hassan et al., 2012). Aqueous extract given at 600 mg/kg/day for 15 days to streptozotocin-induced diabetic rats for 28 days reduced fasting plasma glucose from 320 to 247 mg/dL, total cholesterol from 71 to 40.8 mg/dL, LDL-cholesterol from 12.9 to 3.1 mg/dL, and triglycerides from 198.2 to 131 mg/dL (Ranasinghe et al., 2012). True cinnamon powder fed to rats at 8% of a high-fat diet for 4 weeks decreased total cholesterol by 42.4%, LDL-cholesterol by 77.7%, VLDL-cholesterol by 36.6%, and triglycerides by 36.6% and also attenuated hepatic steatosis (Kassaee et al., 2017).

Aqueous extract given to healthy adults at 85 mg/day for 30 days, then 250 mg/day for 30 days, followed by 500 mg daily for 30 days decreased total plasma cholesterol from 226.4 to 210.2 mg/dL and LDL-cholesterol from 152.8 to 129.8 mg/dL (Ranasinghe et al., 2017).

True cinnamon given at 750 mg twice a day for 12 weeks to patients with fatty livers decreased plasma glucose from 92 to 84.3 mg/dL, cholesterol from 167 to 133 mg/dL, triglycerides from 222.8 to 178.6 mg/dL, hr-CRP, and insulin resistance (Askari et al., 2014).

Cinnamtannin B at the concentration of 0.1 mM promoted intake of glucose by 3T3-L1 adipocytes in a manner comparable with 100 nM of insulin (Taher et al., 2007).

Kidneys: A single oral dose of 3 g of true cinnamon given to rats induced a 114.5% increase in urine secretion within 6 hours (Jayaweera et al., 2018).

Bones and cartilages: Phenolic extract given to rats for 9 days at 200 mg/kg/day reversed CFA-induced edema (Rathi et al., 2013).

Immune system: Extract given orally to rats at 100 mg/kg/day decreased pathogenic bacteria-induced mortality by 17% and increased circulating phagocytosis, neutrophil adhesion, serum immunoglobulin, and antibodies (Niphade et al., 2009).

Brain: Aqueous extract given at the dose of 50 mg/kg/day for 20 weeks to rats poisoned with monosodium glutamate attenuated cognitive dysfunction and increased neurons count in the dente gyrus area of the hippocampus (Madhavadas & Subramanian, 2017).

Warning: Intake of true cinnamon should not be more that 2–4 g/day. Higher doses trigger tachycardia followed by depression and sleepiness (Czygan, 2004).

Comments: Moyse Charas (1619–1698) was the apothecary to King Charles II.

REFERENCES

Alston, C., 1770. *Lectures on the Materia Medica: Containing the Natural History of Drugs, their Virtues and Doses: Also Directions for the Study of the Materia Medica; and an Appendix on the Method of Prescribing*. Edward and Charles Dilly.

Angmor, J.D., Dewick, P.M. and Evans, W.C., 1979. Chemical changes in cinnamon oil during the preparation of the bark; Biosynthesis of cinnamaldehyde and related compounds–studies on Cinnamomum ceylanicum, II1. *Planta Medica*, *35*(4), pp. 342–347.

Askari, F., Rashidkhani, B. and Hekmatdoost, A., 2014. Cinnamon may have therapeutic benefits on lipid profile, liver enzymes, insulin resistance, and high-sensitivity C-reactive protein in nonalcoholic fatty liver disease patients. *Nutrition Research*, *34*(2), pp. 143–148.

Charas, M., 1861. *Pharmacopée Royale Galénique et Chymique*. Chez l' auteur.

Czygan, F.C., 2004. *Herbal Drugs and Phytopharmaceuticals: A Handbook for Practice on a Scientific Basis*. CRC Press.

Farrukh, A., 2019. Upshot outcome of Cinnamomum verum powder relative to hypertension. *International Journal of Innovative Science and Research Technology*, *4*(6), June.

Hassan, S.A., Barthwal, R., Nair, M.S. and Haque, S.S., 2012. Aqueous bark extract of Cinnamomum zeylanicum: A potential therapeutic agent for streptozotocin-induced type 1 diabetes mellitus (T1DM) rats. *Tropical Journal of Pharmaceutical Research*, *11*(3), pp. 429–435.

Jayaweera, G., Makuloluwa, T., Perera, K., Amararatne, J.K., Premakumara, S. and Ratnasooriya, D., 2018. Oral diuretic activity of hot water extract of h-grade quills of Cinnamomum zeylanicum blume in rats. *International Journal of Pharmacy and Pharmaceutical Sciences*, *10*(10).

Kassaee, S.M., Goodarzi, M.T., Roodbari, N.H. and Yaghmaei, P., 2017. The effects of Cinnamomum zeylanicum on lipid profiles and histology via up-regulation of LDL receptor gene expression in hamsters fed a high cholesterol diet. *Jundishapur Journal of Natural Pharmaceutical Products*, *12*(3).

Lémery, N., 1716. *Traité universel des drogues simples, mises en ordre alphabétique. Où l'on trouve leurs différens noms . . . et tout ce qu'il y a de particulier dans les animaux, dans les végétaux, et dans les minéraux*. Au dépend de la Companie.

Madhavadas, S. and Subramanian, S., 2017. Cognition enhancing effect of the aqueous extract of Cinnamomum zeylanicum on non-transgenic Alzheimer's disease rat model: Biochemical, histological, and behavioural studies. *Nutritional Neuroscience*, *20*(9), pp. 526–537.

Nam, J.W., Phansalkar, R.S., Lankin, D.C., McAlpine, J.B., Leme-Kraus, A.A., Bedran-Russo, A.K., Chen, S.N. and Pauli, G.F., 2020. Targeting trimeric and tetrameric Proanthocyanidins of Cinnamomum verum Bark as Bioactives for Dental Therapies. *Journal of Natural Products*, *83*(11), pp. 3287–3297.

Niphade, S.R., Asad, M., Chandrakala, G.K., Toppo, E. and Deshmukh, P., 2009. Immunomodulatory activity of Cinnamomum zeylanicum bark. *Pharmaceutical Biology*, *47*(12), pp. 1168–1173.

Pereira, J., 1843. *The Elements of Materia Medica and Therapeutics*. Lea and Blanchard.

Ranasinghe, P., Jayawardena, R., Pigera, S., Wathurapatha, W.S., Weeratunga, H.D., Premakumara, G.S., Katulanda, P., Constantine, G.R. and Galappaththy, P., 2017. Evaluation of pharmacodynamic properties and safety of Cinnamomum zeylanicum (Ceylon cinnamon) in healthy adults: A phase I clinical trial. *BMC Complementary and Alternative Medicine*, *17*, pp. 1–9.

Ranasinghe, P., Perera, S., Gunatilake, M., Abeywardene, E., Gunapala, N., Premakumara, S., Perera, K., Lokuhetty, D. and Katulanda, P., 2012. Effects of Cinnamomum zeylanicum (Ceylon cinnamon) on blood glucose and lipids in a diabetic and healthy rat model. *Pharmacognosy Research*, *4*(2), p. 73.

Rathi, B., Bodhankar, S., Mohan, V. and Thakurdesai, P., 2013. Ameliorative effects of a polyphenolic fraction of Cinnamomum zeylanicum L. bark in animal models of inflammation and arthritis. *Scientia Pharmaceutica*, *81*(2), pp. 567–590.

Taher, M., Majid, F.A.A. and Sarmidi, M.R., 2007. The effect of cinnamtannin B1 on cell proliferation and glucose uptake of 3T3-L1 cells. *Natural Product Communications*, *2*(1), p. 1934578X0700200112.

46 Watermelon (*Citrullus lanatus* (Thunb.) Matsum. & Nakai)

Citrullus lanatus (**Thunb.**) **Matsum. & Nakai**

Etymology: The word *Citrullus* might derive from the French *citron* and from the Latin *lanatus* = woolly

Family: Cucurbitaceae

Synonyms: *Citrullus edulis* Spach; *Citrullus vulgaris* Schrad.; *Colocynthis citrullus* (L.) Kuntze; *Cucumis citrullus* (L.) Ser.; *Cucurbita citrullus* L.; *Momordica lanata* Thunb.

Common names: Watermelon; pastèque, pepon (Fr.); wassermelone (Ger.); melancia (Port.); арбуз (Rus.); sandía (Spa.)

Part used: Fruit

Constituents: L-arginine, L-citrulline (about 2.5 mg/g), fixed oil (unsaturated fatty acids: oleic, linoleic acid) (Ramazan et al., 2012; Tarazona-Díaz et al., 2017; Hartman et al., 2019).

Medical history: Watermelon is diuretic for Dioscorides and Galen. Matthioli (1572) writes that watermelons were kept by farmers in the barns among the hay and used at his time in Italy as cooling when weather was too hot as well as for fevers. Fusch (1555) defines watermelon as humid and cold to the second degree. Acts on blood and bile (Alston, 1770). Watermelon was used as a refrigerant and a diuretic and given for inflammation and fever in 19th-century Europe. In 19th-century India, watermelon was given as cooling to patients "heated by an acrid and irritating bile" (Ainslie, 1813).

Medicinal uses: Kidney stones (Iran, Pakistan); liver diseases (Bangladesh)

Blood pressure: In post-menopausal hypertensive women, ingestion for 6 weeks of watermelon powder corresponding to L-citrulline/L-arginine 2.7g/1.3g/day decreased brachial systolic blood pressure from 134 to 129 mmHg (Figueroa et al., 2011). Obese patients taking 2 cups of watermelon daily for 4 weeks reduced systolic blood pressure (Connoly et al., 2019). In a subsequent study of post-menopausal women, intake of 360 mL of juice twice a day for 4 weeks had no effects on blood pressure (Ellis et al., 2021). Watermelon juice given at 500 mL/day to young adults prevented impairment of

DOI: 10.1201/9781003301455-46

endothelium-dependent vasoconstriction induced by ingestion of 75 g of glucose (Vincellette et al., 2021).

Plasma lipids and glucose: Juice given freely for 13 weeks to mice decreased fat mass, plasma cholesterol, LDL-cholesterol, and intimal lesions of aortic arches (Poduri et al., 2013). The juice given at 1 g/kg/day for 14 days to alloxan-induced diabetic rats normalized glycemia, total cholesterol, triglycerides, and plasma insulin as well as hepatic glycogen (Ajiboye et al., 2020).

Dyslipidemic patients taking 6 g of watermelon extract for 42 days had decreases in plasma cholesterol and LDL-cholesterol (Massa et al., 2016). Obese patients taking 2 cups of watermelon daily for 4 weeks experienced decreases in plasma triglyceride and LDL-cholesterol and improved HDL-cholesterol as well as total antioxidant capacity, but there were no effects on total cholesterol and plasma glucose (Connoly et al., 2019).

Type 2 diabetic patients ingesting 3 g of L-citrulline before breakfast for 8 weeks had a decrease in fasting blood glucose from 157.9 to 134.9 mg/dL and decreases in both markers of inflammation such as interleukin-6 (Abbaszadeh et al., 2021).

Kidneys: Extract of pulp given to rats poisoned with calcium oxalate decreased the formation of stones in the kidney (Siddiqui et al., 2018). L-citrulline protects from kidney damage in type 1 diabetic mice (Romero et al., 2013).

Comments: (i) Whitelaw Ainslie (1767–1837) was an English surgeon based in Madras.

(ii) With age, decrease in the production and bioavailability of NO (owed to ROS) has been directly implicated in the decline of vascular endothelial function and hypertension. Hyperglycemia, vascular inflammation, and hypercholesterolemia altogether contribute to endothelial NO synthetase uncoupling with the production of superoxide anions that react with vascular NO to yield hydrogen peroxide and peroxynitrite (Katusic, 2001). Argininosuccinate synthase and argininosuccinate lyase catalyze the conversion of L-citrulline to L-arginine which is the substrate for endothelial NO synthetase (Volino-Souza et al., 2022). Increasing L-arginine availability decreases brachial/aortic blood pressure (Figueroa et al., 2011).

(iii) L-citrulline displayed anti-inflammatory effects *in vitro* on peritoneal macrophages of diabetic rats (Breuillard et al., 2015) and as such might be able to mitigate the pro-inflammatory state linked to diabetes and obesity.

(iv) It is clear that intake of fresh watermelon juice is beneficial for the aging cardiovascular system. However, watermelon should be avoided by hyperkaliemic patients.

(v) The glycemic index of watermelon is 72, and therefore, diabetic patients should not eat too much of it (Foster-Powell et al., 2002). These patients need to consume foods with glycemic indices <55 (Atkinson et al., 2008).

REFERENCES

Abbaszadeh, F., Azizi, S., Mobasseri, M. and Ebrahimi-Mameghani, M., 2021. The effects of citrulline supplementation on meta-inflammation and insulin sensitivity in type 2 diabetes: A randomized, double-blind, placebo-controlled trial. *Diabetology & Metabolic Syndrome*, *13*(1), p. 52.

Ainslie, W., 1813. *Materia Indica of Hindoostan*. Government Press.

Ajiboye, B.O., Shonibare, M.T. and Oyinloye, B.E., 2020. Antidiabetic activity of watermelon (Citrullus lanatus) juice in alloxan-induced diabetic rats. *Journal of Diabetes & Metabolic Disorders*, *19*, pp. 343–352.

Alston, C., 1770. *Lectures on the Materia Medica: Containing the Natural History of Drugs, their Virtues and Doses: Also Directions for the Study of the Materia Medica; and an Appendix on the Method of Prescribing*. Edward and Charles Dilly.

Atkinson, F.S., Foster-Powell, K. and Brand-Miller, J.C., 2008. International tables of glycemic index and glycemic load values. *Diabetes Care*, *31*(12), pp. 2281–2283.

Breuillard, C., Bonhomme, S., Couderc, R., Cynober, L. and De Bandt, J.P., 2015. In vitro anti-inflammatory effects of citrulline on peritoneal macrophages in Zucker diabetic fatty rats. *British Journal of Nutrition*, *113*(1), pp. 120–124.

Connolly, M., Lum, T., Marx, A., Hooshmand, S., Kern, M., Liu, C. and Hong, M.Y., 2019. Effect of fresh watermelon consumption on risk factors for cardiovascular disease in overweight and obese adults (P06-102-19). *Current Developments in Nutrition*, *3*(Supplement_1), pp. nzz031–P06.

Ellis, A.C., Mehta, T., Nagabooshanam, V.A., Dudenbostel, T., Locher, J.L. and Crowe-White, K.M., 2021. Daily 100% watermelon juice consumption and vascular function among postmenopausal women: A randomized controlled trial. *Nutrition, Metabolism and Cardiovascular Diseases*, *31*(10), pp. 2959–2968.

Figueroa, A., Sanchez-Gonzalez, M.A., Perkins-Veazie, P.M. and Arjmandi, B.H., 2011. Effects of watermelon supplementation on aortic blood pressure and wave reflection in individuals with prehypertension: A pilot study. *American Journal of Hypertension*, *24*(1), pp. 40–44.

Foster-Powell, K., Holt, S.H. and Brand-Miller, J.C., 2002. International table of glycemic index and glycemic load values: 2002. *The American Journal of Clinical Nutrition*, *76*(1), pp. 5–56.

Fusch, L., 1555. *De Historia Stirpium Commetarii Insignes*. Lugduni Apud Ioan Tornaesium.

Hartman, J.L., Wehner, T.C., Ma, G. and Perkins-Veazie, P., 2019. Citrulline and arginine content of taxa of Cucurbitaceae. *Horticulturae*, *5*(1), p. 22.

Katusic, Z.S., 2001. Vascular endothelial dysfunction: Does tetrahydrobiopterin play a role? *American Journal of Physiology-Heart and Circulatory Physiology*, *281*(3), pp. H981–H986.

Massa, N.M., Silva, A.S., de Oliveira, C.V., Costa, M.J., Persuhn, D.C., Barbosa, C.V. and Goncalves, M.D.C., 2016. Supplementation with watermelon extract reduces total cholesterol and LDL cholesterol in adults with dyslipidemia under the influence of the MTHFR C677T polymorphism. *Journal of the American College of Nutrition*, *35*(6), pp. 514–520.

Matthioli, P.A., 1572. *Commentaires sur les Six Livres de Pedacius Dioscorides Anazarbeen de la matière medicinale*. A l'Escue de Milan.

Poduri, A., Rateri, D.L., Saha, S.K., Saha, S. and Daugherty, A., 2013. Citrullus lanatus 'sentinel'(watermelon) extract reduces atherosclerosis in LDL receptor-deficient mice. *The Journal of Nutritional Biochemistry*, *24*(5), pp. 882–886.

Ramazan, A., Ozcan, M.M., Kanbur, G.S. and Dursun, N., 2012. Some physico-chemical properties of edible and forage watermelon seeds. *Iranian Journal of Chemistry and Chemical Engineering*, *31*(4), pp. 41–47.

Romero, M.J., Yao, L., Sridhar, S., Bhatta, A., Dou, H., Ramesh, G., Brands, M.W., Pollock, D.M., Caldwell, R.B., Cederbaum, S.D. and Head, C.A., 2013. L-citrulline protects from kidney damage in type 1 diabetic mice. *Frontiers in Immunology*, *4*, p. 480.

Siddiqui, W.A., Shahzad, M., Shabbir, A. and Ahmad, A., 2018. Evaluation of anti-urolithiatic and diuretic activities of watermelon (Citrullus lanatus) using in vivo and in vitro experiments. *Biomedicine & Pharmacotherapy*, *97*, pp. 1212–1221.

Tarazona-Díaz, M.P., Martínez-Sánchez, A. and Aguayo, E., 2017. Preservation of bioactive compounds and quality parameters of watermelon juice enriched with L-Citrulline through short thermal treatment. *Journal of Food Quality*, *2017*.

Vincellette, C.M., Losso, J., Early, K., Spielmann, G., Irving, B.A. and Allerton, T.D., 2021. Supplemental watermelon juice attenuates acute hyperglycemia-induced macro- and microvascular dysfunction in healthy adults. *The Journal of Nutrition*, *151*(11), pp. 3450–3458.

Volino-Souza, M., Oliveira, G.V.D., Conte-Junior, C.A., Figueroa, A. and Alvares, T.S., 2022. Current evidence of watermelon (Citrullus lanatus) ingestion on vascular health: A food science and technology perspective. *Nutrients*, *14*(14), p. 2913.

47 Lemon (*Citrus limon* (L.) Osbeck)

Citrus limon (L.) Osbeck

Etymology: From *citron* = a city of Judea and the Persian *limu* = lemon

Family: Rutaceae

Synonyms: *Citrus limonum* Risso; *Citrus medica* subsp. *Limonia* (Risso) Hook. f.; *Citrus medica* var. *limon* L.

Common names: Lemon; citron (Fr.); zitrone (Ger.); limão (Port.); лимон (Rus.); limón (Spa.)

Part used: Fruit

Constituents: Essential oil (limonene, neral) (Paw et al., 2020), flavanones (diosmin), flavanone glycosides (hesperidin, eriocitrin) (Del Río et al., 2004), citric acid (Karadeniz, 2004), ascorbic acid (about 100 mg/mL of fresh juice) (Harris & Ray, 1933; Rekha et al., 2012).

Medical history: Lemon was unknown to the ancient Greeks and Romans. It probably originates from northern India, and Isaac (1959) explains that the Jewish people came in contact with lemon during the Babylonian exile, introduced it in Israel, and brought it with the diaspora to the Mediterranean region. Lemon is said to be mentioned in the Torah: "and you shall take for yourselves on the first day, the fruit of the hadar tree" (Leviticus 23:40).

It was brought to Italy by the Crusaders around the 10th century. Alston (1770) in his lectures describes lemon as antiseptic, for worms, putrefaction, putrid fever, scurvy, bilious acrimony, thirst and citrus peels for heart palpitation, weak stomach, obstruction of viscera, and poisoning. In 19th-century Europe, it was used to make a refreshing drink, used for scurvy, for nausea, to wash the mouth, and for high fever (Pereira, 1843; Moquin-Tandon, 1861). The essential oil has been used in perfumery. In 19th-century India, lemon juice was given as a refrigerant and for scurvy, fever, inflammation, rheumatism, and diarrhea (Watt, 1889).

Medicinal uses: Kidney stones (Morocco); remove blood from urine, diarrhea, prevent hair loss, dandruff (India); fatigue, inhibit formation of bumps and tumors, nausea, dizziness, heaviness (Myanmar); stomachache, seasickness, fish sting (Palau).

DOI: 10.1201/9781003301455-47

Blood pressure: Intake of a lemon daily combined with a daily walk caused a decrease of blood pressure and heart rate in healthy volunteers (Kato et al., 2014). Rats given 5% lemon juice for 90 days had a decrease in blood pressure by about 5 mmHg (Miyake et al., 1998). Intake of ascorbic acid at 500 mg/kg/day for 30 days decreased systolic blood pressure from about 155 to 140 mmHg in hypertensive patients (Duffy et al., 1999).

Blood lipid and glucose: In rabbit on a high-fat diet, ingesting lemon juice at 1 mL/kg/day for 30 days decreased plasma cholesterol from 345.3 to 150.8 mg/dL and LDL-cholesterol from 273 to 122.6 mg/dL (Khan et al., 2010). Rats fed a high-fat diet containing eriocitrin at 0.7% of diet for 21 days had a decrease in cholesterol from about 400 to 350 mg/dL and caused an increase in fecal cholesterol and fecal bile (Miyake et al., 2006).

Gallbladder: Lemon juice stimulates hepatobiliary secretion (Cherng et al., 2006) and by doing so contributes to removing cholesterol (precursor of bile acids). Lemon juice assist in flushing gallstones (Savage et al., 1992).

Kidneys: In rats poisoned with ethylene glycol [v/v] (EG) and ammonium chloride, intake of lemon juice at 6 μL/kg/days for 10 days prevented the deposition of calcium oxalate in kidneys (Touhami et al., 2007). Intake of 30 mL/day of lemon juice for 6 weeks by hyperuricemic patients caused a decrease in plasma uric acid (Wang et al., 2017).

Bones and cartilages: Ethanol extract of peels given to rats orally at 100 mg/kg/day for 18 days decreased hind ankle joint inflammatory cell infiltration (Ahmed et al., 2018).

Warning: Because taking lemon juice can cause gastric reflux and gastric acidity, it must be avoided by patients with sensitive stomachs. Lemon affects the first-pass metabolism of certain drugs (Saito et al., 2005; Sridharan & Sivaramakrishnan, 2016). One alternative could be to drink half a lemon juice in a glass of water after an oily and meaty dish.

Comments: (i) With aging, ascorbic acid content in the myocardium decreases (Schaus, 1957), and ascorbic acid deficiency has been designated a risk factor for ischemic heart diseases (Horsey et al., 1981) and stroke (Gale et al., 1995). In post-menopausal women, ascorbic acid improved large artery elasticity (Moreau et al., 2005). Ascorbic acid decreased incidences of postoperative atrial fibrillation (Baker & Coleman, 2016).

(ii) With aging, dehydration, and decreased urinary pH, uric acid forms stones in the kidneys that can be removed by increasing water intake and increasing urinary pH and citric acid (Gutman, 1968). Citric acid attenuates the formation of kidney stones by forming complexes with calcium ions (Gul & Monga, 2014).

(iii) Habitual exercise delays vascular aging (Seals et al., 2009).

REFERENCES

Ahmed, O.M., Ashour, M.B., Fahim, H.I. and Ahmed, N.A., 2018. Citrus limon and paradisi fruit peel hydroethanolic extracts prevent the progress of complete Freund's adjuvant-induced arthritis in male Wistar rats. *Advances in Animal and Veterinary Sciences*, 6(10), pp. 443–455.

Alston, C., 1770. *Lectures on the Materia Medica: Containing the Natural History of Drugs, their Virtues and Doses: Also Directions for the Study of the Materia Medica; and an Appendix on the Method of Prescribing*. Edward and Charles Dilly.

Baker, W.L. and Coleman, C.I., 2016. Meta-analysis of ascorbic acid for prevention of post-operative atrial fibrillation after cardiac surgery. *American Journal of Health-System Pharmacy*, 73(24), pp. 2056–2066.

Cherng, S.C., Chen, Y.H., Lee, M.S., Yang, S.P., Huang, W.S. and Cheng, C.Y., 2006. Acceleration of hepatobiliary excretion by lemon juice on 99mTc-tetrofosmin cardiac SPECT. *Nuclear Medicine Communications*, 27(11), pp. 859–864.

Del Río, J.A., Fuster, M.D., Gómez, P., Porras, I., Garcıa-Lidón, A. and Ortuño, A., 2004. Citrus limon: A source of flavonoids of pharmaceutical interest. *Food Chemistry*, 84(3), pp. 457–461.

Duffy, S., Gokce, N., Holbrook, M., Huang, A., Frei, B., Keaney, J.F. and Vita, J.A., 1999. Treatment of hypertension with ascorbic acid. *The Lancet*, 354(9195), pp. 2048–2049.

Gale, C.R., Martyn, C.N., Winter, P.D. and Cooper, C., 1995. Vitamin C and risk of death from stroke and coronary heart disease in cohort of elderly people. *BMJ*, 310(6994), pp. 1563–1566.

Gul, Z. and Monga, M., 2014. Medical and dietary therapy for kidney stone prevention. *Korean Journal of Urology*, 55(12), pp. 775–779.

Gutman, A.B., 1968. Uric acid nephrolithiasis. *The American Journal of Medicine*, 45(5), pp. 756–779.

Harris, L.J. and Ray, S.N., 1933. Standardisation of the antiscorbutic potency of ascorbic acid. *Biochemical Journal*, 27(6), p. 2016.

Horsey, J., Livesley, B. and Dickerson, J.W., 1981. Ischaemic heart disease and aged patients: Effects of ascorbic acid on lipoproteins. *Journal of Human Nutrition*, 35(1), pp. 53–58.

Isaac, E., 1959. Influence of religion on the spread of citrus: The religious practices of the Jews helped effect the introduction of citrus to Mediterranean lands. *Science*, 129(3343), pp. 179–186.

Karadeniz, F., 2004. Main organic acid distribution of authentic citrus juices in Turkey. *Turkish Journal of Agriculture and Forestry*, 28(4), pp. 267–271.

Kato, Y., Domoto, T., Hiramitsu, M., Katagiri, T., Sato, K., Miyake, Y., Aoi, S., Ishihara, K., Ikeda, H., Umei, N. and Takigawa, A., 2014. Effect on blood pressure of daily lemon ingestion and walking. *Journal of Nutrition and Metabolism*, 2014.

Khan, Y., Khan, R.A., Afroz, S. and Siddiq, A., 2010. Evaluation of hypolipidemic effect of citrus lemon. *Journal of Basic and Applied Sciences*, 6(1), pp. 39–43.

Miyake, Y., Kuzuya, K., Ueno, C., Katayama, N., Hayakawa, T., Tsuge, H. and Osawa, T., 1998. Suppressive effect of components in lemon juice on blood pressure in spontaneously hypertensive rats. *Food Science and Technology International, Tokyo*, 4(1), pp. 29–32.

Miyake, Y., Suzuki, E., Ohya, S., Fukumoto, S., Hiramitsu, M., Sakaida, K., Osawa, T. and Furuichi, Y., 2006. Lipid-lowering effect of eriocitrin, the main flavonoid in lemon fruit, in rats on a high-fat and high-cholesterol diet. *Journal of Food Science*, 71(9), pp. S633–S637.

Moquin-Tandon, A., 1861. *Element de Botanique Médicale*. J.Baillière et Fils.

Moreau, K.L., Gavin, K.M., Plum, A.E. and Seals, D.R., 2005. Ascorbic acid selectively improves large elastic artery compliance in postmenopausal women. *Hypertension*, 45(6), pp. 1107–1112.

Paw, M., Begum, T., Gogoi, R., Pandey, S.K. and Lal, M., 2020. Chemical composition of Citrus limon L. Burmf peel essential oil from North East India. *Journal of Essential Oil Bearing Plants*, 23(2), pp. 337–344.

Pereira, J., 1843. *The Elements of Materia Medica and Therapeutics*. Lea and Blanchard.

Rekha, C., Poornima, G., Manasa, M., Abhipsa, V., Devi, J.P., Kumar, H.T.V. and Kekuda, T.R.P., 2012. Ascorbic acid, total phenol content and antioxidant activity of fresh juices of four ripe and unripe citrus fruits. *Chemical Science Transactions*, 1(2), pp. 303–310.

Saito, M., Hirata-Koizumi, M., Matsumoto, M., Urano, T. and Hasegawa, R., 2005. Undesirable effects of citrus juice on the pharmacokinetics of drugs: Focus on recent studies. *Drug Safety*, 28, pp. 677–694.

Savage, A.P., O'Brien, T. and Lamont, P.M., 1992. Adjuvant herbal treatment for gallstones. *British Journal of Surgery*, 79, p. 168.

Schaus, R., 1957. The ascorbic acid content of human pituitary, cerebral cortex, heart, and skeletal muscle and its relation to age. *The American Journal of Clinical Nutrition*, 5(1), pp. 39–41.

Seals, D.R., Walker, A.E., Pierce, G.L. and Lesniewski, L.A., 2009. Habitual exercise and vascular ageing. *The Journal of Physiology*, 587(23), pp. 5541–5549.

Sridharan, K. and Sivaramakrishnan, G., 2016. Interaction of citrus juices with cyclosporine: Systematic review and meta-analysis. *European Journal of Drug Metabolism and Pharmacokinetics*, 41, pp. 665–673.

Touhami, M., Laroubi, A., Elhabazi, K., Loubna, F., Zrara, I., Eljahiri, Y., Oussama, A., Grases, F. and Chait, A., 2007. Lemon juice has protective activity in a rat urolithiasis model. *BMC Urology*, 7(1), pp. 1–10.

Wang, H., Cheng, L., Lin, D., Ma, Z. and Deng, X., 2017. Lemon fruits lower the blood uric acid levels in humans and mice. *Scientia Horticulturae*, 220, pp. 4–10.

Watt, G., 1889. *A Dictionary of the Economic Products of India*. Printed by the Superintendent of Government Printing, India.

48 Hawthorn (*Crataegus oxyacantha* L.)

Crataegus oxyacantha L.

Etymology: From the Greek *kratos* = strength, *oxy* = sharp, and *akantho* = thorn
Family: Rosaceae
Synonym: *Crataegus curvisepala* Lindm.
Common names: Hawthorn, white thorn; aubepine, aubespin, épine blanche (Fr.); weißdorn, hagdorn (Ger.); espinheiro (Port.); боярышники (Rus.); espino (Spanish)
Parts used: Flower, leaf
Active principles: Flavanes, flavones, flavone glycosides, tannins (proanthocyanidins) (Kolodziej et al., 1984; Orhan, 2018).
Medical history: Dioscorides asserts that hawthorn stops diarrhea and decreases menstrual flow. In 16th-century Germany, it was considered cold and dry to the second degree (Fusch, 1555). Physicians in 17th-century France used hawthorn as an astringent to check diarrhea and bleeding (Lémery, 1617).
Medicinal uses: For anxiety and related symptoms including palpitations and insomnia, 1–2 g in 150 mL of boiling water as an herbal infusion up to 4 times daily (European Union).

Blood pressure: In type-2 diabetic patients, extract of flowers taken for 16 weeks caused a decrease in systolic blood pressure (Walker et al., 2006). Extracts decreased the blood pressure of people with mild hypertension if taken for at least 12 weeks (Cloud et al., 2020; Tassell et al., 2010).
Heart: Intake of an extract of berries (3 x 30 drops) for 8 weeks by cardiac failure patients for 8 weeks mitigated dyspnea and fatigue (Degenring et al., 2003). Tauchert, 2002 presented evidence that extract of leaves (900 mg) given 16 weeks to congestive heart failure patients improved exercise capacity. In dogs suffering from mitral valve disease, intake of tincture of hawthorn for 120 days decreased systolic blood pressure after 90 days (de Souza Balbueno et al., 2020).
Immune system: Infusion (50 g/L of water) given to rats daily at 100 mg/kg/day for 20 days prevented the increase of leukocytes counts induced by colitis (do Nascimento et al., 2021).

DOI: 10.1201/9781003301455-48

Brain: Extract given to rats at 100 mg/kg/day for 15 days caused some levels of protective effects against transient occlusion of cerebral artery (Paul et al., 2017).

Warning: Careful medical monitoring is needed for patients on cardiac glycosides taking hawthorn (Posadzki et al., 2013). The oral LD 50 of the water -methanol extract is above 10 g/kg in rats. In rats, oral intake of 500 mg of aqueous extract of seeds inhibited implantation in pregnant rats (Gardner & McGuffin, 2013). Up to 1 g day is recommended (Martindale, 1958).

REFERENCES

Cloud, A., Vilcins, D. and McEwen, B., 2020. The effect of hawthorn (Crataegus spp.) on blood pressure: A systematic review. *Advances in Integrative Medicine*, *7*(3), pp. 167–175.

de Souza Balbueno, M.C., Junior, K.D.C.P. and de Paula Coelho, C., 2020. Evaluation of the efficacy of Crataegus oxyacantha in dogs with early-stage heart failure. *Homeopathy*, *109*(04), pp. 224–229.

Degenring, F.H., Suter, A., Weber, M. and Saller, R., 2003. A randomised double blind placebo controlled clinical trial of a standardised extract of fresh Crataegus berries (Crataegisan) in the treatment of patients with congestive heart failure NYHA II. *Phytomedicine*, *10*, pp. 363–369.

do Nascimento, R.D.P., da Fonseca Machado, A.P., Lima, V.S., Moya, A.M.T.M., Reguengo, L.M., Junior, S.B., Leal, R.F., Cao-Ngoc, P., Rossi, J.C., Leclercq, L. and Cottet, H., 2021. Chemoprevention with a tea from hawthorn (Crataegus oxyacantha) leaves and flowers attenuates colitis in rats by reducing inflammation and oxidative stress. *Food Chemistry: X*, *12*, p. 100139.

Elango, C. and Devaraj, S.N., 2010. Immunomodulatory effect of Hawthorn extract in an experimental stroke model. *Journal of Neuroinflammation*, *7*(1), pp. 1–13.

Fusch, L., 1555. *De Historia Stirpium Commetarii Insignes*. Lugduni Apud Ioan Tornaesium.

Gardner, Z. and McGuffin, M. eds., 2013. *American Herbal Products Association's Botanical Safety Handbook*. CRC Press.

Kolodziej, H., Ferreira, D. and Roux, D.G., 1984. Synthesis of condensed tannins. Part 12. Direct access to [4, 6]-and [4, 8]-all-2, 3-cis-procyanidin derivatives from (−)-epicatechin: Assessment of bonding positions in oligomeric analogues from Crataegus oxyacantha L. *Journal of the Chemical Society, Perkin Transactions*, *1*, pp. 343–350.

Lémery, N., 1716. *Traité universel des drogues simples, mises en ordre alphabétique. Où l'on trouve leurs différens noms . . . et tout ce qu'il y a de particulier dans les animaux, dans les végétaux, et dans les minéraux*. Au dépend de la Companie.

Orhan, I.E., 2018. Phytochemical and pharmacological activity profile of Crataegus oxyacantha L. (hawthorn)-A cardiotonic herb. *Current Medicinal Chemistry*, *25*(37), pp. 4854–4865.

Paul, S., Sharma, S., Paliwal, S.K. and Kasture, S., 2017. Role of Crataegus oxyacantha (Hawthorn) on scopolamine induced memory deficit and monoamine mediated behaviour in rats. *Oriental Pharmacy and Experimental Medicine*, *17*, pp. 315–324.

Posadzki, P., Watson, L. and Ernst, E., 2013. Contamination and adulteration of herbal medicinal products (HMPs): An overview of systematic reviews. *European Journal of Clinical Pharmacology*, *69*, pp. 295–307.

Tassell, M.C., Kingston, R., Gilroy, D., Lehane, M. and Furey, A., 2010. Hawthorn (Crataegus spp.) in the treatment of cardiovascular disease. *Pharmacognosy Reviews*, *4*(7), p. 32.

Tauchert, M., 2002. Efficacy and safety of *crataegus* extract WS 1442 in comparison with placebo in patients with chronic stable New York Heart Association class-III heart failure. *American Heart Journal, 143*, pp. 910–915.

Walker, A.F., Marakis, G., Simpson, E., Hope, J.L., Robinson, P.A., Hassanein, M. and Simpson, H.C., 2006. Hypotensive effects of hawthorn for patients with diabetes taking prescription drugs: A randomised controlled trial. *British Journal of General Practice, 56*(527), pp. 437–443.

49 Coriander (*Coriandrum sativum* L.)

Coriandrum sativum L.

Etymology: From the Greek *koris* = a bug and *andros* = male and the Latin *sativum* = cultivated

Family: Apiaceae

Synonym: *Selinum coriandrum* Krause

Common names: Coriander; coriandre (Fr.); koriander (Ger.); coentro (Port.) кориандр (Rus.); cilantro (Spanish)

Parts used: Leaf, seed

Constituent: Essential oil (linalool) (Mandal & Mandal, 2015).

Medical history: Dioscorides says about coriander that it is anti-inflammatory, sedative, and able to induce "delirium with undecent talk if taken in large amounts" (!). For Galen, it is a tonic. Alston (1770) describes coriander as carminative and good for vertigo. Coriander was known to the Indian physician Sushruta (6th century BC) and used in 19th-century India as a tonic, to facilitate digestion (Ainslie, 1813), for the heart, thirst, to induce urination and as an aphrodisiac (Udoy Chand Dutt, 1877). In 19th-century Europe, coriander was used for fatigue and added to purgative medicines (Guibourt, 1836).

Medicinal uses: Loss of appetite, flatulence, digestive discomfort (1–3 g or one teaspoon of crushed seeds in 150 mL of boiling water, 1–3 cups a day before meals) (European Union); abdominal pain (Turkey); headache, diuretic (Iraq); jaundice (Iran; India; Pakistan); indigestion (Bangladesh); pain in the chest (Myanmar); stomach diseases (China); indigestion (Korea); diabetes, hypertension, urinary tract problems (the Philippines)

Blood pressure: Coriandrum given at 2 g/day to rats on a high-salt and high-fructose diet decreased blood pressure and increased plasma NO (Wang et al., 2022).

Plasma lipids and glucose: Coriander seeds are beneficial for blood lipids (Chithra, & Leelamma, 1997). Coriander seeds given to rats as 10% of a high-fat diet for 75 days decreased total plasma cholesterol from 157.3 to 85.1 mg/dL (Dhanapakiam et al., 2007). An extract given orally at 20 mg/kg to rats decreased glycemia (Aissaoui et al., 2011).

Two capsules containing 500 mg of fruit powder given daily to 50 patients with type-2 diabetes for 6 weeks decreased glycemia from 172.7 to 86 mg/dL, reduced LDL-cholesterol from 137.8 to 67.5 mg/dL, and increased

DOI: 10.1201/9781003301455-49

HDL-cholesterol (Parsaeyan, 2012). Coriander seed powder given at 1 g/day to diabetic type-2 patients for 6 weeks decreased plasma cholesterol from 161.5 to 145.2 mg/dL, triglycerides from 171.6 to 130.4 mg/dL, and LDL-cholesterol from 133 to 111.4 mg/dL (Zamany et al., 2022).

Heart: Coriander has cardioprotective effects experimentally in rats. Methanol extract given prophylactically to rats at 300 mg/kg orally for 30 days prevented myocardial infarction induced by isoproterenol (Patel et al., 2012). Aqueous extract given orally for 15 days at 1 g/kg protected rats against isoproterenol-induced heart failure (Dhyani et al., 2020). Rats fed 300 mg/kg of coriander for 12 days were protected against potassium chloride-induced arrythmia prophylactically and curatively (Rehman et al., 2016).

Gallbladder: Rats given coriander seeds as 10% of a diet enriched with fats increased fecal bile acids from 26.4 to 57.5 mg/rat/day and fecal neutral sterols from 8.2 to 105.3 mg/rat/day (Dhanapakiam et al., 2007).

Kidneys: Coriandrum given at 2 g/day to rats on a high-salt and high-fructose diet increased the urinary excretion of sodium ions by inhibiting renal Na^+/H^+ exchangers (Wang et al., 2022).

Bones and cartilages: Patients with arthritis consuming 5 g of seeds per day for 60 days experienced decreased plasma MDA from 150.3 to 95.3 nmol/dL and decreased alkaline phosphatase (Rai & Andallu, no year of publication available).

Skin and hair: Essential oil applied topically twice a day for 5 weeks protected rodents against ultraviolet B as evidenced by decreases in dermal MDA, matrix metalloproteinase-1, and AP-1 and increase in transforming growth factor β (TGFβ) and collagen (Hwang et al., 2012).

Brain: Mice fed a diet containing 15% coriander for 45 days were protected against scopolamine-induced dementia with decreases cerebral acetylcholine esterase activity (Mani et al., 2011).

Warning: Coriander seed consumption should not exceed a daily dose of 4 g (Czygan, 2004).

Comments: (i) Ultraviolet B and high ROS concentration in the dermis inhibit the expression of TGFβ and the subsequent synthesis of collagen (Shin et al., 2019).

(ii) Alzheimer's disease involves the decreased inability of cholinergic neurons to produce acetylcholine, resulting in irreversible memory loss. Scopolamine is a plant tropane alkaloid used to mimic Alzheimer's disease in rodents as it binds to and block post-synaptic neurons in the brain and increases acetylcholinesterase activity. Acetylcholinesterase is an enzyme found in the synapses of brain neurons that catalyzes the conversion of acetylcholine to acetic acid and choline (Yadang et al., 2020). Intake of coriander regularly at a dietary dose could potentially have some beneficial effects on memory during aging.

(iii) Regular consumption of coriander at a normal dietary dose (a few seeds) could potentially have beneficial effects on aging individuals.

REFERENCES

Aissaoui, A., Zizi, S., Israili, Z.H. and Lyoussi, B., 2011. Hypoglycemic and hypolipidemic effects of Coriandrum sativum L. in Meriones shawi rats. *Journal of Ethnopharmacology*, *137*(1), pp. 652–661.

Ainslie, W., 1813. *Materia Indica of Hindoostan*. Government Press.

Alston, C., 1770. *Lectures on the Materia Medica: Containing the Natural History of Drugs, their Virtues and Doses: Also Directions for the Study of the Materia Medica; and an Appendix on the Method of Prescribing*. Edward and Charles Dilly.

Chithra, V. and Leelamma, S., 1997. Hypolipidemic effect of coriander seeds (Coriandrum sativum): Mechanism of action. *Plant Foods for Human Nutrition*, *51*(2), pp. 167–172.

Czygan, F.C., 2004. *Herbal Drugs and Phytopharmaceuticals: A Handbook for Practice on a Scientific Basis*. CRC Press.

Dhanapakiam, P., Joseph, J.M., Ramaswamy, V.K., Moorthi, M. and Kumar, A.S., 2007. The cholesterol lowering property of coriander seeds (Coriandrum sativum): Mechanism of action. *Journal of Environmental Biology*, *29*(1), p. 53.

Dhyani, N., Parveen, A., Siddiqi, A., Hussain, M.E. and Fahim, M., 2020. Cardioprotective efficacy of Coriandrum sativum (L.) seed extract in heart failure rats through modulation of endothelin receptors and antioxidant potential. *Journal of Dietary Supplements*, *17*(1), pp. 13–26.

Dutt, U.C., 1877. *The Materia Medica of the Hindus*. Thacker, Spink, & Co.

Guibourt, N.J.B.G., 1836. *Histoire abrégée des drogues simples*. Méquignon-Marvis Père et fils.

Hwang, J.Y., Park, T.S., Kim, D.H., Hwang, E.Y., Lee, J.N., young Lee, J., Lee, G.T., Lee, K. and Son, J.H., 2012. Anti-wrinkle Compounds Isolated from the Seeds of Arctium lappa L. 생명과학회지, *22*(8), pp. 1092–1098.

Mandal, S. and Mandal, M., 2015. Coriander (Coriandrum sativum L.) essential oil: Chemistry and biological activity. *Asian Pacific Journal of Tropical Biomedicine*, *5*(6), pp. 421–428.

Mani, V., Parle, M., Ramasamy, K. and Abdul Majeed, A.B., 2011. Reversal of memory deficits by Coriandrum sativum leaves in mice. *Journal of the Science of Food and Agriculture*, *91*(1), pp. 186–192.

Parsaeyan, N., 2012. The effect of coriander seed powder consumption on atherosclerotic and cardioprotective indices of type 2 diabetic patients. *Iranian Journal of Diabetes and Obesity*, *4*(2, Summer). pp. 86–90.

Patel, D.K., Desai, S.N., Gandhi, H.P., Devkar, R.V. and Ramachandran, A.V., 2012. Cardio protective effect of Coriandrum sativum L. on isoproterenol induced myocardial necrosis in rats. *Food and Chemical Toxicology*, *50*(9), pp. 3120–3125.

Rai, S.R. and Andallu, B., Efficacy of coriander (Coriandrum sativum L.) seeds in combating oxidative stress in arthritis patient. *Nutritional Deprivation in the Midst of Plenty*, pp. 57–61.

Rehman, N., Jahan, N., Khan, K.M. and Zafar, F., 2016. Anti-arrhythmic potential of Coriandrum sativum seeds in salt induced arrhythmic rats. *Pakistan Veterinary Journal*, *36*(4).

Salem, M.A., Manaa, E.G., Osama, N., Aborehab, N.M., Ragab, M.F., Haggag, Y.A., Ibrahim, M.T. and Hamdan, D.I., 2022. Coriander (Coriandrum sativum L.) essential oil and oil-loaded nano-formulations as an anti-aging potentiality via TGFβ/SMAD pathway. *Scientific Reports*, *12*(1), p. 6578.

Shin, J.W., Kwon, S.H., Choi, J.Y., Na, J.I., Huh, C.H., Choi, H.R. and Park, K.C., 2019. Molecular mechanisms of dermal aging and antiaging approaches. *International Journal of Molecular Sciences*, *20*(9), p. 2126.

Wang, X., Liu, Y., Wang, Y., Dong, X., Wang, Y., Yang, X., Tian, H. and Li, T., 2022. Protective effect of Coriander (Coriandrum sativum L.) on high-fructose and high-Salt diet-induced hypertension: Relevant to improvement of renal and intestinal function. *Journal of Agricultural and Food Chemistry*, *70*(12), pp. 3730–3744.

Yadang, F.S.A., Nguezeye, Y., Kom, C.W., Betote, P.H.D., Mamat, A., Tchokouaha, L.R.Y., Taiwé, G.S., Agbor, G.A. and Bum, E.N., 2020. Scopolamine-induced memory impairment in mice: Neuroprotective effects of Carissa edulis (Forssk.) Valh (Apocynaceae) aqueous extract. *International Journal of Alzheimer's Disease*, *2020*.

Zamany, S., Mahdavi, A.M., Pirouzpanah, S. and Barzegar, A., 2022. The effects of coriander seed supplementation on serum glycemic indices, lipid profile and parameters of oxidative stress in patients with type 2 diabetes mellitus: A randomized double-blind placebo-controlled clinical trial. https://doi.org/10.21203/rs.3.rs-262149/v2

50 Saffron (*Crocus sativus* L.)

Crocus sativus L.

Etymology: From the Greek *Krokos*, a young man of Greek mythology transformed into a flower, a string, or a filament, and the Latin *sativus* = cultivated

Family: Iridaceae

Common names: Saffron; safran (Fr.; Ger.); açafrão (Port.); шафр ан (Rus.); azafrán (Spa.)

Part used: Stigma

Constituents: Carotenoids (crocetin and its glycoside crocin), monoterpenes (safranal and its glycoside: picrocrocin) (Xi & Qian, 2006).

Medical history: Saffron is mentioned in the Shir Hashirim or Songs of Songs (4:14) of King Solomon (10th century BC). Saffron was known to Hippocrates, who advocated it's use for uterine diseases. Saffron is diuretic and aphrodisiac for Dioscorides and anti-inflammatory for Pliny the Elder. Fusch (1555) defines saffron as hot to the second degree and dry to the first. In his 18th-century lectures, Alston (1770) advocates the use of saffron for "lowness of the spirit", palpitations of the heart, vertigo, asthma, jaundice, and malignant fevers. During the 19th century in Europe, it was used for asthma and stomach cramps, to induce uterine contractions, for hypochondriacs, as a dye, and in theriacs (Guibourt, 1836).

Medicinal uses: Tonic (Turkey), antispasmodic (Iraq, Iran)

Blood pressure: Tablets of dried powdered stigma given to healthy volunteers at 400 mg/day for 7 days caused a decrease in standing systolic blood pressure from 126.5 to 115.5 mmHg and standing systolic blood pressure from 75 to 73 mmHg (Modaghegh et al., 2008). Intake of dried powder of stigmas given at 100 mg/day for 12 weeks to type-2 diabetic patients caused a decrease in systolic blood pressure from 132.7 to 124.5 mmHg and in diastolic blood pressure from 79.5 to 76.7 mmHg (Ebrahimi et al., 2019).

Heart: Crocetin protected rats against isoproterenol-induced myocardial injuries (Liu & Qian, 1994)

Plasma lipid and glucose: Tablets of dried powdered stigma given to healthy volunteers at 400 mg/day for 7 days caused a decrease in cholesterol from 167.5 to 158.5 mg/dL and in triglycerides from 125.9 to 111.4 mg/dL (Modaghegh et al., 2008).

Crocetin given to rabbits or rats poisoned with a high-fat diet prevented the formation of aortic atheromatous plaques (Gainer & Jones, 1975; Yuanxiong et al., 2004). In quails on a high-fat diet, crocetin given prophylactically

DOI: 10.1201/9781003301455-50

prevented the formation of atheromatous plaques. *In vitro*, crocin decreased vascular endothelial cell apoptosis, the proliferation of vascular smooth muscle cells, and the formation of foam cells induced by oxidized LDL (He et al., 2005). Crocin given to type-2 diabetic patients induced a decrease in fasting blood glucose (Sepahi et al., 2022).

Kidneys: Aqueous extracts of saffron given to rats orally at 240 mg/kg increased urine excretion from about 10 to 30 mL within 5 hours and increased sodium ions excretion from 65 meq/L/5h to 120 meq/L/5h while decreasing creatinine (Shariatifar et al., 2014). Hydroalcoholic extract of saffron given to type-2 diabetic patients at 15 mg/day for 8 weeks decreased plasma uric acid from 4.8 to 4.1 mg/dL and blood urea nitrogen from 28.9 to 24.4 mg/dL (Milajerdi et al., 2017).

Bones and cartilages: Crocin given orally at 20 mg/kg/day for 12 weeks to ovariectomized rats increased the bone mineral density of L4 vertebrae from about 0.1 to 0.2 g/cm^2 and improved trabecular microarchitecture, and normalized bone MDA (Algandaby, 2019).

Brain: Saffron given at 15 mg twice per day for 22 weeks to elderly patients suffering from dementia caused improvements in cognitive function (Akhondzadeh et al., 2010).

Warnings: The maximum daily dose of saffron is 1.5 g (Czygan et al., 2004). Use cautiously in patients using anticoagulants or antiplatelet agents, hormonal agents, antidepressants, or antihypertensives (Posadzki et al., 2013).

REFERENCES

Akhondzadeh, S., Shafiee Sabet, M., Harirchian, M.H., Togha, M., Cheraghmakani, H., Razeghi, S., Hejazi, S.S., Yousefi, M.H., Alimardani, R., Jamshidi, A. and Rezazadeh, S.A., 2010. A 22-week, multicenter, randomized, double-blind controlled trial of Crocus sativus in the treatment of mild-to-moderate Alzheimer's disease. *Psychopharmacology, 207*, pp. 637–643.

Algandaby, M.M., 2019. Crocin attenuates metabolic syndrome-induced osteoporosis in rats. *Journal of Food Biochemistry, 43*(7), p. e12895.

Alston, C., 1770. *Lectures on the Materia Medica: Containing the Natural History of Drugs, their Virtues and Doses: Also Directions for the Study of the Materia Medica; and an Appendix on the Method of Prescribing.* Edward and Charles Dilly.

Czygan, F.C., 2004. *Herbal Drugs and Phytopharmaceuticals: A Handbook for Practice on a Scientific Basis.* CRC Press.

Ebrahimi, F., Aryaeian, N., Pahlavani, N., Abbasi, D., Hosseini, A.F., Fallah, S., Moradi, N. and Heydari, I., 2019. The effect of saffron (Crocus sativus L.) supplementation on blood pressure, and renal and liver function in patients with type 2 diabetes mellitus: A double-blinded, randomized clinical trial. *Avicenna Journal of Phytomedicine, 9*(4), p. 322.

Gainer, J.L. and Jones, J.R., 1975. The use of crocetin in experimental atherosclerosis. *Experientia, 31*, pp. 548–549.

Guibourt, N.J.B.G., 1836. *Histoire abrégée des drogues simples.* Méquignon-Marvis Père et fils.

He, S.Y., Qian, Z.Y., Tang, F.T., Wen, N., Xu, G.L. and Sheng, L., 2005. Effect of crocin on experimental atherosclerosis in quails and its mechanisms. *Life Sciences, 77*(8), pp. 907–921.

Liu, T. and Qian, Z., 1994. Protective effect of crocetin on isoproterenol-induced myocardial injury in rats. *Chinese Traditional and Herbal Drugs*.

Milajerdi, A., Jazayeri, S., Bitarafan, V., Hashemzadeh, N., Shirzadi, E., Derakhshan, Z., Mahmoodi, M., Rayati, A., Djazayeri, A. and Akhondzadeh, S., 2017. The effect of saffron (Crocus sativus L.) hydro-alcoholic extract on liver and renal functions in type 2 diabetic patients: A double-blinded randomized and placebo control trial. *Journal of Nutrition & Intermediary Metabolism*, *9*, pp. 6–11.

Modaghegh, M.H., Shahabian, M., Esmaeili, H.A., Rajbai, O. and Hosseinzadeh, H., 2008. Safety evaluation of saffron (Crocus sativus) tablets in healthy volunteers. *Phytomedicine*, *15*(12), pp. 1032–1037.

Posadzki, P., Watson, L. and Ernst, E., 2013. Contamination and adulteration of herbal medicinal products (HMPs): An overview of systematic reviews. *European Journal of Clinical Pharmacology*, *69*, pp. 295–307.

Sepahi, S., Golfakhrabadi, M., Bonakdaran, S., Lotfi, H. and Mohajeri, S.A., 2022. Effect of crocin on diabetic patients: A placebo-controlled, triple-blinded clinical trial. *Clinical Nutrition ESPEN*, *50*, pp. 255–263.

Shariatifar, N., Shoeibi, S., Sani, M.J., Jamshidi, A.H., Zarei, A., Mehdizade, A. and Dadgarnejad, M., 2014. Study on diuretic activity of saffron (stigma of Crocus sativus L.) Aqueous extract in rat. *Journal of Advanced Pharmaceutical Technology & Research*, *5*(1), p. 17.

Xi, L. and Qian, Z., 2006. Pharmacological properties of crocetin and crocin (digentiobiosyl ester of crocetin) from saffron. *Natural Product Communications*, *1*(1), p. 1934578X0600100112.

Yuanxiong, D., Zhiyu, Q. and Futian, T., 2004. Effects of crocetin on experimental atherosclerosis in rats. *Zhong cao yao= Chinese Traditional and Herbal Drugs*, *35*(7), pp. 777–781.

51 Melon (*Cucumis melo* L.)

Cucumis melo L.

Etymology: From the Latin *cucumis* = a cucumber and *melo* = an apple

Family: Cucurbitaceae

Synonym: *Melo sativus* Sageret

Common names: Melon; melon, pompon (Fr.); pfeben (Ger.); melão (Port.); дыня (Rus.); melaon (Sp.)

Part used: Fruit

Constituents: L-citrulline (about 300 mg/kg), ascorbic acid (Hartman et al., 2019), fixed oil (unsaturated fatty acids: linoleic acid) (de Melo et al., 2000).

Medical history: Dioscorides advocates melon as diuretic and anti-inflammatory and for beautifying the face. Matthioli in 16th-century Italy advocates the use of melon for the liver and as a diuretic. Fusch (1555) says it is cold and humid. In 18th-century French medical practice, the seeds of melon were a cold remedy used in skin cosmetics, for fever, to abrogate sexual appetite, and to induce sleep. Alston (1770) recommends melon seeds to treat urinary tract infections, inflammation, fever, and urinary stones, and recommends the fruits for individuals with a hot temperament.

Medicinal uses: Constipation, difficult urination (Pakistan); fever, thirst, diuresis (Korea); emetic, urinary problems, eczema (the Philippines)

Plasma lipids and cholesterol: Ethanol extract given daily at 100 mg/kg tohamsters on a high-fat diet daily for 10 days decreased total plasma cholesterol from about 800 to 400 mg/dL and plasma glucose from about 150 to 100 mg/dL. This regimen caused a decrease in LDL-cholesterol and an elevation in HDL-cholesterol (Shankar et al., 2015). Seed oil given to rats at 9.5% of a high-fat diet reduced plasma cholesterol by 24% and enhanced the excretion of fecal bile acids by 150% (Hao et al., 2020).

Heart: Intake of melon at 500 mg/kg/day for 30 days by rats with myocardial infarction induced by isoprenaline normalized lipid peroxidation in myocardial tissues and protected the myocardium against myofibril degeneration (Asdaq et al., 2021).

Kidneys: Hydro-alcoholic and methanolic extracts of seeds protected rats against the formation of kidney stones induced by ethylene glycol (Saleem et al., 2021; Eidi & Ashjazadeh, 2023).

Warning: With a glycemic index of 65, melon is not advised in diabetic patients (Foster-Powell et al., 2002; Atkinson et al., 2008).

Comment: Cold remedies as defined by ancient physicians are often anti-inflammatory and/or cardioprotective.

REFERENCES

Asdaq, S.M.B., Venna, S., Mohzari, Y., Alrashed, A., Alajami, H.N., Aljohani, A.O., Mushtawi, A.A.A., Alenazy, M.S., Alamer, R.F., Alanazi, A.K. and Nayeem, N., 2021. Cucumis melo enhances enalapril mediated Cardioprotection in Rats with Isoprenaline Induced Myocardial Injury. *Processes*, *9*(3), p. 557.

Atkinson, F.S., Foster-Powell, K. and Brand-Miller, J.C., 2008. International tables of glycemic index and glycemic load values. *Diabetes Care*, *31*(12), pp. 2281–2283.

de Melo, M.L.S., Narain, N. and Bora, P.S., 2000. Characterisation of some nutritional constituents of melon (Cucumis melo hybrid AF-522) seeds. *Food Chemistry*, *68*(4), pp. 411–414.

Eidi, M. and Ashjazadeh, L., 2023. Anti-urolithiatic effect of Cucumis melo L. var inodorous in male rats with kidney stones. *Urolithiasis*, *51*(1), pp. 1–9.

Foster-Powell, K., Holt, S.H. and Brand-Miller, J.C., 2002. International table of glycemic index and glycemic load values: 2002. *The American Journal of Clinical Nutrition*, *76*(1), pp. 5–56.

Fusch, L., 1555. *De Historia Stirpium Commetarii Insignes*. Lugduni Apud Ioan Tornaesium.

Hao, W., Zhu, H., Chen, J., Kwek, E., He, Z., Liu, J., Ma, N., Ma, K.Y. and Chen, Z.Y., 2020. Wild melon seed oil reduces plasma cholesterol and modulates gut microbiota in hypercholesterolemic hamsters. *Journal of Agricultural and Food Chemistry*, *68*(7), pp. 2071–2081.

Hartman, J.L., Wehner, T.C., Ma, G. and Perkins-Veazie, P., 2019. Citrulline and arginine content of taxa of Cucurbitaceae. *Horticulturae*, *5*(1), p. 22.

Matthioli, P.A., 1572. *Commentaires sur les Six Livres de Pedacius Dioscorides Anazarbeen de la matière medicinale*. A l'Escue de Milan.

Saleem, A., Islam, M., Saeed, H. and Iqtedar, M., 2021. In-vivo evaluation of anti-urolithiatic activity of different extracts of peel and pulp of cucumis melo L. in mice model of kidney stone formation. *Pakistan Journal of Zoology*, *53*(4).

Shankar, K., Singh, S.K., Kumar, D., Varshney, S., Gupta, A., Rajan, S., Srivastava, A., Beg, M., Srivastava, A.K., Kanojiya, S. and Mishra, D.K., 2015. Cucumis melo ssp. agrestis var. agrestis ameliorates high fat diet induced dyslipidemia in Syrian golden hamsters and inhibits adipogenesis in 3T3-L1 adipocytes. *Pharmacognosy Magazine*, *11*(Suppl 4), p. S501.

52 Cucumber (*Cucumis sativus* L.)

Cucumis sativus L.

Etymology: From the Latin *cucumis* = a cucumber and *sativus* = cultivated

Family: Cucurbitaceae

Common names: Cucumber; concombre (Fr.); gurke (Ger.); pepino (Port., Spa.); огурéц (Rus.)

Part used: Fruit

Constituents: L-citrulline (200 mg/kg), L-arginine (Hartman et al., 2019), linolenic acid (Fishwick et al., 1977).

Medical history: Dioscorides recommends cucumber for the stomach, to refresh, for jaundice, and as a diuretic, while Pliny the Elder considers it as beneficial for arteries and angina. In the 16th century, Fusch (1555) defines cucumber as cold and dry. Lémery (1716) says of cucumber that it attenuates "excessive movement of blood". In 19th-century Europe, it was used in skin cosmetics (Guibourt, 1836), while in India it was cooling, diuretic, and tonic (Dymock, 1884).

Medicinal uses: Typhoid (Iran. Iraq); fever, general debility, indigestion (Pakistan); diuretic, tonic, cooling (India); sore throat (Bangladesh); diuretic (Myanmar); dysentery (Cambodia, Laos, Vietnam); cooling, diuretic (Korea); gallstone (Indonesia)

Blood pressure: Healthy volunteers (19–23 years old) given 180 mL of cucumber juice, expressed from 200 gr fresh cucumber, every day for 7 days had decreases in systolic blood pressure from 115.5 to 111.3 mmHg and diastolic blood pressure from 79.3 to 77.2 mmHg (Fadilah et al., 2020). Intake of cucumber juice at the single dose of 400 g in fasting volunteers decreased systolic blood pressure from 130 to 118.6 mmHg and diastolic blood pressure from 78.7 to 75.3 mmHg dL after 2 hours (Bartimaeus et al., 2016).

Plasma lipids and glucose: The oral administration of pectin extracted from the fruit at 5 g/kg evoked hypolipidemic action in normal as well as cholesterol-fed experimental animals (Sudheesh & Vijayalakshmi, 1999). Ethanol extract of fruits given to rats at 400 mg/kg 30 min after the oral administration of 2 g of glucose caused a reduction in post-prandial glycemia from 95.4 to 75 mg/dL after 90 min. The extract given daily for 12 days to streptozotocin-induced diabetic rats decreased fasting plasma glucose from 236.6 to 96.4 mg/dL (Glibenclamide 250 mg/kg: 80.9 mg/dL), total cholesterol from 154.4 to 60.7 mg/dL, LDL-cholesterol from 93.6 to 38.5 mg/dL, and triglycerides from 182.5 to 90.9 mg/dL (Karthiyayini et al., 2015).

DOI: 10.1201/9781003301455-52

Intake of cucumber juice at the single dose of 400 g in fasting volunteers decreased fasting glycemia from 5.3 to 4.3 mg/dL after 2 hours (Bartimaeus et al., 2016).

Kidneys: Ethanol extract of cucumber given orally to rats at 500 mg/kg/day for 28 days protected kidneys against alloxan-induced diabetes with a reduction in plasma creatinine from 0.9 to 0.3 mg/dL (normal value 0.3 mg/dL) and prevented glomerular and tubular necrosis (Ofoego et al., 2020). Ethanol extract given orally at 400 mg/kg/day for 28 days protected rats against ethylene-glycol-induced kidney stone formation (Pethakar et al., 2017).

Bones and cartilages: Rats receiving daily aqueous extract at 400 mg/kg for 14 days were protected against formaldehyde-induced arthritis as evidence by a 60% decrease in edema and a decrease in hr-CRP (Fonkoua et al., 2022).

Skin and hair: Juice inhibited the activity of hyaluronidase and elastase with IC_{50} of 20.9 and 6.1 μg/mL, respectively (Fiume et al., 2014; Nema et al., 2011).

Brain: Fresh cucumber paste given orally to mice at the daily dose of 6 g/kg for 15 days attenuated amnesia induced by diazepam (Kumar et al., 2014).

Comments: (i) Rats receiving 4 mL/kg of fresh cucumber homogenate were protected against carrageenan-induced paw edema (Agatemor et al., 2015).

(ii) With aging and especially in diabetic patients, obese individuals, smokers, and alcoholics, the vascular endothelium is in a state of chronic low-grade inflammation, and that state prevents the physiological activity of NO and other regulators of blood pressure, resulting, at least in part, in arterial stiffness and increased blood pressure.

(iii) Cold medicines as defined by ancient physicians are often beneficial for inflammation and ROS accumulation.

(iv) When keratinocytes are exposed to ultraviolet B irradiations, they secrete pro-inflammatory cytokines that stimulate the secretion of elastase by fibroblasts. Elastase catalyse the decomposition of elastin in the dermis. In parallel, mesenchymal cells secrete hyaluronidase that decompose hyaluronic acid (Garg, 2017). Cucumber contain substantial amounts of potassium ions and must be avoided in hyperkaliemic patients.

REFERENCES

Agatemor, U., Nwodo, O. and Anosike, C., 2015. Anti-inflammatory activity of *Cucumis sativus* L. *British Journal of Pharmaceutical Research*, 8(2), pp. 1–8.

Bartimaeus, E.A.S., Echeonwu, J.G. and Ken-Ezihuo, S.U., 2016. The effect of Cucumis sativus (cucumber) on blood glucose concentration and blood pressure of apparently healthy individuals in Port Harcourt. *European Journal of Biomedical and Pharmaceutical Sciences*, 3(12), pp. 108–114.

Dymock, W., 1884. *The Vegetable Materia Medica of Western India*. Education Society Press.

Fadlilah, S., Sucipto, A. and Judha, M., 2020. Cucumber (Cucumis sativus) and tomato (Solanum lycopersicum) juice effective to reduce blood pressure. *GSC Biological and Pharmaceutical Sciences*, 10(1), pp. 001–008.

Fishwick, M.J., Wright, A.J. and Galliard, T., 1977. Quantitative composition of the lipids of cucumber fruit (Cucumis sativus). *Journal of the Science of Food and Agriculture*, 28(4), pp. 394–398.

Fiume, M.M., Bergfeld, W.F., Belsito, D.V., Hill, R.A., Klaassen, C.D., Liebler, D.C., Marks Jr, J.G., Shank, R.C., Slaga, T.J., Snyder, P.W. and Andersen, F.A., 2014. Safety assessment of Cucumis sativus (cucumber)-derived ingredients as used in cosmetics. *International Journal of Toxicology*, 33(Suppl 2), pp. 47S–64S.

Fonkoua, M., Lambou, G.F., Nguemto, G.T., Ngassa, D.N., Youovop, J., Edoa, C.A., Ngondi, J.L. and Enyong, J.O., 2022. Aqueous extract of Cucumis sativus fruit attenuates rheumatoid arthritis associated disorders on animal model of rheumatoid arthritis induced by formaldehyde. *Arabian Journal of Medicinal and Aromatic Plants*, 8(3), pp. 131–148.

Fusch, L., 1555. *De Historia Stirpium Commetarii Insignes*. Lugduni Apud Ioan Tornaesium.

Garg, C., 2017. Molecular mechanisms of skin photoaging and plant inhibitors. *International Journal of Green Pharmacy (IJGP)*, 11(02).

Guibourt, N.J.B.G., 1836. *Histoire abrégée des drogues simples*. Méquignon-Marvis Père et fils.

Hartman, J.L., Wehner, T.C., Ma, G. and Perkins-Veazie, P., 2019. Citrulline and arginine content of taxa of Cucurbitaceae. *Horticulturae*, 5(1), p. 22.

Karthiyayini, T., Kumar, R., Kumar, K.S., Sahu, R.K. and Roy, A., 2015. Evaluation of antidiabetic and hypolipidemic effect of Cucumis sativus fruit in streptozotocin-induced-diabetic rats. *Biomedical and Pharmacology Journal*, 2(2), pp. 351–355.

Kumar, M., Garg, A. and Parle, M., 2014. Amelioration of diazepam induced memory impairment by fruit of *Cucumis sativus* L in aged mice by using animal models of Alzheimer's disease. *International Journal of Pharmacy and Pharmaceutical Research*, 6(4), pp. 1015–1023.

Lémery, N., 1716. *Traité universel des drogues simples, mises en ordre alphabétique. Où l'on trouve leurs différens noms . . . et tout ce qu'il y a de particulier dans les animaux, dans les végétaux, et dans les minéraux*. Au dépend de la Companie.

Nema, N.K., Maity, N., Sarkar, B. and Mukherjee, P.K., 2011. Cucumis sativus fruit-potential antioxidant, anti-hyaluronidase, and anti-elastase agent. *Archives of Dermatological Research*, 303, pp. 247–252.

Ofoego, U.C., Nweke, E.O. and Nzube, O.M., 2020. Ameliorative effect of ethanolic extract of Cucumis sativus (Cucumber) pulp on alloxan induced kidney toxicity in male adult wistar rats ameliorative effect of ethanolic extract of Cucumis sativus (Cucumber) pulp on alloxan induced kidney toxicity. *Journal of Natural Sciences Research*, 9, pp. 12–22.

Pethakar, S.R., Hurkadale, P.J. and Hiremath, R.D., 2017. Evaluation of antiurolithiatic potentials of hydro-alcoholic extract of Cucumis sativus L. *Indian Journal of Pharmaceutical Education and Research*, 51(4S).

Sudheesh, S. and Vijayalakshmi, N.R., 1999. Lipid-lowering action of pectin from Cucumis sativus. *Food Chemistry*, 67(3), pp. 281–286.

53 Pumpkin (*Cucurbita pepo* L.)

Cucurbita pepo L.

Etymology: From the Latin *cucurbita* = a gourd and *pepo* = pumpkin

Family: Cucurbitaceae

Synonym: *Cucurbita maxima* Duschesne

Common names: Marrow, pumpkin, squash; citrouille, pastèque (Fr.); kürbis (Ger.); abóbora (Port.); тыква (Rus.); calbaza (Spa.)

Part used: Fruit

Constituents: Trigonelline (23 mg/100g), nicotinic acid (Yoshinari et al., 2009; Dong et al., 2021; Li et al., 2021), pectin (Salima et al., 2022), fixed oil (unsaturated fatty acids: oleic acid, linoleic acid) (Murkovic et al., 1996).

Medical history: Pumpkin was unknown to ancient Greek and Roman physicians, as well as to Middle Ages physicians in Europe. Brought from North America, it is described in the 16th century by the French physician Lobel under the name *"Pepo maximus Indicus compressus"* and mentioned by Daléchamps (1615) under the name of *"grand pompon plat d'Indie"*. A description is found in the *Universalis plantarum historiae, Tomus II* (1651) of Bauhin on page 226, one can read *"provenit anno 1605"*, suggesting the introduction of the plant in Europe in the year 1605. The seeds were used as diuretic, cooling, inflammation, and fevers in 18th-century Scotland (Alston, 1770) and for tapeworm in 19th-century Europe and North America (Peirera, 1843)

Medicinal uses: Benign prostatic hyperplasia, overactive bladder (2.5–7.5 g, 2 times daily) (European Union); bronchitis, leprosy, vermifuge, burn portion, jaundice (India); toothache (Bangladesh);

Blood pressure: Pumpkin seed oil given orally to rats at 100 mg/kg once daily for 6 weeks reduced L-NAME-induced elevation of blood pressure, mitigated cardiac pathological alterations and ECG changes, decreased MDA, and normalized NO (El-Mosallamy et al., 2012).

Oil of seeds given at 2 g/day for 12-weeks to post-menopausal women decreased diastolic blood pressure from 81.1 to 75.6 mmHg (Gossell-Williams et al., 2011). Volunteers taking oil of seeds at 1 g/day for 90 days experienced decreases in systolic blood pressure from 145.6 to 130 mmHg and in diastolic blood pressure from 84.3 to 81 mmHg (Majid et al., 2020). Five grams of pumpkin seeds per day for 60 days given to hyperlipidemic, hyperglycemic, and hypertensive women reduced systolic blood pressure from 132.5 to 122 mmHg and diastolic blood pressure from 81.9 to 75.7 mmHg (Monica et al., 2022).

DOI: 10.1201/9781003301455-53

Plasma lipids and glucose: Pumpkin paste given at 1% of diet to diabetic rats for 49 days had no effect on insulin but decreased total cholesterol from 109 to 93.6 mg/dL (Yoshinari et al., 2009).

Oil of seeds given at 2 g/day for 12-weeks to post-menopausal women increased HDL-cholesterol from 0.9 to 1 mmol/L (Gossell-Williams et al., 2011). Five grams of pumpkin seeds per day for 60 days to hyperlipidemic, hyperglycemic, and hypertensive women decreased fasting glucose from 94.7 to 89.6 mg/dL, total plasma cholesterol from 188.5 to 184.2 mg/dL, and LDL-cholesterol from 127.8 to 120.2 mg/dL (Monica et al., 2022).

Skin and hair: Oil of seeds applied externally protected the skin of rodent against ultraviolet B irradiation (Nakavoua et al., 2021). Oil of seeds (5%) applied to mice topically 6 days a week for 3 weeks maintained 75% of hair follicles in the anagen phase in the presence of testosterone (Hajhashemi et al., 2019).

Brain: Ethanol extract of seeds given orally at 200 mg/kg/day for 28 days to rats caused some levels of protection against aluminium chloride-induced cognitive dysfunction (Yadav et al., 2023).

Warning: Eating pumpkin seeds without chewing can potentially provoke rectal impaction (Chandrasekhara, 1983). Pumpkin seeds are rich in potassium and should be avoided in hyperkaliemic patients.

Comment: (i) Matthias de Lobel (1538–1616) was a French botanist and physician author of *Plantarum seu stirpium historia* (1576).

(ii) Jean Bauhin (1551–1613) was a Swiss botanist and physician author of *Universalis plantarum historiae* (1651).

(iii) At the base of a hair follicle is the dermal papilla, comprising cells expressing receptors to testosterone that inhibit insulin-like growth factor and hair growth (anagen phase) (Grymowicz et al., 2020).

REFERENCES

Alston, C., 1770. *Lectures on the Materia Medica: Containing the Natural History of Drugs, their Virtues and Doses: Also Directions for the Study of the Materia Medica; and an Appendix on the Method of Prescribing.* Edward and Charles Dilly.

Chandrasekhara, K.L., 1983. Pumpkin-seed impaction. *Annals of Internal Medicine*, 98(5_Part_1), pp. 675–675.

Daléchamps, 1615. *De l' histoire generale des plantes simples.* Chez Heritier Guillaume Rouille.

Dong, X.J., Chen, J.Y., Chen, S.F., Li, Y. and Zhao, X.J., 2021. The composition and anti-inflammatory properties of pumpkin seeds. *Journal of Food Measurement and Characterization*, 15, pp. 1834–1842.

El-Mosallamy, A.E., Sleem, A.A., Abdel-Salam, O.M., Shaffie, N. and Kenawy, S.A., 2012. Antihypertensive and cardioprotective effects of pumpkin seed oil. *Journal of Medicinal Food*, 15(2), pp. 180–189.

Gossell-Williams, M., Hyde, C., Hunter, T., Simms-Stewart, D., Fletcher, H., McGrowder, D. and Walters, C.A., 2011. Improvement in HDL cholesterol in postmenopausal women supplemented with pumpkin seed oil: Pilot study. *Climacteric*, 14(5), pp. 558–564.

Grymowicz, M., Rudnicka, E., Podfigurna, A., Napierala, P., Smolarczyk, R., Smolarczyk, K. and Meczekalski, B., 2020. Hormonal effects on hair follicles. *International Journal of Molecular Sciences*, *21*(15), p. 5342.

Hajhashemi, V., Rajabi, P. and Mardani, M., 2019. Beneficial effects of pumpkin seed oil as a topical hair growth promoting agent in a mice model. *Avicenna Journal of Phytomedicine*, *9*(6), p. 499.

Li, Y., Ramaswamy, H.S., Li, J., Gao, Y., Yang, C., Zhang, X., Irshad, A. and Ren, Y., 2021. Nutrient evaluation of the seed, pulp, flesh, and peel of spaghetti squash. *Food Science and Technology*, *42*.

Lobel, M., 1576. *Plantarum seu Stirpium Historia*. Ex officina Christophori Plantini.

Majid, A.K., Ahmed, Z. and Khan, R., 2020. Effect of pumpkin seed oil on cholesterol fractions and systolic/diastolic blood pressure. *Food Science and Technology*, *40*, pp. 769–777.

Monica, S.J., John, S., Madhanagopal, R., Sivaraj, C., Khusro, A., Arumugam, P., Gajdacs, M., Lydia, D.E., Sahibzada, M.U.K., Alghamdi, S. and Almehmadi, M., 2022. Chemical composition of pumpkin (Cucurbita maxima) seeds and its supplemental effect on Indian women with metabolic syndrome. *Arabian Journal of Chemistry*, *15*(8), p. 103985.

Murkovic, M., Hillebrand, A., Winkler, J., Leitner, E. and Pfannhauser, W., 1996. Variability of fatty acid content in pumpkin seeds (Cucurbita pepo L.). *Zeitschrift für Lebensmittel-Untersuchung und Forschung*, *203*(3), pp. 216–219.

Nakavoua, A.H., Enoua, G.C., Manhan-Iniangas, S., Chalard, P. and Figuérédo, G., 2021. Use of Cucurbita pepo Oil to fight against the UV action on the skin. *Green and Sustainable Chemistry*, *11*(2), pp. 49–58.

Pereira, J., 1843. *The Elements of Materia Medica and Therapeutics*. Lea and Blanchard.

Salima, B., Seloua, D., Djamel, F. and Samir, M., 2022. Structure of pumpkin pectin and its effect on its technological properties. *Applied Rheology*, *32*(1), pp. 34–55.

Yadav, R.S.P., Shenoy, B.V., Kumar, N., Kumar, G.P. and Kumar, S.N., 2023. In vivo acetylcholinesterase activity and Antioxidant property of Cucurbita pepo ethanolic extract in Alzheimer's disease induced by Aluminium chloride in Sprague Dawley rat model. *Research Journal of Pharmacy and Technology*, *16*(3), pp. 1065–1071.

Yoshinari, O., Sato, H. and Igarashi, K., 2009. Anti-diabetic effects of pumpkin and its components, trigonelline and nicotinic acid, on Goto-Kakizaki rats. *Bioscience, Biotechnology, and Biochemistry*, *73*(5), pp. 1033–1041.

54 Cumin (*Cuminum cyminum* L.)

Cuminum cyminum L.

Etymology: From the Hebrew *kammôn* = cumin and the Greek *kyminon* = cumin
Common name: Cumin; kreuzkümmel (Ger.); cominho (Port.); comino (Spa.)
Part used: Fruit
Constituent: Essential oil (cuminaldehyde, cuminol) (Bettaieb et al., 2011; Ali & Jumma, 2019).
Medical history: Greeks and Romans used cumin to treat inflammation (Dioscorides), for bitter eructations (Galen), and for tumors (Pliny the Elder). In 12th-century Byzantium, cumin was recommended for digesting meat and for intestinal worms, but excess causes delirium (Simeon Seth). In 16th-century Italy, Matthioli recommends cumin for ergotism, ulcers, and tumors, and he asserts that the seeds drank with cooked wine increase sperm. For Alston (1770), it induces perspiration and is used for vertigo and for weak and windy stomach. In 19th-century Europe, cumin was used as a stimulant and carminative. It was used as a stomachic and carminative and for milk secretion and nausea in pregnancy in 19th-century India (Watt, 1889).
Medicinal uses: Facilitate digestion (Iran); colitis, regulate menstrual cycle, promote lactation, carminative (Iraq)
Blood pressure: The oral administration of an aqueous extract of cumin to renovascular hypertensive rats at 200_ mg/kg for 9_ weeks caused a decrease of systolic blood pressure, NO bioavailability, and vascular inflammation (Kalaivani et al., 2013). Essential oil given to obese patients at 75 mg 2 times per day for 8 weeks caused a decrease in diastolic blood pressure from 84 to 81.4 mmHg (Morovati et al., 2019)
Plasma lipids and glucose: Cumin given orally at 250 mg/kg/day for 6 weeks to alloxan-induced diabetic rats decreased glycemia and total cholesterol from 153.4 to 98.2 mg/dL (control value: 75 mg/dL) (Dhandapani et al., 2002). Oral administration by prediabetic subjects of 75 mg of cumin essential oil per day for 10 weeks improved lipid profiles except for cholesterol (Jafari et al., 2018). Type-2 diabetic patients given 5 g of cumin per day for 60 days had a decrease in glycaemia as well as a 47% fall in total cholesterol (Andallu & Ramya, 2007).
Seeds given orally at 250 mg/kg for 6 weeks to alloxan-induced diabetic rats increased the levels of glutathione in the pancreas, liver, intestine, and aorta (Surya et al., 2005). Cuminol and cuminaldehyde at the concentration of 25 μg/mL boosted glucose-dependently insulin secretion from pancreatic

DOI: 10.1201/9781003301455-54

β-cells due ATP-sensitive potassium ions channel blockade and an increase in intracellular calcium ions concentration (Patil et al., 2013).

Bones and cartilages: Methanol extract given orally at 1 g/kg/day for 10 days to ovariectomized rats sustained the calcium ions content and mechanical strength of bones (Shirke et al., 2008).

Immune system: Extract given orally at 200 mg/kg/day safeguarded the immune system of mice against cyclosporine A poisoning as evidenced by by increased lymphocytes count (Chauhan et al., 2010).

Brain: Aqueous extract given orally at 300 mg/kg/day for 12 days attenuated scopolamine-induced dementia (Koppula & Choi, D.K., 2011).

Warnings: Individuals deficient in ADAMTS13 are at risk of developing thrombotic thrombocytopenic purpura when ingesting cumin seeds (Naing et al., 2016). The oral LD_{50} of essential oil is 0.7 mL/kg (Ceylan et al., 2003). Some individuals develop anaphylactic shock with cumin (Boxer et al., 1997). Must be avoided in pregnancy. In general, intake of essential oils orally should be avoided at all cost.

REFERENCES

Ali, A.M. and Jumma, H.J.A., 2019. Yield, quality and composition of cumin essential oil as affected by storage period. *International Journal of Analytical Mass Spectrometry and Chromatography*, 7(1), pp. 9–17.

Alston, C., 1770. *Lectures on the Materia Medica: Containing the Natural History of Drugs, their Virtues and Doses: Also Directions for the Study of the Materia Medica; and an Appendix on the Method of Prescribing*. Edward and Charles Dilly.

Andallu, B. and Ramya, V., 2007. Antihyperglycemic, cholesterol-lowering and HDL-raising effects of cumin (Cuminum cyminum) seeds in type-2 diabetes. *Journal of Natural Remedies*, pp. 142–149.

Bettaieb, I., Bourgou, S., Sriti, J., Msaada, K., Limam, F. and Marzouk, B., 2011. Essential oils and fatty acids composition of Tunisian and Indian cumin (Cuminum cyminum L.) seeds: A comparative study. *Journal of the Science of Food and Agriculture*, 91(11), pp. 2100–2107.

Boxer, M., Roberts, M. and Grammer, L., 1997. Cumin anaphylaxis: A case report. *Journal of Allergy and Clinical Immunology*, 99(5), pp. 722–723.

Ceylan, E., Özbek, H. and Ağaoğlu, Z., 2003. Cuminum cyminum L. (kimyon) meyvesi uçucu yağının median lethal doz (LD50) düzeyi ve sağlıklı ve diyabetli farelerde hipoglisemik etkisinin araştırılması. *Van Tıp Dergisi*, 10(2), pp. 29–35.

Chauhan, P.S., Satti, N.K., Suri, K.A., Amina, M. and Bani, S., 2010. Stimulatory effects of Cuminum cyminum and flavonoid glycoside on Cyclosporine-A and restraint stress induced immune-suppression in Swiss albino mice. *Chemico-Biological Interactions*, 185(1), pp. 66–72.

Dhandapani, S., Subramanian, V.R., Rajagopal, S. and Namasivayam, N., 2002. Hypolipidemic effect of Cuminum cyminum L. on alloxan-induced diabetic rats. *Pharmacological Research*, 46(3), pp. 251–255.

Jafari, T., Mahmoodnia, L., Tahmasebi, P., Memarzadeh, M.R., Sedehi, M., Beigi, M. and Fallah, A.A., 2018. Effect of cumin (Cuminum cyminum) essential oil supplementation on metabolic profile and serum leptin in pre-diabetic subjects: A randomized double-blind placebo-controlled clinical trial. *Journal of Functional Foods*, 47, pp. 416–422.

Kalaivani, P., Saranya, R.B., Ramakrishnan, G., Ranju, V., Sathiya, S., Gayathri, V., Thiya-garajan, L.K., Venkhatesh, J.R., Babu, C.S. and Thanikachalam, S., 2013. Cuminum cyminum, a dietary spice, attenuates hypertension via endothelial nitric oxide synthase and NO pathway in renovascular hypertensive rats. *Clinical and Experimental Hypertension*, *35*(7), pp. 534–542.

Koppula, S. and Choi, D.K., 2011. Cuminum cyminum extract attenuates scopolamine-induced memory loss and stress-induced urinary biochemical changes in rats: A noninvasive biochemical approach. *Pharmaceutical Biology*, *49*(7), pp. 702–708.

Matthioli, P.A., 1572. *Commentaires sur les Six Livres de Pedacius Dioscorides Anazarbeen de la matière medicinale*. A l'Escue de Milan.

Morovati, A., Pourghassem Gargari, B. and Sarbakhsh, P., 2019. Effects of cumin (Cuminum cyminum L.) essential oil supplementation on metabolic syndrome components: A randomized, triple-blind, placebo-controlled clinical trial. *Phytotherapy Research*, *33*(12), pp. 3261–3269.

Naing, T.W., Rahman, E.U., Abdulfattah, O. and Schmidt, F., 2016. Thrombotic thrombocytopenic purpura with cardiac involvement after an allergic reaction to turmeric and cumin: A case report. *Journal of Cardiology & Current Research*, *7*(3), p. 00247.

Patil, S.B., Takalikar, S.S., Joglekar, M.M., Haldavnekar, V.S. and Arvindekar, A.U., 2013. Insulinotropic and β-cell protective action of cuminaldehyde, cuminol and an inhibitor isolated from Cuminum cyminum in streptozotocin-induced diabetic rats. *British Journal of Nutrition*, *110*(8), pp. 1434–1443.

Shirke, S.S., Jadhav, S.R. and Jagtap, A.G., 2008. Methanolic extract of Cuminum cyminum inhibits ovariectomy-induced bone loss in rats. *Experimental Biology and Medicine*, *233*(11), pp. 1403–1410.

Surya, D., Vijayakumar, R.S. and Nalini, N., 2005. Oxidative stress and the role of cumin (Cuminum cyminum Linn.) in alloxan-induced diabetic rats. *Journal of Herbs, Spices & Medicinal Plants*, *11*(3), pp. 127–139.

Watt, G., 1889. *A Dictionary of the Economic Products of India*. Printed by the Superintendentof Government Printing, India.

55 Turmeric (*Curcuma longa* L.)

Curcuma longa L.

DOI: 10.1201/9781003301455-55

Etymology: From the Arabic *kurkum* = turmeric and the Latin *longa* = long

Family: Zingiberaceae

Synonym: *Curcuma domestica* Valeton

Common names: Turmeric, Indian saffron, long-rooted turmeric; curcuma, safran d'Inde, souchet des Indes (Fr.); gerbwulzel (Ger.); cúrcuma (Port.; Spa); куркумы (Rus.)

Part used: Rhizome

Constituents: Arylheptanoids (curcumin, desmethoxycurcumin), sesquiterpenes (Czygan, 2004).

Medical history: Turmeric was known to Sushruta (6th century BC) as *haridra* and used for inflammation, disorders of the blood, and jaundice as well as in numerous medicinal formulations (Dutt, 1877; Watt, 1889; Drury, 1873). Turmeric was known to Dioscorides and Pliny as *cyperus indicus*, which indicates that spice trade already existed between the West and the East during antiquity.

    In 16th-century Europe, turmeric was called *terra merita* because turmeric powder looks somewhat like some sort of potting soil. Daléchamps (1615) asserts that turmeric is useful for "opening internal parts" and for dropsy and cachexia. In 18-th century France, Lémery recommends intake of turmeric for liver and pancreatic obstructions, wounds, dropsy, diabetes, jaundice, and kidney stones and as a diuretic. For Alston in 18th-century Scotland, turmeric is a diuretic and is used for jaundice and for dropsy. It was used as a tonic in 19th-century North America.

Medicinal uses: In the European Union, turmeric is used for dyspepsia (0.5–1.0 g in 150 mL of boiling water as an infusion, 2–3 times daily); hepatitis (Turkey); tonic (Iran); sore throat, cough, fever (Sikkim); blood diseases, eczema (Bangladesh); sprain (India); diabetes (the Philippines); stomach ulcers, poison by black magic (!) (Papua New Guinea)

Blood pressure: Turmeric is more effective at reducing blood pressure than curcumin. Intake of turmeric at 50 mg/kg/day to rat on a high-fat and high-fructose diet decreased systolic blood pressure and prevented ventricular dysfunction (Hasimun et al., 2021).

Curcumin given at 500 mg/kg/day for 7 days to healthy individuals after breakfast caused a mild decrease in blood pressure (Soni & Kutian, 1992).

    Turmeric given daily to type-2 diabetic patients for 3 months caused a decrease in arterial stiffness (Srinivasan et al., 2019).

Plasma lipids and glucose: Curcumin given at 500 mg/kg/day for 7 days to healthy individuals after breakfast caused a decrease in total cholesterol of 11.6%, increased HDL-cholesterol by 29%, attenuated triglycerides, and decreased plasma MDA by 33% (Soni & Kutian, 1992).

    Two capsules containing 180 mg each of an extract (with more than 20% arylheptanoids) given to healthy volunteers before breakfast decreased postprandial glycemia by 15.7% after 30 min and insulinemia by 26.5% (Thota et al., 2018).

Heart: Turmeric given to rats orally at 100 mg/kg/day attenuated myocardial infarction induced by occlusion of the left ascending coronary artery (Mohanty et al., 2004).

Gallbladder: Curcumin given to rats as 0.5% of a high-cholesterol diet increased the fecal excretion of bile acids and cholesterol (Rao et al., 1970).

Kidneys: Curcumin given orally to rats at 30 mg/kg/day for 2 weeks to streptozotocin-induced diabetic rats increased creatinine and urea clearance and decreased proteinuria (Sharma et al., 2006).

Bones and cartilages: Arylheptanoids given to ovariectomized rats at 60 mg 3 times a week for 2 months prevented the loss of trabecular bones by about 50% (Wright et al., 2010). Aqueous extract given to patients with osteoarthritis at 500 mg twice a day for 42 days caused clinical improvement (Madhu et al., 2013).

Skin and hair: An extract given orally to mice at 1 g/kg twice a day for 19 weeks prevented increased skin thickness, the formation of wrinkles, matrix metalloprotein proteinase-2 expression, and decreased skin elasticity induced by ultraviolet B irradiation (Sumiyoshi & Kimura, 2009).

Brain: Curcumin given orally at 300 mg/kg/day for 60 days to transgenic Alzheimer's disease model of mice decreased the symptoms of dementia by preventing hypothalamic synaptic degradation and amyloid β proteins deposition in the cortex and hypothalamus and downregulated the expression of β-Site amyloid precursor protein cleavage enzyme 1 in the cortex (Zheng et al., 2017).

Warnings: Patients with obstruction of the biliary ducts as well as those with gallstones should avoid taking turmeric or curcumin (Czygan, 2004). Curcumin can induce bleeding, and turmeric should not be taken by patients on warfarin and other anticoagulants (Gronich et al., 2022). Curcumin inhibits cytochrome P450 3A4 and could affect the bioavailability of drugs taken orally and metabolized by this enzyme in the intestines (Appiah-Opong et al., 2007). A maximum of 9 g/day of rhizomes must not be exceeded. Use cautiously in patients using beta blockers or those with increased risk of bleeding (Posadzki et al., 2013).

Comments: (i) Sushruta (6th century BC) was an Indian ayurvedic physician and author of a book titled *Sushruta Samhita*. He is probably the first to define and propose a treatment for diabetes. His contribution to medical science is immense.

(ii) John Michael Maisch (1831–1893) was a pharmacist, professor of materia medica and botany at the Philadelphia College of Pharmacy, and author of a book titled *A manual of organic materia medica* (1885).

(iii) About 75% of 1 g/kg of curcumin ingested orally is excreted in the feces of rats, and traces remain of curcumin in the urine (Wahlström & Blennow, 1978). Curcumin that escapes first-pass metabolism reduces plasma markers of lipid peroxidation. Curcumin given at 500 mg/kg/day for 7 days to healthy individuals after breakfast caused a decrease in plasma MDA by 33% (Soni & Kutian, 1992).

(iv) Curcumin at high concentration is an adsorbent or quenching agent (Bisson et al., 2016; Nelson et al., 2017), and it induces the intestinal aggregation and fecal excretion of cholesterol, fatty acids, and biliary sterols, thereby decreasing hepatic and plasma cholesterol.

(v) β-Site amyloid precursor protein cleavage enzyme 1 (BACE 1) is a β-secretase enzyme that catalyzes the cleavage of transmembrane amyloid precursor protein (APP) into precursors of amyloid β proteins aggregates (Read, 2015).

(vi) Intake of turmeric at a dietary dose (mixed with olive oil and black pepper) for salad dressing or as a light infusion could potentially be beneficial for elderly people.

REFERENCES

Alston, C., 1770. *Lectures on the Materia Medica: Containing the Natural History of Drugs, their Virtues and Doses: Also Directions for the Study of the Materia Medica; and an Appendix on the Method of Prescribing.* Edward and Charles Dilly.

Appiah-Opong, R., Commandeur, J.N., van Vugt-Lussenburg, B. and Vermeulen, N.P., 2007. Inhibition of human recombinant cytochrome P450s by curcumin and curcumin decomposition products. *Toxicology*, 235(1–2), pp. 83–91.

Bisson, J., McAlpine, J.B., Friesen, J.B., Chen, S.N., Graham, J. and Pauli, G.F., 2016. Can invalid bioactives undermine natural product-based drug discovery? *Journal of Medicinal Chemistry*, 59(5), pp. 1671–1690.

Czygan, F.C., 2004. *Herbal Drugs and Phytopharmaceuticals: A Handbook for Practice on a Scientific Basis.* CRC Press.

Daléchamps, 1615. *De l'histoire generale des plantes simples.* Chez Heritier Guillaume Rouille.

Dutt, U.C., 1877. *The Materia Medica of the Hindus.* Thacker, Spink, & Co.

Gronich, N., Hurani, H., Weisz, I. and Halabi, S., 2022. Spontaneous bleeding and curcumin: Case report. *Journal of Clinical Images and Medical Case Reports*, 3(11), p. 2141.

Hasimun, P., Sulaeman, A., Hidayatullah, A. and Mulyani, Y., 2021. Effect of Curcuma longa L. extract on noninvasive cardiovascular biomarkers in hypertension animal models. *Journal of Applied Pharmaceutical Science*, 11(8), pp. 085–089.

Madhu, K., Chanda, K. and Saji, M.J., 2013. Safety and efficacy of Curcuma longa extract in the treatment of painful knee osteoarthritis: A randomized placebo-controlled trial. *Inflammopharmacology*, 21, pp. 129–136.

Mohanty, I., Arya, D.S., Dinda, A., Joshi, S., Talwar, K.K. and Gupta, S.K., 2004. Protective effects of Curcuma longa on ischemia-reperfusion induced myocardial injuries and their mechanisms. *Life Sciences*, 75(14), pp. 1701–1711.

Nelson, K.M., Dahlin, J.L., Bisson, J., Graham, J., Pauli, G.F. and Walters, M.A., 2017. The essential medicinal chemistry of curcumin: Miniperspective. *Journal of Medicinal Chemistry*, 60(5), pp. 1620–1637.

Posadzki, P., Watson, L. and Ernst, E., 2013. Contamination and adulteration of herbal medicinal products (HMPs): An overview of systematic reviews. *European Journal of Clinical Pharmacology*, 69, pp. 295–307.

Rao, D.S., Sekhara, N.C., Satyanarayana, M.N. and Srinivasan, M., 1970. Effect of curcumin on serum and liver cholesterol levels in the rat. *The Journal of Nutrition*, 100(11), pp. 1307–1315.

Read, J., 2015. *β-Site Amyloid Precursor Protein Cleaving Enzyme 1 (BACE1) as a Novel Alzheimer's Disease Intervention Target* (Doctoral dissertation, Deakin University).

Sharma, S., Kulkarni, S.K. and Chopra, K., 2006. Curcumin, the active principle of turmeric (Curcuma longa), ameliorates diabetic nephropathy in rats. *Clinical and Experimental Pharmacology and Physiology*, *33*(10), pp. 940–945.

Soni, K. and Kutian, R., 1992. Effect of oral curcumin administranon on serum peroxides and cholesterol levels in human volunteers. *Indian Journal of Physiology and Pharmacology*, *36*(4), pp. 273–275.

Srinivasan, A., Selvarajan, S., Kamalanathan, S., Kadhiravan, T., Prasanna Lakshmi, N.C. and Adithan, S., 2019. Effect of Curcuma longa on vascular function in native Tamilians with type 2 diabetes mellitus: A randomized, double-blind, parallel arm, placebo-controlled trial. *Phytotherapy Research*, *33*(7), pp. 1898–1911.

Sumiyoshi, M. and Kimura, Y., 2009. Effects of a turmeric extract (Curcuma longa) on chronic ultraviolet B irradiation-induced skin damage in melanin-possessing hairless mice. *Phytomedicine*, *16*(12), pp. 1137–1143.

Thota, R.N., Dias, C.B., Abbott, K.A., Acharya, S.H. and Garg, M.L., 2018. Curcumin alleviates postprandial glycaemic response in healthy subjects: A cross-over, randomized controlled study. *Scientific Reports*, *8*(1), p. 13679.

Wahlström, B. and Blennow, G., 1978. A study on the fate of curcumin in the rat. *Acta pharmacologica et toxicologica*, *43*(2), pp. 86–92.

Watt, G., 1889. *A Dictionary of the Economic Products of India*. Printed by the Superintendentof Government Printing, India.

Wright, L.E., Frye, J.B., Timmermann, B.N. and Funk, J.L., 2010. Protection of trabecular bone in ovariectomized rats by turmeric (Curcuma longa L.) is dependent on extract composition. *Journal of Agricultural and Food Chemistry*, *58*(17), pp. 9498–9504.

Zheng, K., Dai, X., Xiao, N.A., Wu, X., Wei, Z., Fang, W., Zhu, Y., Zhang, J. and Chen, X., 2017. Curcumin ameliorates memory decline via inhibiting BACE1 expression and β-Amyloid pathology in 5× FAD transgenic mice. *Molecular Neurobiology*, *54*, pp. 1967–1977.

56 Quince (*Cydonia oblonga* Mill.)

Cydonia oblonga Mill.

Etymology: From *Cydonia*, a city in the Island of Crete and the Latin *oblonga* = oblong

Family: Rosaceae

Synonyms: *Cydonia vulgaris* Pers.; *Pyrus cydonia* L.

Common names: Quince; coing (Fr.); kitten (Ger.); marmelo (Port.); айв á (Rus.); menbrillo, marmellos (Sp.)

Part used: Fruit

Constituents: Hydroxycinnamic acid derivatives (3-*O*-caffeoylquinic acid, 5-*O*-caffeoylquinic = chlorogenic acid), flavonol glycosides (quercetin 3-galactoside, rutin) (Silva et al., 2005), pectin (about 2 g/100g) (Forni et al., 1994), ascorbic acid (about 70 mg/100 g) (Rop et al., 2011).

Medical history: Hippocrates used quince as an astringent for diarrhea. It was known to Dioscorides as *malum cotoneum* and used "for those who canot breath if their head are not straight" (suggesting heart problems). Pliny the Elder asserts that quince is diuretic and useful for dysentery. Ibn al Wahshiya in the Middle Ages recommends quince as an antidote for poisons. In 16th-century Germany, Fusch (1555) defines quince as cold to the first degree and dry to the second. For bilious diseases, kidney stones, and ophthalmia (seeds only) in 18th-century Sweden (Bergius, 1782), while astringent and antiseptic in Scotland (Alston, 1770). The fruit juice was used as astringent and cooling, and the seed mucilage was used externally for cracked lips and painful hemorrhoids in 19th-century Europe and North America (Pereira, 1843)

Medicinal uses: Nipple wounds, diarrhea, cough (Turkey); cough, diarrhea (Afghanistan); dysentery (India)

Blood pressure: Aqueous extract of seeds at 600 mg/kg/day for 21 days prevented rise in blood pressure in rats fed a high-cholesterol, high-glucose diet (Rahman et al., 2021).

Plasma lipids and glucose: Streptozotocin-induced diabetic rats given an aqueous extract of fruits orally at the daily dose of 240 mg/kg for 6 weeks had normalized plasma triglycerides from about 80 to 60 mg/dL and total cholesterol from about 90 to 65 mg/dL, decreased LDL-cholesterol from about 22 to 16 mg/dL, and increased HDL-cholesterol (Mirmohammadlu et al., 2015). Ethanol extract of fruits given orally at 200 mg/kg/day for 8 weeks to mice on a high-fat diet caused decreases in body weight from 41.9 to 36.5 g, total plasma cholesterol from 188.4 to 171.8 mg/dL, plasma glucose from 196.8 to 185.2 mg/dL, triglycerides from 90.4 to 77.3 mg/dL, and plasma insulin from 9.1 to 5.6 ng/mL and increased HDL-cholesterol from 126.9 to 150.9 mg/dL. In epidydimal adipose tissues, the regimen activated AMP-activated protein kinase (AMPK) (Lee et_al., 2022).

Kidneys: Streptozotocin-induced diabetic rats given an aqueous extract of fruits orally at the daily dose of 240 mg/kg for 6 weeks showed decreased plasma urea (from about 55 to 35 mg/dL) as well as plasma creatinine (Mirmohammadlu et al., 2015).

Immune system: Aqueous extract given to rats for about 60 days prevented the development of atopic dermatitis and decreased plasma immunoglobulin E (Shinomiya et al., 2009).

Comments: (i) Ibn al Wahshiya was an Iraqi physician (900 AD) who wrote a book titled *Kitab al-sumum* (*The book of poisons*).

(ii) The Swedish botanist Petrus Jonas Bergius (1730–1790) professor at the Medical College of Stockholm wrote a book titled *Materia medica e regno Vegetabili* (1782).

(iii) AMPK is an energy sensor in cells that once activated triggers the intake of plasmatic glucose and fatty acids. It is the target of antidiabetic drugs such as metformin (Musi, 2006). It is also activated by reactive oxygen species (Hinchy et al., 2018).

(iv) Quince is edible when cooked in the form of delicious compotes and recommended for the aging cardiovascular system.

REFERENCES

Alston, C., 1770. *Lectures on the Materia Medica: Containing the Natural History of Drugs, their Virtues and Doses: Also Directions for the Study of the Materia Medica; and an Appendix on the Method of Prescribing*. Edward and Charles Dilly.

Bergius, P.J., 1782. *Materia medica e regno vegetabili, sistens simplicia officinalia, pariter atque culinaria. Secundum systema sexuale, ex autopsia et experientia, fideliter digessit Petrus Jonas Bergius . . . Editio secunda correctior. Cum privilegio s.r: majest. Sueciæ, & elect. Saxon. Stockholmiæ, typis Petri Hesselberg*.

Forni, E., Penci, M. and Polesello, A., 1994. A preliminary characterization of some pectins from quince fruit (Cydonia oblonga Mill.) and prickly pear (Opuntia ficus indica) peel. *Carbohydrate Polymers*, 23(4), pp. 231–234.

Fusch, L., 1555. *De Historia Stirpium Commetarii Insignes*. Lugduni Apud Ioan Tornaesium.

Hinchy, E.C., Gruszczyk, A.V., Willows, R., Navaratnam, N., Hall, A.R., Bates, G., Bright, T.P., Krieg, T., Carling, D. and Murphy, M.P., 2018. Mitochondria-derived ROS activate AMP-activated protein kinase (AMPK) indirectly. *Journal of Biological Chemistry*, 293(44), pp. 17208–17217.

Lee, H.S., Jeon, Y.E., Jung, J.I., Kim, S.M., Hong, S.H., Lee, J., Hwang, J.S., Hwang, M.O., Kwon, K. and Kim, E.J., 2022. Anti-obesity effect of Cydonia oblonga Miller extract in high-fat diet-induced obese C57BL/6 mice. *Journal of Functional Foods*, 89, p. 104945.

Mirmohammadlu, M., Hosseini, S.H., Kamalinejad, M., Gavgani, M.E., Noubarani, M. and Eskandari, M.R., 2015. Hypolipidemic, hepatoprotective and renoprotective effects of Cydonia oblonga Mill. fruit in streptozotocin-induced diabetic rats. *Iranian Journal of Pharmaceutical Research: IJPR*, 14(4), p. 1207.

Musi, N., 2006. AMP-activated protein kinase and type 2 diabetes. *Current Medicinal Chemistry*, *13*(5), pp. 583–589.

Pereira, J., 1843. *The Elements of Materia Medica and Therapeutics*. Lea and Blanchard.

Rahman, U., Saleem, M., Mahnashi, M.H., Alqahtani, Y.S., Alqarni, A.O., Alyami, B.A., Mushtaq, M.N. and Qasim, S., 2021. Antihypertensive and safety studies of Cydonia oblonga M. *Pakistan Journal of Pharmaceutical Sciences*, *34*(2 (Supplementary)), pp. 687–691.

Rop, O., Balik, J., Řezníček, V., Jurikova, T., Škardová, P., Salaš, P., Sochor, J., Mlček, J. and KramáŘová, D., 2011. Chemical characteristics of fruits of some selected quince (Cydonia oblonga Mill.) cultivars. *Czech Journal of Food Sciences*, *29*(1), pp. 65–73.

Shinomiya, F., Hamauzu, Y. and Kawahara, T., 2009. Anti-allergic effect of a hot-water extract of quince (Cydonia oblonga). *Bioscience, Biotechnology, and Biochemistry*, *73*(8), pp. 1773–1778.

Silva, B.M., Andrade, P.B., Martins, R.C., Valentão, P., Ferreres, F., Seabra, R.M. and Ferreira, M.A., 2005. Quince (Cydonia oblonga Miller) fruit characterization using principal component analysis. *Journal of Agricultural and Food Chemistry*, *53*(1), pp. 111–122.

57 Artichoke (*Cynara scolymus* L.)

Cynara scolymus L.

Etymology: From the Greek *cyna* = dog and *scolymos* = artichoke

Family: Asteraceae

Synonym: *Cynara cardunculus* L.

Common names: Artichoke; artichaut (Fr.); artischockenblätter (Ger.); alcachofra (Port.); артишок (Rus.); alcachofera (Spa.)

Parts used: Flower receptacle, leaf

Constituents: Hydroxycinnamic acid derivatives (chlorogenic acid, cynarin), flavonol glycosides, sesquiterpene lactones (cynaropicrin) (Fritsche et al., 2002).

Medical history: Dioscorides called artichoke *scolymus* and used it as diuretic (as did Galen and Pliny the Elder) while defining it as hot and dry to the second degree. Galen warns that eating artichoke makes blood melancholic. In 17th-century France, physicians used artichoke for jaundice and water retention and as aphrodisiac (Daléchamps, 1615). It was used as diuretic in 19th-century France (Moquin-Tandon, 1861).

Medicinal uses: Enhances biliary excretion, dyspepsia (1.5 g of leaves in 150 mL of boiling water, 4 times a day) (European Union); diuretic (Turkey); jaundice, liver tonic, digestive (Iran).

Plasma lipids and glucose: Artichoke has beneficial effects on the lipid profile of hamsters (Qiang et al., 2012). Powder given to rats as part of a high-fat diet at 0.005% for 16 weeks attenuated the increased mass of adipose tissues, increased fecal cholesterol, decreased plasma glucose and plasma insulin, and brought total cholesterol close to normal (8.7 to 6.9 mg/dL) (Kwon et al., 2018). Ethanol extract of leaves given orally at 400 mg/kg/day for 60 days to rats on high-fat diet decreased body weight, plasma total cholesterol, and triglycerides (Ben Salem et al., 2022).

Two capsules each containing 1.8 g of aqueous extract of leaves given to hyperlipidemic patients morning and evening each day for 42 days decreased total plasma cholesterol from 7.7 to 6.3 mmol/L and LDL-cholesterol from 5.2 to 4.2 mmol/L (Englisch et al., 2000). Aqueous extract of leaves given at 1280 mg/day for 12 weeks to hypercholesterolemic patients caused a decrease in total cholesterol from 7.1 to 6.8 mmol/L (Bundy et al., 2008).

Heart: Extract of leaves given orally at 400 mg/kg/day for 60 days to rats on a high-fat diet prevented cardiac tissue injuries (Ben Salem et al., 2022).

      DOI: 10.1201/9781003301455-57

Gallbladder: Aqueous extract of leaves given at the single dose of 1.9 g to volunteers induced a 94.3% increase in bile secretion after 60 min (Kirchhoff et al., 1994).

Adipose tissue: Powder of leaves (500 g/capsule) given to hypertensive patients twice daily for 8 weeks caused a decrease in weight (Ardalani et al., 2020).

Kidneys: Ethanol extract of leaves given orally at 400 mg/kg/day for 60 days to rats on a high-fat diet caused renoprotective effects (Ben Salem et al., 2022).

Skin and hair: Cynaropicrin in rodents prevented the hyperproliferation of keratinocytes and melanocytes induced by ultraviolet B irradiation. *In vitro*, cyanopicrin inhibited the NF-κB-mediated transactivation of matrix metalloproteinase-1 by fibroblasts (Tanaka et al., 2013).

Brain: Methanol extract of leaves given to rats orally at 1.6 g/kg daily evoked some protection against diethylnitrosamine-induced locomotor dysfunction. Brains tissues of treated animals contained lower levels of caspase-3 (Cicek et al., 2022).

Warning: Contra-indicated for patients with obstruction of bile duct, cholangitis, or gallstones.

Comments: (i) When fibroblasts in the dermis are exposed to proinflammatory cytokines, NF-kB or MAPK are activated and this results in the production of matrix metalloproteinases, of matrix metalloproteinase-1, which catalyzes the degradation of collagen I and III, and matrix metalloproteinase-12, which degrades elastin, contributing to skin aging (Pittayapruek et al., 2016).

(ii) Artichoke consumption could potentially be beneficial for elderly people. In case of hyperkaliemia it must be avoided.

REFERENCES

Ardalani, H., Jandaghi, P., Meraji, A. and Moghadam, M.H., 2020. The effect of Cynara scolymus on blood pressure and BMI in hypertensive patients: A randomized, double-blind, placebo-controlled, clinical trial. *Complementary Medicine Research*, 27(1), pp. 40–46.

Ben Salem, M., Affes, H., Dhouibi, R., Charfi, S., Turki, M., Hammami, S., Ayedi, F., Sahnoun, Z., Zeghal, K.M. and Ksouda, K., 2022. Effect of Artichoke (cynara scolymus) on cardiac markers, lipid profile and antioxidants levels in tissue of HFD-induced obesity. *Archives of Physiology and Biochemistry*, 128(1), pp. 184–194.

Bundy, R., Walker, A.F., Middleton, R.W., Wallis, C. and Simpson, H.C., 2008. Artichoke leaf extract (Cynara scolymus) reduces plasma cholesterol in otherwise healthy hypercholesterolemic adults: A randomized, double blind placebo controlled trial. *Phytomedicine*, 15(9), pp. 668–675.

Cicek, B., Genc, S., Yeni, Y., Kuzucu, M., Cetin, A., Yildirim, S., Bolat, I., Kantarci, M., Hacimuftuoglu, A., Lazopoulos, G. and Tsatsakis, A., 2022. Artichoke (Cynara Scolymus) methanolic leaf extract alleviates Diethylnitrosamine-induced toxicity in BALB/c mouse brain: Involvement of oxidative stress and Apoptotically related Klotho/PPARγ signaling. *Journal of Personalized Medicine*, 12(12), p. 2012.

Daléchamps, 1615. *De l' histoire generale des plantes simples*. Chez Heritier Guillaume Rouille.

Englisch, W., Beckers, C., Unkauf, M., Ruepp, M. and Zinserling, V., 2000. Efficacy of artichoke dry extract in patients with hyperlipoproteinemia. *Arzneimittelforschung*, *50*(03), pp. 260–265.

Fritsche, J., Beindorff, C.M., Dachtler, M., Zhang, H. and Lammers, J.G., 2002. Isolation, characterization and determination of minor artichoke (Cynara scolymus L.) leaf extract compounds. *European Food Research and Technology*, *215*, pp. 149–157.

Kirchhoff, R., Beckers, C.H., Kirchhoff, G.M., Trinczek-Gärtner, H., Petrowicz, O. and Reimann, H.J., 1994. Increase in choleresis by means of artichoke extract. *Phytomedicine*, *1*(2), pp. 107–115.

Kwon, E.Y., Kim, S.Y. and Choi, M.S., 2018. Luteolin-enriched artichoke leaf extract alleviates the metabolic syndrome in mice with high-fat diet-induced obesity. *Nutrients*, *10*(8), p. 979.

Moquin-Tandon, A., 1861. *Element de Botanique Médicale*. J.Baillière et Fils.

Pittayapruek, P., Meephansan, J., Prapapan, O., Komine, M. and Ohtsuki, M., 2016. Role of matrix metalloproteinases in photoaging and photocarcinogenesis. *International Journal of Molecular Sciences*, *17*(6), p. 868.

Qiang, Z., Lee, S.O., Ye, Z., Wu, X. and Hendrich, S., 2012. Artichoke extract lowered plasma cholesterol and increased fecal bile acids in Golden Syrian hamsters. *Phytotherapy Research*, *26*(7), pp. 1048–1052.

Tanaka, Y.T., Tanaka, K., Kojima, H., Hamada, T., Masutani, T., Tsuboi, M. and Akao, Y., 2013. Cynaropicrin from Cynara scolymus L. suppresses photoaging of skin by inhibiting the transcription activity of nuclear factor-kappa B. *Bioorganic & Medicinal Chemistry Letters*, *23*(2), pp. 518–523.

58 Lesser Cardamom (*Elettaria cardamomum* (L.) Maton)

Elettaria cardamomum (L.) Maton

Etymology: From the Sanskrit *ela* = cardamom and the Latin *cardamomum* = cardamom

Family: Zingiberaceae

Synonyms: *Amomum cardamomum* L.; *Cardamomum officinale* (L.) Salisb.

Common names: Cardamom, lesser cardamon; cardamome (Fr.); kardamom (Ger.); cardamomo (Port.; Spa.); кардамон (Rus.)

Part used: Seed

Constituents: Essential oil (1,8-cineole, terpinyl acetate) (Savan & Küçükbay, 2013), chalcones (cardamonin) (Makhija et al., 2022).

Medical history: In ancient India, Sushruta used lesser cardamom for cold, asthma, piles, and diseases of the bladder and kidneys. Dioscorides says it is good for sciatica and recommends it as diuretic. The plant is described in the *Hortus malabaricus* of Hendrik van Rheede, governor of Dutch Malabar from 1669 to 1676, under the name *elatteri*. Lémery in 18th-century France writes that people of Malabar chewed the seeds to cool the mouth in case of hot weather and for fever and asserts that it is good for flatulence and to promote urination. It is a hot and dry carminative for vertigo and for "windy stomach" in the lectures of Alston (1770). It is a tonic in 19th-century India (O'Shaughnessy, 1842) and Europe (Pereira, 1843).

Medicinal uses: Abdominal pain (Bangladesh); cough, headache, heart diseases, urinary disorders (Myanmar)

Blood pressure: Lesser cardamom given to hypertensive patients at 1.5 g twice a day for 12 weeks decreased systolic blood pressure from 154.2 to 134.8 mmHg and diastolic blood pressure from 91.8 to 79.6 mmHg and enhanced plasmatic fibrinolytic activity (Verma et al., 2009).

Plasma lipids and glucose: Lesser cardamom given to hypertensive patients at 1.5 g twice a day for 12 weeks decreased total cholesterol, triglycerides, and LDL-cholesterol by 19, 15, and 25%, respectively (Verma et al., 2009). Essential oil given to rats on a high-fat diet at 3 g/kg for 8 weeks caused reductions of plasma total cholesterol by 31%, LDL-cholesterol by 44%, and triglycerides by 42% (Nagashree et al., 2017).

DOI: 10.1201/9781003301455-58

Obese patients given cardamom for 2 months showed decreased total cholesterol, LDL-cholesterol, and triglycerides (Fatemeh et al., 2017).

Heart: Essential oil given to rats on a high-fat diet at 3 g/kg for 8 weeks decreased the content of cardiac cholesterol by 39% and increased cardiac SOD activity (Nagashree et al., 2017).

Brain: Ethanol extract given to rats at 1 g/kg/day for 15 days attenuated scopolamine-induced dementia (Kunwar et al., 2015).

Warning: Do not exceed 3 g per day (Singletary, 2022).

REFERENCES

Alston, C., 1770. *Lectures on the Materia Medica: Containing the Natural History of Drugs, their Virtues and Doses: Also Directions for the Study of the Materia Medica; and an Appendix on the Method of Prescribing*. Edward and Charles Dilly.

Fatemeh, Y., Siassi, F., Rahimi, A., Koohdani, F., Doostan, F., Qorbani, M. and Sotoudeh, G., 2017. The effect of cardamom supplementation on serum lipids, glycemic indices and blood pressure in overweight and obese pre-diabetic women: A randomized controlled trial. *Journal of Diabetes & Metabolic Disorders, 16*, pp. 1–9.

Kunwar, T., Kumar, N. and Kothiyal, P., 2015. Effect of Elettaria cardamomum hydroethanolic extract on learning and memory in Scopolamine induced amnesia. *World Journal of Pharmaceutical Sciences*, pp. 75–85.

Lémery, N., 1716. *Traité universel des drogues simples, mises en ordre alphabétique. Où l'on trouve leurs différens noms . . . et tout ce qu'il y a de particulier dans les animaux, dans les végétaux, et dans les minéraux*. Au dépend de la Companie.

Makhija, P., Handral, H.K., Mahadevan, G., Kathuria, H., Sethi, G. and Grobben, B., 2022. Black cardamom (Amomum subulatum Roxb.) fruit extracts exhibit apoptotic activity against lung cancer cells. *Journal of Ethnopharmacology, 287*, p. 114953.

Nagashree, S., Archana, K.K., Srinivas, P., Srinivasan, K. and Sowbhagya, H.B., 2017. Antihypercholesterolemic influence of the spice cardamom (Elettaria cardamomum) in experimental rats. *Journal of the Science of Food and Agriculture, 97*(10), pp. 3204–3210.

O'Shaughnessy, W.B., 1842. *The Bengal Dispensatory and Companion to the Pharmacopœia . . . Chiefly Compiled . . . by W. B. O'Shaughnessy, Etc*. Calcutta printed.

Pereira, J., 1843. *The Elements of Materia Medica and Therapeutics*. Lea and Blanchard.

Savan, E.K. and Küçükbay, F.Z., 2013. Essential oil composition of Elettaria cardamomum Maton. *Journal of Applied Biological Sciences, 7*(3), pp. 42–45.

Singletary, K., 2022. Cardamom: Potential health benefits. *Nutrition Today, 57*(1), pp. 38–49.

Verma, S.K., Jain, V. and Katewa, S.S., 2009. Blood pressure lowering, fibrinolysis enhancing and antioxidant activities of cardamom (Elettaria cardamomum). *Indian Journal of Biochemistry and Biophysics, 46*, pp. 503–506.

59 Horsetail (*Equisetum arvense* L.)

Equisetum arvense L.

Etymology: From the Latin *equus* = horse, *seta* = bristle, and *arvense* = from the fields

Family: Equisetaceae

Synonyms: *Allostelites arvensis* (L.) Börner; *Equisetum campestre* Schultz; *Equisetum saxicola* Suksd.

Common names: Horsetail; prêles (Fr.) schachtelhalmkraut (Ger.); cavalinha (Port.); хвощ (Rus.); cola de caballo (Spa.)

Part used: Aerial part

Constituents: Flavones, flavone glycosides, phenolic acids, silicon (about 20 g/kg) (Carnat et al., 1991; Labun et al., 2013).

Medical history: Horsetail was known to Dioscorides as *hippuris* and used as a diuretic: "*urinam cit*". For Pliny the Elder, horsetail is useful for orthopnea. It is astringent for 16th-century German physicians (Fusch, 1555). In 18th-century Scotland, it was used for wounds and ulcers of the gut, kidneys, and bladder (Alston, 1770).

Medicinal uses: Diuretic (European Union, Turkey); kidney stones (Pakistan); cough (China)

Blood pressure: Powder given at 900 mg/day to hypertensive patients for 3 months decreased systolic and diastolic blood pressure of 12.6 and 8.1 mmHg, respectively (Carneiro et al., 2022).

Blood lipids and glucose: Methanol extract given orally at 100 mg/kg/day for about 40 days to streptozotocin-induced diabetic rats attenuated weight loss, decreased plasma glucose from about 35 to 30 mmol/L (normal about 8 mmol/L), improved glucose tolerance, and reduced insulinemia (Hegedűs et al., 2020).

Heart: Methanol extract given orally at 100 mg/kg/day about 40 days to streptozotocin-induced diabetic rats increased the expression of left ventricular SIRT1 (Hegedűs et al., 2020).

Kidneys: Powder given at 900 mg/day caused diuretic effects similar to those from hydrochlorothiazide at 50 mg/day (Carneiro et al., 2022).

Bones and cartilages: Ethanol extract given to rats at 120 mg/kg/day for 30 days resulted in increased mandibular bone mineral density (Arbabzadegan et al., 2019).

Warning: The dose of 12 g/day must not be exceeded (European Union). A 52-year old man was hospitalized for liver poisoning after consuming 500 mL of a decoction daily for 2 weeks (Klnçalp et al., 2012).

Comments: (i) Pliny asserts that the plant is of use for orthopnea, difficulty breathing when lying down, a symptom of heart weakness.

(ii) SIRT1 is a NAD⁺-dependent deacetylase that protects the cardiovascular system against ROS by inducing SOD and CAT expression, but it decreases with aging. In aging rats, although SIRT1 activity decreases, moderate exercise promotes its activity (Russomanno et al., 2017).

(iii) Yingmao et al. (2001) demonstrated that *Equisetum hiemale* L. was able to improve systole-diastolic performance and coronary flow and attenuate heart rate.

REFERENCES

Alston, C., 1770. *Lectures on the Materia Medica: Containing the Natural History of Drugs, their Virtues and Doses: Also Directions for the Study of the Materia Medica; and an Appendix on the Method of Prescribing.* Edward and Charles Dilly.

Arbabzadegan, N., Moghadamnia, A.A., Kazemi, S., Nozari, F., Moudi, E. and Haghanifar, S., 2019. Effect of equisetum arvense extract on bone mineral density in Wistar rats via digital radiography. *Caspian Journal of Internal Medicine*, 10(2), p. 176.

Carnat, A., Petitjean-Freytet, C., Muller, D. and Lamaison, J.L., 1991. Main constituents of the sterile fronds of Equisetum arvense L. *Plantes Médicinales et Phytothérapie*, 25(1), pp. 32–38.

Carneiro, D.M., Jardim, T.V., Araújo, Y.C.L., Arantes, A.C., de Sousa, A.C., Barroso, W.K.S., Sousa, A.L.L., de Carvalho Cruz, A., da Cunha, L.C. and Jardim, P.C.B.V., 2022. Antihypertensive effect of Equisetum arvense L.: A double-blind, randomized efficacy and safety clinical trial. *Phytomedicine*, 99, p. 153955.

Fusch, L., 1555. *De Historia Stirpium Commetarii Insignes.* Lugduni Apud Ioan Tornaesium.

Hegedűs, C., Muresan, M., Badale, A., Bombicz, M., Varga, B., Szilágyi, A., Sinka, D., Bácskay, I., Popoviciu, M., Magyar, I. and Szarvas, M.M., 2020. SIRT1 activation by Equisetum arvense L. (Horsetail) modulates insulin sensitivity in streptozotocin induced diabetic rats. *Molecules*, 25(11), p. 2541.

Klnçalp, S., Ekiz, F., Basar, Ö., Çoban, S. and Yüksel, O., 2012. Equisetum arvense (Field Horsetail)-induced liver injury. *European Journal of Gastroenterology & Hepatology*, 24(2), pp. 213–214.

Labun, P., Grulova, D., Salamon, I. and Serseň, F., 2013. Calculating the silicon in horsetail (Equisetum arvense L.) during the vegetation season. *Food and Nutrition Sciences*, 4(5), pp. 510–514.

Russomanno, G., Corbi, G., Manzo, V., Ferrara, N., Rengo, G., Puca, A.A., Latte, S., Carrizzo, A., Calabrese, M.C., Andriantsitohaina, R. and Filippelli, W., 2017. The anti-ageing molecule sirt1 mediates beneficial effects of cardiac rehabilitation. *Immunity & Ageing*, 14, pp. 1–9.

Yingmao, C., Yu, L., You, Z., Suzhen, H., Shuzhi, C., Qiushi, Z. and Cheng, Z., 2001. Effect of Equisetum hiemale L on the physiological cardiac performances of rat. *Journal of Chengde Medical College*, 18(3), pp. 184–187.

60 Philippines Wax Flower (*Etlingera elatior* (Jack) R.M. Sm.)

Etlingera elatior (Jack) **R.M. Sm.**

DOI: 10.1201/9781003301455-60

Etymology: After A.E. Etlinger, an 18th-century botanist, and the Latin *eliator* = tall

Family: Zingiberaceae

Synonyms: *Alpinia elatior* Jack; *Alpinia speciosa* (Blume) D. Dietr.; *Elettaria speciosa* Blume; *Nicolaia speciosa* (Blume) Horan.; *Phaeomeria magnifica* (Roscoe) K. Schum.; *Phaeomeria imperialis* (Roscoe) Lindl.; *Phaeomeria speciosa* (Blume) Koord.

Common names: Philippines wax flower, torch ginger; gingembre torche, rose de porcelaine (Fr.); fackelingwer (Ger.); bastão do imperador (Port.); имбирь Факел (Rus.); antorcha de jengibre (Spa.)

Part used: Inflorescence

Constituents: Flavone glycosides, hydroxycinnamic acid derivatives (3-*O*-caffeoylquinic acid, chlorogenic acid), essential oils (Chan et al., 2011).

Medical history: Philippines wax flower has been used as medicinal food in Thailand, Malaysia, Indonesia, and the Philippines since the dawn of time. The Scottish botanist and physician William Jack (1795–1822) describes for the first time the plant he found on the west coast of Sumatra and informs us that the natives called it *Bunga kenchong*. He notes that "it is a remarkable species" (Malayan Miscellanies, 1822). The German botanist Karl Ludwig von Blume (1796–1862), in his *Enumeratio plantarum Javae* (1827), gives a description and notes that it grows in the primary rainforest of Java: "*crescit in sylvis humidis Java insulae*".

Medicinal uses: Body ache (Malaysia); hypertension (Aceh).

Blood pressure: Ethanol extract of fruits given at 800 mg/kg/day to streptozotocin-induced diabetic rats for 4 weeks decreased blood pressure from about 140 to 120 mmHg and reduced plasma levels of hr-CRP (Widyarini et al., 2022).

Plasma lipids and glucose: Aqueous extract of flowers given orally at 1 g/kg/day for 6 weeks to rats decreased the glycemia of streptozotocin-induced diabetic rats from about 25 to 10 mmol/L (normal: about 5 mmol/L) and decreased cholesterol, triglycerides, LDL-cholesterol, urea, and creatinine (Nor et al., 2020). Ethanol extract of fruits given at 800 mg/kg/day to streptozotocin-induced diabetic rats for 4 weeks prevented weight loss and decreased plasma glucose from 262.8 to 89.9 mg/dL (Widyarini et al., 2022).

Kidneys: Aqueous extract of flowers given orally at 1 g/kg/day for 6 weeks decreased plasma urea and creatinine in rats (Nor et al., 2020).

Comment: The flower buds are common sights in almost all Southeast Asian wet markets, and their consumption, as traditionally made in Malaysia and Indonesia, could potentially be beneficial for elderly people.

REFERENCES

Al-Mansoub, M.A., Asif, M., Revadigar, V., Hammad, M.A., Chear, N.J.Y., Hamdan, M.R., Majid, A.M.S.A., Asmawi, M.Z. and Murugaiyah, V., 2021. Chemical composition, antiproliferative and antioxidant attributes of ethanolic extract of resinous sediment from Etlingera elatior (Jack.) inflorescence. *Brazilian Journal of Pharmaceutical Sciences*, *57*.

Chan, E.W., Lim, Y.Y. and Wong, S.K., 2011. Phytochemistry and pharmacological properties of Etlingera elatior: A review. *Pharmacognosy Journal*, *3*(22), pp. 6–10.

Juwita, T., Puspitasari, I.M. and Levita, J., 2018. Torch ginger (Etlingera elatior): A review on its botanical aspects, phytoconstituents and pharmacological activities. *Pakistan Journal of Biological Sciences*, *21*(4), pp. 151–165.

Nor, N.A.M., Noordin, L., Bakar, N.H.A. and Ahmad, W.A.N.W., 2020. Evaluation of antidiabetic activities of Etlingera elatior flower aqueous extract in vitro and in vivo. *Journal of Applied Pharmaceutical Science*, *10*(8), pp. 043–051.

Widyarini, T., Indarto, D. and Purwanto, B., 2022. Modulation effects of Etlingera elatior ethanol extract as anti-inflammatory on chronic kidney disease in mice with hypertension and diabetes. *Journal of Population Therapeutics and Clinical Pharmacology*, *29*(04), pp. 140–149.

61 Japanese Horseradish (*Eutrema japonicum* (Miq.) Koidz.)

Eutrema japonicum (Miq.) Koidz.

Etymology: From the Greek *eu* = perfect and *trema* = hole and the Latin *japonicum* = from Japan

Family: Brassicaceae

Synonyms: *Cochlearia wasabi* Sieb.; *Eutrema wasabi* (Siebold) Maxim.; *Lunaria japonica* Miq.; *Wasabia japonica* (Miq.) Matsum.

Common names: Japanese horseradish, wasabi

Part used: Root

Constituents: Glucosinolates (sinigrin), isothicyanates (6-methyl(sulfinyl)hexyl isothiocyanate) (Park et al., 2006).

Medical history: Wasabi was not known to Europeans until probably the end of the 19th century, but it has been used in Japan since the dawn of time as an extremely spicy condiment. We had to wait for the observations of the German botanist Philipp Franz von Siebold (1796–1866), who lived in Japan where he married a Japanese woman and had a daughter.

Blood pressure: The addition of 5% wasabi powder into a high-fat, high-carbohydrate diet given to rats for 8 weeks decreased body weight from 460 to 416 g, systolic blood pressure from 146 to 132 mmHg, triglycerides from 1.7 to 0.9 mmol/L, and cholesterol from 1.5 to 1 mmol/L. This regimen decreased heart inflammation (Thomaz et al., 2022).

Aqueous extract (from decoction of leaves) given orally to rats (genetically modified to mimic hyperlipidemia) at 4 g/kg/day for 4 weeks reduced systolic blood pressure from 240 to 196.6 mmHg, plasma triglycerides from 496 to 443.1 mg/dL, and insulinemia from 11.6 to 7.6 ng/mL (Oowatari et al., 2016).

Skin and hair: 6-Methyl(sulfinyl)hexyl isothiocyanate induced the growth of dermal papilla cells *in vitro* and upregulated the expression of vascular endothelial growth factor and adenosine 2AB receptor (Yamada-Kato et al., 2018).

Brain: An extract (corresponding to 9.6 mg of 6-methyl(sulfinyl)hexyl isothiocyanate/day) given for 12 weeks to patients with myalgic encephalomyelitis improved cognitive function (Oka et al., 2022).

DOI: 10.1201/9781003301455-61

Warnings: A Japanese woman was admitted to hospital for a left ventricular dysfunction after eating enormous quantities of wasabi (Finkel-Oron et al., 2019). Wasabi should be avoided by patients with thyroid dysfunction (Yamada-Kato et al., 2018).

Comments: (i) Glucosinolates such as sinigrin decompose into isothiocyanates such as allyl isothiocyanate which are ROS scavengers (Haina et al., 2010).

(ii) Adenosine 2AB receptor once activated induces the formation of blood vessels (angiogenesis) by promoting the expression of VEGF and endothelial NO synthetase (Du et al., 2015).

REFERENCES

Du, X., Ou, X., Song, T., Zhang, W., Cong, F., Zhang, S. and Xiong, Y., 2015. Adenosine A2B receptor stimulates angiogenesis by inducing VEGF and eNOS in human microvascular endothelial cells. *Experimental Biology and Medicine*, 240(11), pp. 1472–1479.

Finkel-Oron, A., Olchowski, J., Jotkowitz, A. and Barski, L., 2019. Takotsubo cardiomyopathy triggered by wasabi consumption: Can sushi break your heart? *BMJ Case Reports CP*, 12(9), p. e230065.

Haina, Y.U.A.N., Shanjing, Y.A.O., Yuru, Y.O.U., Gongnian, X.I.A.O. and Qi, Y., 2010. Antioxidant activity of isothiocyanate extracts from broccoli. *Chinese Journal of Chemical Engineering*, 18(2), pp. 312–321.

Oka, T., Yamada, Y., Lkhagvasuren, B., Nakao, M., Nakajima, R., Kanou, M., Hiramatsu, R. and Nabeshima, Y.I., 2022. Clinical effects of wasabi extract containing 6-MSITC on myalgic encephalomyelitis/chronic fatigue syndrome: An open-label trial. *BioPsychoSocial Medicine*, 16(1), pp. 1–12.

Oowatari, Y., Ogawa, T., Katsube, T., Iinuma, K., Yoshitomi, H. and Gao, M., 2016. Wasabi leaf extracts attenuate adipocyte hypertrophy through PPARγ and AMPK. *Bioscience, Biotechnology, and Biochemistry*, 80(8), pp. 1594–1601.

Park, Y., Cho, M., Park, S., Lee, Y., Jeong, B. and Chung, J., 2006. Sinigrin contents in different tissues of wasabi and antimicrobial activity of their water extracts. *Korean Journal of Horticultural Science & Technology*, 24(4), pp. 480–487.

Thomaz, F.S., Tan, Y.P., Williams, C.M., Ward, L.C., Worrall, S. and Panchal, S.K., 2022. Wasabi (Eutrema japonicum) reduces obesity and blood pressure in diet-induced metabolic syndrome in rats. *Foods*, 11(21), p. 3435.

Yamada-Kato, T., Okunishi, I., Fukamatsu, Y., Tsuboi, H. and Yoshida, Y., 2018. Stimulatory effects of 6-methylsulfinylhexyl isothiocyanate on cultured human follicle dermal papilla cells. *Food Science and Technology Research*, 24(3), pp. 567–572.

62 Asafetida (*Ferula assa-foetida* L.)

Ferula assa-foetida L.

Etymology: From the Latin *ferula* = a rod, the Persian *aza* = mastic, and the Latin *foetida* = fetid

Family: Apiaceae

Synonyms: *Ferula foetida* St.-Lag.; *Narthex assafoetida* (L.) Falc. *Scorodosma foetida* Bunge

Common names: Asafetida, devil's dung; ase fétide (Fr.); teufelsdreck (Ger.); assa-fétida (Port.); асафетида (Rus.); asa fétida (Spa.)

Part used: Resin

Constituents: Sesquiterpene coumarins (foetidine, saradaferin), alkyl sulfurs (Iranshahy & Iranshahi, 2011; Asghari et al., 2016).

Medical history: Asafetida was known to Sushruta. Pliny the Elder calls asafetida *laserpitium* and Avicenna *andjudaan*. The German naturalist Engelbert Kaempfer in his *Amoenitatum exoticarum* (1712) gave a first description of asafetida from a collection he personally observed in Iran, where the plant was called *Hingiseb*. In 19th-century India, it was a stimulant, antispasmodic, carminative, tonic, and given for flatulence (O'Shaughnessy, 1842), heart diseases, and a favoured spice for Brahmins (Ainslie, 1813). In 19th-century Europe and North America, it was used for convulsions, as an antispasmodic, and for flatulent colic (Pereira, 1843)

Medicinal uses: Wounds (Iran); asthma, stomach problems, ulcer, cough, anxiety, fever (Pakistan); flatulence (India); forbidden to Buddhist monks (Korea)

Blood pressure: Aqueous extract at the concentration of 3 mg/mL relaxed guinea-pig ileum rings challenged with acetylcholine, histamine, or potassium chloride and decreased blood pressure in rats when injected at doses ranging from 0.3 to 2.2 mg/100 g (Fatehi et al., 2004).

Plasma lipids and glucose: Ethanol extract given orally at 150 mg/kg/day for 6 weeks to streptozotocin-induced diabetic rats decreased insulinemia, total cholesterol, triglycerides, LDL-cholesterol and glycemia from 528.2 to 307.8 mg/dL (normal value: 88 mg/dL) (Iranshahi & Alizadeh, 2012).

Adipose tissues: Asafetida given at 50 mg/kg/day for 8 weeks to rats on a high-fructose diet prevented increases in abdominal fat mass (Azizian et al., 2012).

Gallbladder: Intake of asafetida (250 mg in daily chow) for 4 weeks increased the secretion of cholesterol in bile from 0.2 to 0.3 µmol/hr (Sambaiah & Srinivasan, 1991).

DOI: 10.1201/9781003301455-62

Kidneys: Ethanol extract given orally at 150 mg/kg/day for 6 weeks to strep-tozotocin-induced diabetic rats normalized plasma creatinine and urea (Iranshahi & Alizadeh, 2012). Asafetida at 25 mg/kg/day for 28 days prevented the growth of urinary stones induced by ethylene glycol poisoning in rodents (Bagheri et al., 2018).

Bones and cartilages: Hydroalcoholic extract given to rats at 100 mg/kg/day for 28 days prevented collagen-induced arthritis (Sarab et al., 2020).

Warnings: Asafetida should never be given to infants as it causes methemoglobin-emia (Al-Qahtani et al., 2020). Contact dermatitis can happen in susceptible subjects (Tempark et al., 2016). None of the plants listed in this handbook are intended for pediatric or veterinary use.

REFERENCES

Ainslie, W., 1813. *Materia Indica of Hindoostan*. Government Press.

Al-Qahtani, S., Abusham, S. and Alhelali, I., 2020. Severe methemoglobinemia secondary to Ferula asafetida ingestion in an infant: A case report. *Saudi Journal of Medicine & Medical Sciences*, 8(1), p. 56.

Asghari, J., Atabaki, V., Baher, E. and Mazaheritehrani, M., 2016. Identification of sesquiterpene coumarins of oleo-gum resin of Ferula assa-foetida L. from the Yasuj region. *Natural Product Research*, 30(3), pp. 350–353.

Azizian, H., Rezvani, M.E., Esmaeili, D.M. and Bagheri, S.M., 2012. Anti-obesity, fat lowering and liver steatosis protective effects of *Ferula asafetida* gum in type 2 diabetic rats: Possible involvement of leptin. *Iranian Journal of Diabetes and Obesity*, 4(3, Autumn), pp. 120–126.

Bagheri, S.M., Yadegari, M., Behpur, M. and Javidmehr, D., 2018. Antilithiatic and hepatoprotective effects of Ferula assa-foetida oleo-gum-resin on ethylene glycol-induced lithiasis in rats. *Urological Science*, 29(4), p. 180.

Fatehi, M., Farifteh, F. and Fatehi-Hassanabad, Z., 2004. Antispasmodic and hypotensive effects of Ferula asafetida gum extract. *Journal of Ethnopharmacology*, 91(2–3), pp. 321–324.

Iranshahi, M. and Alizadeh, M., 2012. Antihyperglycemic effect of Asafetida (Ferula assafoetida Oleo-Gum-Resin) in streptozotocin-induced diabetic rats. *World Applied Sciences Journal*, 17(2), pp. 157–162.

Iranshahy, M. and Iranshahi, M., 2011. Traditional uses, phytochemistry and pharmacology of asafetida (Ferula assa-foetida oleo-gum-resin)—A review. *Journal of Ethnopharmacology*, 134(1), pp. 1–10.

O'Shaughnessy, W.B., 1842. *The Bengal Dispensatory and Companion to the Pharmacopœia . . . Chiefly Compiled . . . by W. B. O'Shaughnessy, Etc*. Calcutta printed.

Pereira, J., 1843. *The Elements of Materia Medica and Therapeutics*. Lea and Blanchard.

Sambaiah, K. and Srinivasan, K., 1991. Secretion and composition of bile in rats fed diets containing spices. *Journal of Food Science and Technology*, 28(1), pp. 35–38.

Sarab, G.A., Mohammadi, M., Shahmirzady, R.Y. and Gheini, M.H., 2020. Effect of hydroalcoholic extract of Ferula assa-foetida L. resin on rheumatoid arthritis symptoms in the collagen-induced animal model. *Journal of Birjand University of Medical Sciences*, 27(1), pp. 33–43.

Tempark, T., Chatproedprai, S. and Wananukul, S., 2016. Localized contact dermatitis from Ferula assa-foetida oleo-gum-resin essential oil, a traditional topical preparation for stomach ache and flatulence. *Indian Journal of Dermatology, Venereology and Leprology*, 82, p. 467.

63 Figs (*Ficus carica* L.)

Ficus carica L.

Etymology: From the Latin *ficus* = a fig and *carica* = a fig

Family: Moraceae

Synonym: *Ficus kopetdagensis* Pachom.

Common names: Figs; figues (Fr.); feigen (Ger.); figos (Port.); инж ир (Rus.); higos (Sp.)

Part used: Fruit

Constituents: Fixed oil (unsaturated fatty acids: γ-linolenic acid, linoleic acid, oleic acid), phytosterols, anthocyanins, magnesium, potassium, zinc (Nakilcioğlu-Taş, 2019; Kamiloglu & Akgun, 2023).

Medical history: Dioscorides asserts that eating fresh figs induce sweating and acne while eating dry figs is warming, fortifying, good for the kidneys, and good for those with short breath, ascites, and epilepsy. Galen says they treat kidney stones. The 95th chapter of the Coran is titled "The Fig tree" or "*At-Tin*". The Medical School of Salerno (10th century) says that eating figs in excess induces fornication, that figs are a male aphrodisiac, laxative, and good for tumors. In Alston's lectures (1770), figs are laxative, good for nephritic pain, small pox, tumors, and childbirth. In 19th-century North America, figs were considered good for the lungs and to treat nephritic affections (Pereira, 1843).

Medicinal uses: Diarrhea, latex for warts (Turkey); laxative (Pakistan); diseases of the blood, laxative (India)

Blood pressure: Aqueous extract given orally at 1g/kg/day for 21 days to glucose-induced hypertensive rats caused a decrease in mean arterial blood pressure from about 170 to 80 mmHg as well as in heart rate from about 400 to 200 bpm. This regimen was also effective in reducing blood pressure in healthy rats (Alamgeer et al., 2017).

Heart: In isolated hearts of rabbits, ethanol extract of fruits evoked negative inotropic and chronotropic effects (Alamgeer et al., 2017).

Blood lipids and glucose: In streptozotocin-induced diabetic rats, aqueous extract of fruits taken for 4 weeks at 500 mg/kg/day decreased plasma glucose from about 225 to 170 mg/dL and plasma cholesterol from about 175 to 125 mg/dL and increased HDL-cholesterol from about 40 to 60 mg/dL (Arafa et al., 2020).

Bones and cartilages: Aqueous extract given at 50 mg/kg/day for 8 weeks to ovariectomized rats prevented osteoporotic deterioration and the trabecular separation of the femur (Mohammad et al., 2018).

DOI: 10.1201/9781003301455-63

Brain: Hydroalcoholic extract given orally at 250 mg/kg/day for improved the cognitive function of rats (Komaki et al., 2014).

Warnings: Allergic individuals are at risk of anaphylactic shock (Dechamp et al., 1995). With a glycemic index of 61, figs must be avoided by diabetic patients (Foster-Powell et al., 2002; Atkinson et al., 2008).

REFERENCES

Alamgeer, Iman, S., Asif, H. and Saleem, M., 2017. Evaluation of antihypertensive potential of Ficus carica fruit. *Pharmaceutical Biology*, *55*(1), pp. 1047–1053.

Alston, C., 1770. *Lectures on the Materia Medica: Containing the Natural History of Drugs, their Virtues and Doses: Also Directions for the Study of the Materia Medica; and an Appendix on the Method of Prescribing*. Edward and Charles Dilly.

Arafa, E.S.A., Hassan, W., Murtaza, G. and Buabeid, M.A., 2020. Ficus carica and Sizigium cumini regulate glucose and lipid parameters in high-fat diet and streptozocin-induced rats. *Journal of Diabetes Research*, *2020*.

Atkinson, F.S., Foster-Powell, K. and Brand-Miller, J.C., 2008. International tables of glycemic index and glycemic load values: 2008. *Diabetes Care*, *31*(12), pp. 2281–2283.

Dechamp, C., Bessot, J.C., Pauli, G. and Deviller, P., 1995. First report of anaphylactic reaction after fig (Ficus carica) ingestion. *Allergy*, *50*(6), pp. 514–516.

Foster-Powell, K., Holt, S.H. and Brand-Miller, J.C., 2002. International table of glycemic index and glycemic load values: 2002. *The American Journal of Clinical Nutrition*, *76*(1), pp. 5–56.

Kamiloglu, S. and Akgun, B., 2023. Bioactive compounds of fig (Ficus carica). In *Fig (Ficus carica): Production, Processing, and Properties* (pp. 479–512). Springer International Publishing.

Komaki, A., Rastegarmanesh, M. and Sarihi, A., 2014. The effect of hydro-alcoholic extract of Ficus Carica on passive avoidance learning and memory in male rats. *Avicenna Journal of Clinical Medicine*, *20*(4), pp. 312–319.

Mohammad, A., Razaly, N.I., Rani, M.D.M., Aris, M.S.M., Dom, S.M. and Effendy, N.M., 2018. A micro-computed tomography (micro-CT) analysis of postmenopausal osteoporotic rat models supplemented with Ficus carica. *Journal of Applied Pharmaceutical Science*, *8*(6), pp. 39–45.

Nakilcioğlu-Taş, E., 2019. Biochemical characterization of fig (Ficus carica L.) seeds. *Journal of Agricultural Sciences*, *25*(2), pp. 232–237.

Pereira, J., 1843. *The Elements of Materia Medica and Therapeutics*. Lea and Blanchard.

64 Fennel (*Foeniculum vulgare* Mill.)

Foeniculum vulgare Mill.

Etymology: From the Latin *faenum* = hay and *vulgare* = common

Family: Apiaceae

Synonyms: *Anethum foeniculum* L.; *Foeniculum officinale* All.; *Selinum foeniculum* (L.) E.H.L. Krause

Common names: Fennel; fenouil (Fr.); fenchel (Ger.); funcho (Port.); ф енхель (Rus.); hinojo (Sp.)

Part used: Seed

Constituents: Essential oil (anethole, methyl chavicol, fenchone) (Damayanti & Setyawan, 2012).

Medical history: Ancient Greeks called fennel *marathro*. Dioscorides asserts that the seeds are galactagogue and of use for kidney pains and to induce urination. Pliny the Elder advocated the use of fennel for kidney stones. Galen and Simeon Seth describe fennel as diuretic, emmenagogue, hot to the third degree, and dry to the first. In Alston's lectures (1770), fennel is diuretic, nephritic, and carminative and treats lead poisoning. In 19th-century Europe and North America, fennel was used as a carminative and diuretic.

Medicinal uses: In the European Union, fennel is used as a carminative, diuretic, and for menstrual spams (1.5 g of seeds in 250 mL of boiling water steeped for 15 minutes 3 times daily); abdominal pain, sedative (Turkey); carminative, dysentery, cough (Afghanistan); cold, dysentery (Iran, Iraq); diuretic (Iran); cardiac problems, flatulence (Pakistan); antispasmodic (India); carminative, diuretic (Bangladesh); flatulence (Myanmar); inflammation, urinary problems (the Philippines)

Blood pressure: Aqueous extract given orally to hypertensive rats decreased systolic blood pressure (El Bardai et al., 2001). Anethole at 125 mg/kg/day for 20 days given to rats with nicotine- and stress-induced hypertension decreased systolic blood pressure from 135.6 to 121.9 mmHg (Seo et al., 2018). Anethole relaxed isolated rat aorta by opening voltage-dependent calcium ions channels (Soares et al., 2007).

Plasma lipids and glucose: Aqueous extract of seeds given at 300 mg/kg/day orally for 35 days decreased glycemia from 339.3 to 110.4 mg/dL (Anitha et al., 2014). Methanol extract of seeds given orally at 200 mg/kg/day for 11 days to dexamethasone-induced diabetic rats decreased cholesterol from 108.4 to 88.7 mg/dL, and triglycerides from 201.7 to 149.2 mg/dL, and these

 DOI: 10.1201/9781003301455-64

effects were similar to those for pioglitazone at 30 mg/kg/day (Dongare et al., 2010).

Kidneys: Aqueous extract given orally to hypertensive rats induced urination (El Bardai et al., 2001). Hydroalcoholic extract increased urine flow and urinary sodium ions excretion in rats (Beaux et al., 1997).

Platelets: Essential oil or anethole orally administered to mice at 30 mg/kg for 5 days prevented thromboembolism induced by injection of collagen and epinephrine (Tognolini et al., 2007).

Bones and cartilages: Methanol extract of seeds given orally to mice at 200 mg/kg/day for 7 days evoked some protection against formaldehyde-induced arthritis (Choi & Hwang, 2004).

Brain: Methanol extract given orally for 8 days to mice at the daily dose of 100 mg/kg attenuated scopolamine-induced dementia (Joshi & Parle, 2007).

Warnings: Intake of 50 g/kg of seeds per day for 10 days caused kidney injuries in rats (Luaibi et al., 2017). In rats, long-term ingestion of anethole at 550 mg/kg/day caused damage to the liver including the formation of tumors (Newberne et al., 1999).

REFERENCES

Alston, C., 1770. *Lectures on the Materia Medica: Containing the Natural History of Drugs, their Virtues and Doses: Also Directions for the Study of the Materia Medica; and an Appendix on the Method of Prescribing.* Edward and Charles Dilly.

Anitha, T., Balakumar, C., Ilango, K.B., Jose, C.B. and Vetrivel, D., 2014. Antidiabetic activity of the aqueous extracts of Foeniculum vulgare on streptozotocin-induced diabetic rats. *International Journal of Advances in Pharmacy, Biology and Chemistry, 3*(2), pp. 487–494.

Beaux, D., Fleurentin, J. and Mortier, F., 1997. Diuretic action of hydroalcohol extracts of Foeniculum vulgare var dulce (DC) roots in rats. *Phytotherapy Research: An International Journal Devoted to Medical and Scientific Research on Plants and Plant Products, 11*(4), pp. 320–322.

Choi, E.M. and Hwang, J.K., 2004. Antiinflammatory, analgesic and antioxidant activities of the fruit of Foeniculum vulgare. *Fitoterapia, 75*(6), pp. 557–565.

Damayanti, A. and Setyawan, E., 2012. Essential oil extraction of fennel seed (Foeniculum vulgare) using steam distillation. *International Journal of Science and Engineering, 3*(2), pp. 12–14.

Dongare, V.R., Arvindekar, A.U. and Magadum, C.S., 2010. Hypoglycemic effect of Foeniculum vulgare Mill. fruit on dexamethasone induced insulin resistance rats. *Research Journal of Pharmacognosy and Phytochemistry, 2*(2), pp. 163–165.

El Bardai, S., Lyoussi, B., Wibo, M. and Morel, N., 2001. Pharmacological evidence of hypotensive activity of Marrubium vulgare and Foeniculum vulgare in spontaneously hypertensive rat. *Clinical and Experimental Hypertension (New York, NY: 1993), 23*(4), pp. 329–343.

Joshi, H. and Parle, M., 2007. Antiamnesic potentials of Foeniculum vulgare Linn. in mice. *Advances in Traditional Medicine, 7*(2), pp. 182–190.

Luaibi, N.M., Al-Tamimi, A.A. and Shafiq, A.A., 2017. Physiological and histological effects of fennel seeds (Foeniculum vulgare) on kidneys in male rats. *Journal of Pharmaceutical and Biological Sciences, 5*(1), pp. 45–55.

Newberne, P., Smith, R.L., Doull, J., Goodman, J.I., Munro, I.C., Portoghese, P.S., Wagner, B.M., Weil, C.S., Woods, L.A., Adams, T.B. and Lucas, C.D., 1999. The FEMA GRAS assessment of trans-anethole used as a flavouring substance. *Food and Chemical Toxicology*, *37*(7), pp. 789–811.

Seo, E., Kang, P. and Seol, G.H., 2018. Trans-anethole prevents hypertension induced by chronic exposure to both restraint stress and nicotine in rats. *Biomedicine & Pharmacotherapy*, *102*, pp. 249–253.

Soares, P.M.G., Lima, R.F., de Freitas Pires, A., Souza, E.P., Assreuy, A.M.S. and Criddle, D.N., 2007. Effects of anethole and structural analogues on the contractility of rat isolated aorta: Involvement of voltage-dependent Ca2+-channels. *Life Sciences*, *81*(13), pp. 1085–1093.

Tognolini, M., Ballabeni, V., Bertoni, S., Bruni, R., Impicciatore, M. and Barocelli, E., 2007. Protective effect of Foeniculum vulgare essential oil and anethole in an experimental model of thrombosis. *Pharmacological Research*, *56*(3), pp. 254–260.

65 Asam Gelugur (*Garcinia atroviridis* Griff. ex T. Anderson)

Garcinia atroviridis Griff. ex T. Anderson

Etymology: After the French botanist Laurent Garcin (1683–1752) and the Latin *atroviridis* = blackish green

Family: Clusiaceae

Common name: Asam gelugur (Malay)

Part used: Fruit

Constituents: (-) Hydroxycitric acid (Lewis, 1969), prenylated benzophenones (garcinol), prenylated xanthones (camboginol), and flavanones (naringenin) (Shahid et al., 2022).

Medical history: Asam gelugur has been used since the dawn of time to flavor dishes in Thailand, Malaysia, and Indonesia.

   The English botanist and surgeon William Griffith (1810–1845) provides a first description of the plant in the *The flora of British India* (1875). Henry Nicholas Ridley (1855–1956), director of Gardens and Forests in Straits Settlements, informs us in his *Malay materia medica* (1897) that Malays called the plant *asam gelugur* and states that the "fruits are used in curries" and given after childbirth "*ubat barut.*"

Medicinal uses: Stomach ache (Indonesia)

Blood lipids and glucose: Methanol extract given orally at 50 mg/day for 8 weeks to guinea pigs on a high-cholesterol diet decreased plasma cholesterol from 4.5 to 3.6 mmol/L (normal: 1.2 mmol/L) and mitigated aortic atherosclerosis (Amran et al., 2009). Aqueous extract of fruits decreased total cholesterol, triglycerides, LDL-cholesterol, and the formation of atheromas in a manner comparable with that of atorvastatin (Al-Mansoub et al., 2014).

Comment: (-) Hydroxycitric acid inhibits the synthesis of fatty acids by blocking the formation of acetyl-CoA (Tomar et al., 2019).

REFERENCES

Al-Mansoub, M.A., Asmawi, M.Z. and Murugaiyah, V., 2014. Effect of extraction solvents and plant parts used on the antihyperlipidemic and antioxidant effects of Garcinia atroviridis: A comparative study. *Journal of the Science of Food and Agriculture*, 94(8), pp. 1552–1558.

Amran, A.A., Zaiton, Z., Faizah, O. and Morat, P., 2009. Effects of Garcinia atroviridis on serum profiles and atherosclerotic lesions in the aorta of guinea pigs fed a high cholesterol diet. *Singapore Medical Journal*, *50*(3), p. 295.

Lewis, Y.S., 1969. [77] Isolation and properties of hydroxycitric acid. In *Methods in Enzymology* (Vol. 13, pp. 613–619). Academic Press.

Shahid, M., Law, D., Azfaralariff, A., Mackeen, M.M., Chong, T.F. and Fazry, S., 2022. Phytochemicals and Biological Activities of Garcinia atroviridis: A Critical Review. *Toxics*, *10*(11), p. 656.

Tomar, M., Rao, R.P., Dorairaj, P., Koshta, A., Suresh, S., Rafiq, M., Kumawat, R., Paramesh, R. and Venkatesh, K.V., 2019. A clinical and computational study on anti-obesity effects of hydroxycitric acid. *RSC Advances*, *9*(32), pp. 18578–18588.

66 Tournefort's Gundelia (*Gundelia tournefortii* L.)

Gundelia tournefortii L.

Etymology: After Andreas von Gundelsheimer (1668–1715), German botanist and the first European to meet the plant, and the Latin *tourneforti* after the French botanist Joseph Pitton de Tournefort (1656–1708)

Family: Asteraceae

Common names: Tournefort's gundelia; gundele du Levant, gundelia de Tournefort (Fr.)

Part used: Whole plant

Constituents: Hydroxycinnamic acid derivatives (chlorogenic acid) (around 900 mg/100 g at flowering stage) (Haghi et al., 2011), coumarins (scopoletin) (Halabi et al., 2005), fixed oil (unsaturated fatty acids: oleic acid, linoleic acid) (Abdul et al., 2012), essential oil (terpinyl acetate) (Halabi et al., 2005).

Medical history: Archeological evidence suggests that the plant was used for food in Iraq (Savard et al., 2003) and medicinally in Turkey and Iraq during the Neolithic. Joseph Pitton de Tournefort visited Armenia and in his *Relations d'un voyage du Levant* (1717) describes a plant eaten by natives that he called "*Gundelia orientalis*" in honor of his travel companion Gundelsheimer. The plant was reputed to be an emetic and laxative by the French physicians of the 18th century (Lémery, 1716).

Medicinal uses: Enlarged spleen, diabetes, diuretic, gastrointestinal disorders, stroke, toothache (Turkey); chest pain, stroke, liver diseases (Jordan, Iran); hypertension (Iran)

Blood lipids and glucose: Aqueous decoction of roots given orally at 300 mg/kg/day for 22 days to mice poisoned by dexamethasone decreased plasma glucose from 261.76 to 184.8 mg/dL (normal: 179.9 mg/dL) and normalized triglycerides and cholesterol (Azeez & Kheder, 2012). Methanol extract given to rats at 200 mg/kg improved plasma lipids (Mansi et al., 2020). Aqueous extract of leaves given orally and daily at 40 mg/kg to alloxan-induced diabetic rats for 20 days decreased fasting glycemia (Mohammadi et al., 2018). Powder of stems added as 0.2% of a high-cholesterol diet given to rabbits for 2 months decreased cholesterol and triglycerides (Rafiee et al., 2017).

An extract of aerial parts given for 8 weeks to patients with coronary artery disease decreased total cholesterol and LDL-cholesterol (Hajizadeh-Sharafabad et al., 2016).

DOI: 10.1201/9781003301455-66

Heart: Methanol extract given to rats at 200 mg/kg improved creatinine phosphokinase and lactate dehydrogenase (Mansi et al., 2020). Powder of stems added as 0.2% of a high-cholesterol diet given to rabbits for 2 months decreased the formation of coronary atheroma (Rafiee et al., 2017).

Kidneys: Aqueous extract of leaves given orally and daily at 40 mg/kg to alloxan-induced diabetic rats for 20 days decreased plasma urea and protected kidney against hypertrophy (Mohammadi et al., 2018).

Comments: (i) Tournefort's gundelia could potentially be beneficial as a functional food for elderly. More experiments are needed.

(ii) Although Tournefort's gundelia is used for food in the Middle East, the side effects and maximum daily amount are not known. It is in fact the case for most medicinal plants because schools of pharmacy in most developed countries do not teach any more materia medica. It is time to reverse the steam as people are left with herbal products sold on the internet of doubtful origin. Again, the manufacture and delivery of medicinal plants should be the responsibility of pharmacists. Alternatively, national schools of herbalism could train professional herbalists with a license to deliver medicinal plants.

REFERENCES

Abdul, D.A., Hamd, N.S. and Hassan, H.G., 2012. Characteristics of fatty acids content in Gundelia L. oil extract. *Iraqi National Journal of Chemistry*, *45*, pp. 144–148.

Azeez, O.H. and Kheder, A.E., 2012. Effect of Gundelia tournefortii on some biochemical parameters in dexamethasone-induced hyperglycemic and hyperlipidemic mice. *Iraqi Journal of Veterinary Sciences*, *26*(2).

Haghi, G., Hatami, A. and Arshi, R., 2011. Distribution of caffeic acid derivatives in Gundelia tournefortii L. *Food Chemistry*, *124*(3), pp. 1029–1035.

Hajizadeh-Sharafabad, F., Alizadeh, M., Mohammadzadeh, M.H.S., Alizadeh-Salteh, S. and Kheirouri, S., 2016. Effect of Gundelia tournefortii L. extract on lipid profile and TAC in patients with coronary artery disease: A double-blind randomized placebo controlled clinical trial. *Journal of Herbal Medicine*, *6*(2), pp. 59–66.

Halabi, S., Battah, A.A., Aburjai, T. and Hudaib, M., 2005. Phytochemical and Antiplatelet Investigation of Gundelia tournifortii. *Pharmaceutical Biology*, *43*(6), pp. 496–500.

Lémery, N., 1716. *Traité universel des drogues simples, mises en ordre alphabétique. Où l'on trouve leurs différens noms . . . et tout ce qu'il y a de particulier dans les animaux, dans les végétaux, et dans les minéraux*. Au dépend de la Companie.

Mansi, K., Tabaza, Y. and Aburjai, T., 2020. The iron chelating activity of Gundelia tournefortii in iron overloaded experimental rats. *Journal of Ethnopharmacology*, *263*, p. 113–114.

Mohammadi, G., Zangeneh, M.M., Rashidi, K. and Zangeneh, A., 2018. Evaluation of nephroprotective and antidiabetic effects of gundelia tournefortii aqueous extract on diabetic nephropathy in male mice. *Research Journal of Pharmacognosy*, *5*(4), pp. 65–73.

Rafiee, L., Keshvari, M., Atar, A.M., Hamidzadeh, Z., Dashti, G.R., Rafieian-Kopaei, M. and Asgary, S., 2017. Effect of Gundelia tournefortii L. on some cardiovascular risk factors in an animal model. *Journal of Herbmed Pharmacology*, *6*(4), pp. 191–195.

Savard, M., Nesbitt, M. and Gale, R., 2003. Archaeobotanical evidence for early Neolithic diet and subsistence at M'lefaat (Iraq). *Paléorient*, pp. 93–106.

67 Okra (*Hibiscus esculentus* L.)

Hibiscus esculentus L.

Etymology: From the Latin *esculentus* = edible

Synonym: *Abelmoschus esculentus* (L.) Moench

Family: Malvaceae

Common names: Okra; gombo (French); essbarer eibisch (Ger.); quiabo (Port.); бамия (Rus.); quimbombó (Sp.)

Part used: Fruit

Constituents: Flavone glycosides (Liao et al., 2012), pectin (Chen et al., 2014), fixed oil (unsaturated fatty acids: linoleic acid) (Gemede et al., 2015).

Medical history: According to the Andalusian physician Abu Abbas el Nebbati (1166–1239), who travelled in the Middle East, okra was used as a vegetable in Egypt. Similarly, the Italian physician Prospero Alpini (1553–1617) observed okra in Egypt. In the 19th century, okra was used as a diuretic and for sore throats in India (Dymock, 1884). In 19th century France okra was used for the preparation of a cough syrups "syrop de nafé

Medicinal uses: Diabetes (Turkey); syphilis (Iraq); cuts (Sikkim); diuretic (Myanmar)

Blood pressure: Ethanol extract given orally at 150 mg/kg/day for 20 days to rats on a high-fructose diet decreased systolic blood pressure from 174.1 to 102.2 mmHg and diastolic blood pressure from 165.7 to 94.2 mmHg (Mondal et al., 2019).

Plasma lipids and glucose: Ethanol extract given orally at 150 mg/kg/day for 20 days to rats on a high-fructose diet decreased plasma cholesterol from 74.6 to 46.4 mg/dL and triglycerides from 29.4 to 16.6 mg/dL (Mondal et al., 2019).

Okra powder given (10 g in 150 g of yogurt) for 8 weeks to type-2 diabetic patients caused a decrease in fasting plasma glucose and in triglycerides from 187.5 to 165.2 mg.dL, total cholesterol from 191 to 180.7 mg/dL, and LDL-cholesterol from 125.4 to 117.2 mg/dL (Moradi et al., 2020).

In type-2 diabetic patients, 1 g of okra given every 6 hours for 8 weeks decreased fasting blood glucose (Saatchi et al., 2022). Oil of seeds given to rats at 10% of diet for 90 days was able to decrease plasma cholesterol (Srinivasa Rao et al., 1991).

Heart: Feeding rats a high-fructose diet for 6 weeks induced a decrease in P wave and QRS complex duration (indicating enlargement of the atria)

DOI: 10.1201/9781003301455-67

that was normalized by treatment with a methanol extract of seeds given at 150 mg/kg/day for 20 days (Mondal et al., 2019).

Gallbladder: 100 mg of okra dry matter binds 1.6 µmol of bile acids (Kahlon et al., 2007).

Comment: It is clear that eating okra, preferably boiled, is beneficial for the aging cardiovascular system. However, since it contains high amounts of potassium it should be avoided in hyperkaliemic patients.

REFERENCES

Chen, Y., Zhang, J.G., Sun, H.J. and Wei, Z.J., 2014. Pectin from Abelmoschus esculentus: Optimization of extraction and rheological properties. *International Journal of Biological Macromolecules*, 70, pp. 498–505.

Dymock, W., 1884. *The Vegetable Materia Medica of Western India*. Education Society Press.

Gemede, H.F., Ratta, N., Haki, G.D., Woldegiorgis, A.Z. and Beyene, F., 2015. Nutritional quality and health benefits of okra (Abelmoschus esculentus): A review. *Journal of Food Processing & Technology*, 6(458), p. 2.

Kahlon, T.S., Chapman, M.H. and Smith, G.E., 2007. In vitro binding of bile acids by okra, beets, asparagus, eggplant, turnips, green beans, carrots, and cauliflower. *Food Chemistry*, 103(2), pp. 676–680.

Liao, H., Liu, H. and Yuan, K., 2012. A new flavonol glycoside from the Abelmoschus esculentus Linn. *Pharmacognosy Magazine*, 8(29), p. 12.

Mondal, K., KP, S.G. and Manandhar, S., 2019. Anti-hypertensive effect of Abelmoschus esculentus (okra) seed extracts in fructose-induced hypertensive rats. *Indian Journal of Physiology and Pharmacology*, 63(2), pp. 175–181.

Moradi, A., Tarrahi, M.J., Ghasempour, S., Shafiepour, M., Clark, C.C. and Safavi, S.M., 2020. The effect of okra (Abelmoschus esculentus) on lipid profiles and glycemic indices in Type 2 diabetic adults: Randomized double blinded trials. *Phytotherapy Research*, 34(12), pp. 3325–3332.

Saatchi, A., Aghamohammadzadeh, N., Beheshtirouy, S., Javadzadeh, Y., Afshar, F.H. and Ghaffary, S., 2022. Anti-hyperglycemic effect of Abelmoschus culentesus (Okra) on patients with diabetes type 2: A randomized clinical trial. *Phytotherapy Research*, 36(4), pp. 1644–1651.

Srinivasa Rao, P., Udayasekhara Rao, P. and Sesikeran, B., 1991. Serum cholesterol, triglycerides and total lipid fatty acids of rats in response to okra (hibiscus escuientus) seed oil. *Journal of the American Oil Chemists' Society*, 68(6), pp. 433–435.

68 Kangkong (*Ipomoea aquatica* Forssk.)

Ipomoea aquatica Forssk.

Etymology: From the Greek *ips* = worm and *homoios* = similar and the Latin
  aquatica = aquatic
Family: Convolvulaceae
Synonyms: *Convolvulus repens* Vahl; *Ipomoea repens* Roth; *Ipomoea reptans*
  Poir.
Common names: Kangkong, swamp cabbage, water spinach; patate aquatique
  (Fr.); wasserspinat (Ger.); espinafre de água (Port.); espinaca de agua (Spa.)
Parts used: Leaf, stem
Constituents: Flavone glycosides, hydroxycinnamic derivatives (4,5-di-*O*-caf-
  feoylquinic acid) (Lawal et al., 2017).
Medical history: A description of the plant is available in *Flora Aegyptiaco-
  Arabica* (1775) written by the Finish naturalist Peter Forsskål (1732–1763),
  who at the cost of his life explored Egypt and Yemen. He notes the aquatic
  habit "*caulis in rivulis repens*". In 19th-century China, the plant was laxative, anti-inflammatory, and cooling (Porter Smith, 1871).
Medicinal uses: Kidney stones (Myanmar; Thailand); diabetes (the Philippines); galactagogue (Bangladesh; India)

Plasma lipids and cholesterol: Ethanol extract of leaves given orally to streptozotocin-induced diabetic mice at 200 mg/kg/day for 3 weeks decreased glycemia from 22.8 to 7.3 mmol/L, and this effect was somewhat comparable
  with the effects of metformin (Hamid et al., 2011). Methanol extract given
  orally at 400 mg/kg for 30 days to rats poisoned with a cholesterol-enriched
  diet decreased heart cholesterol from 6.7 to 3.8 mg/dL and kidney cholesterol from 10.8 to 5.8 mg/dL (Sivaraman & Muralidaran, 2010).
Brain: Hydroalcoholic extract given orally to mice at 400 mg/kg/day for
  4 weeks attenuated dementia induced by intracerebrovascular injection of
  amyloid protein β (Sivaraman et al., 2016).

Comments: (i) With aging and lack of hydration, cholesterol tends to form aggregates in renal tubules (nephrons), resulting in inflammation and immune
  response that contributes to kidney injuries (Mulay et al., 2014).

DOI: 10.1201/9781003301455-68

(ii) It is clear that eating boiled kangkong is beneficial for elderly people. However, since it contains high amounts of potassium it should be avoided in hyperkaliemic patients.

REFERENCES

Hamid, K., Ullah, M.O., Sultana, S., Howlader, M.A., Basak, D., Nasrin, F. and Rahman, M.M., 2011. Evaluation of the leaves of *Ipomoea aquatica* for its hypoglycemic and antioxidant activity. *Journal of Pharmaceutical Sciences and Research*, *3*(7), p. 1330.

Lawal, U., Leong, S.W., Shaari, K., Ismail, I.S., Khatib, A. and Abas, F., 2017. α-glucosidase inhibitory and antioxidant activities of Different Ipomoea aquatica cultivars and LC–MS/MS profiling of the active cultivar. *Journal of Food Biochemistry*, *41*(2), p. e12303.

Mulay, S.R., Evan, A. and Anders, H.J., 2014. Molecular mechanisms of crystal-related kidney inflammation and injury. Implications for cholesterol embolism, crystalline nephropathies and kidney stone disease. *Nephrology Dialysis Transplantation*, *29*(3), pp. 507–514.

Porter Smith, F., 1871. *Contributions Towards the Materia Medica and Natural History of China for the Use of Medical Missionaries and Students*. American Presbytarian Mission Press.

Sivaraman, D. and Muralidaran, P., 2010. Hypolipidemic activity of Ipomoea aquatica Forsk. Leaf extracts on lipid profile in hyperlipidemic rats. *International Journal of Pharmaceutical and Biological Archive*, *1*, pp. 175–179.

Sivaraman, D., Panneerselvam, P. and Muralidharan, P., 2016. Memory and brain neurotransmitter restoring potential of hydroalcoholic extract of ipomoea aquatica Forsk on amyloid beta A beta (25–35) induced cognitive deficits in Alzheimer's mice. *International Journal of Pharmacology*, pp. 52–65.

69 Sweet Potato (*Ipomoea batatas* (L.) Lam)

Ipomoea batatas (L.) Lam

Etymology: From the Greek *ips* = worm and *homoios* = similar and the Spanish *batata* = sweet potato

Family: Convolvulaceae

Synonyms: *Batatas edulis* (Thunb. ex Murray) Choisy; *Convolvulus batatas* L.; *Convolvulus edulis* Thunb. ex Murray

Common names: Sweet potato, camote; patate douce (Fr.); süßkartoffel (Ger.); batata doce (Port.); batata (Spa.)

Part used: Root

Constituents: Anthocyanins (Harada et al., 2004), hydroxycinnamic acid derivatives (chlorogenic acid) (Kojima & Uritani, 1972), fibers, polysaccharides (Takamine et al., 2005).

Medical history: According to the French botanist Alphonse de Candolle (1806–1893) in his book *Origin of cultivated plants* (1884), Christopher Columbus may have brought to Spain samples of sweet potato that were then disseminated throughout Portuguese and Spanish colonies. It was used in 19th-century India as a laxative (Watt, 1889).

Medicinal uses: Swelling (Indonesia); stomachache (Papua New Guinea); give strength to farmers (North Borneo); hypertension (the Philippines); constipation (Korea)

Plasma lipids and glucose: Ethanol extract of peeled sweet potato given to alloxan-induced diabetic rats at 100 mg/kg for 35 days reduced blood, liver, and kidney MDA as well as creatinine from 3.1 to 0.7 mg/dL, and systolic blood pressure from 139.1 to 106.5 mmHg (Herawati et al., 2020). In rats, intake of sweet potato for 16 weeks attenuated high-fat-diet-induced aortic stiffness (Garner et al., 2017).

Type-2 diabetic patients ingesting 4 tablets (containing 336 mg of powdered skinned sweet potatoes) per day for 6 weeks experienced a decrease in insulin resistance (Ludvik et al., 2003). Skinned sweet potatoes given to type-2 diabetic patients at 4 g per day for 3 months decreased cholesterol from 248.7 to 214.6 mg/dL and fasting glucose from 143.7 to 128.5 mg/dL (Ludvik et al., 2004).

Bones and cartilages: An extract given daily for 4 weeks to ovariectomized mice decreased plasma levels of collagen terminal telopeptide (Melissa et al., 2018).

DOI: 10.1201/9781003301455-69

Immune system: A polysaccharide given orally at 50 mg/kg/day to mice for 7 days increased lymphocytes count (Zhao et al., 2005).

Brain: An extract given orally to mice evoked some protection against amyloid protein β-induced dementia (Kim et al., 2011).

Comments: (i) Excess of glucose in the plasma favors protein glycation, which results in the production of low-molecular-mass aminated compounds, collectively termed advanced glycation end products or AGE (Nakamura & Kawaharada, 2021). AGE reacts with proteins to generate cellular dysfunction, glomerulosclerosis, and interstitial fibrosis of the kidneys (Vallon & Komers, 2011).

(ii) A cholesterol-rich diet progressively increases arterial stiffness over time. This process can be reverted by decreasing plasma cholesterol (Wilkinson & Cockcroft, 2007).

(iii) Intake of yogurt is necessary to maintain a healthy intestinal bacterial flora. Fiber of sweet potatoes given to rats mixed with diet induced the production of propionic acid by gut bacteria. Propionic acid is an inhibitor of cholesterol synthesis that targets β-hydroxy β-methylglutaryl-CoA synthetase (Takamine et al., 2005).

(iv) The glycemic index of sweet potatoes is about 61, so diabetics should not eat too much of them (Foster-Powell et al., 2002; Atkinson et al., 2008).

(v) Collagen terminal telopeptide is a breakdown product from the degradation of bone matrix collagen by osteoclasts, and its presence in plasma indicates bone loss (Christy et al., 2014).

REFERENCES

Atkinson, F.S., Foster-Powell, K. and Brand-Miller, J.C., 2008. International tables of glycemic index and glycemic load values. *Diabetes Care*, *31*(12), pp. 2281–2283.

Christy, A.L., D'Souza, V., Babu, R.P., Takodara, S., Manjrekar, P., Hegde, A. and Rukmini, M.S., 2014. Utility of C-terminal telopeptide in evaluating levothyroxine replacement therapy-induced bone loss. *Biomarker Insights*, *9*, pp. BMI-S13965.

Foster-Powell, K., Holt, S.H. and Brand-Miller, J.C., 2002. International table of glycemic index and glycemic load values: 2002. *The American Journal of Clinical Nutrition*, *76*(1), pp. 5–56.

Garner, T., Ouyang, A., Berrones, A.J., Campbell, M.S., Du, B. and Fleenor, B.S., 2017. Sweet potato (Ipomoea batatas) attenuates diet-induced aortic stiffening independent of changes in body composition. *Applied Physiology, Nutrition, and Metabolism*, *42*(8), pp. 802–809.

Harada, K., Kano, M., Takayanagi, T., Yamakawa, O. and Ishikawa, F., 2004. Absorption of acylated anthocyanins in rats and humans after ingesting an extract of Ipomoea batatas purple sweet potato tuber. *Bioscience, Biotechnology, and Biochemistry*, *68*(7), pp. 1500–1507.

Herawati, E.R.N., Santosa, U., Sentana, S. and Ariani, D., 2020. Protective effects of anthocyanin extract from purple sweet potato (Ipomoea batatas L.) on blood MDA levels, liver and renal activity, and blood pressure of hyperglycemic rats. *Preventive Nutrition and Food Science*, *25*(4), p. 375.

Kim, J.K., Choi, S.J., Cho, H.Y., Kim, Y.J., Lim, S.T., Kim, C.J., Kim, E.K., Kim, H.K., Peterson, S. and Shin, D.H., 2011. Ipomoea batatas attenuates amyloid β peptide-induced neurotoxicity in ICR mice. *Journal of Medicinal Food*, *14*(3), pp. 304–309.

Kojima, M. and Uritani, I., 1972. Elucidation of the structure of a possible intermediate in chlorogenic acid biosynthesis in sweet potato root tissue. *Plant and Cell Physiology*, *13*(6), pp. 1075–1084.

Ludvik, B., Neuffer, B. and Pacini, G., 2004. Efficacy of Ipomoea batatas (Caiapo) on diabetes control in type 2 diabetic subjects treated with diet. *Diabetes Care*, *27*(2), pp. 436–440.

Ludvik, B., Waldhäusl, W., Prager, R., Kautzky-Willer, A. and Pacini, G., 2003. Mode of action of Ipomoea batatas (Caiapo) in type 2 diabetic patients. *Metabolism*, *52*(7), pp. 875–880.

Melissa, T., Suyasa, I.K., Lestari, A.A.W. and Astawa, P., 2018. CTx serum level in mice post-ovariectomy after administration of sweet potato extract (Ipomoea batatas) is lower than without administration of sweet potato extract (Ipomoea batatas). *Indonesia Journal of Biomedical Science*, *12*(2).

Nakamura, A. and Kawaharada, R., 2021. Advanced glycation end products and oxidative stress in a Hyperglycaemic environment. *Fundamentals of Glycosylation*. IntechOpen.

Takamine, K., Hotta, H., Degawa, Y., Morimura, S. and Kida, K., 2005. Effects of dietary fiber prepared from sweet potato pulp on cecal fermentation products and microflora in rats. *Journal of Applied Glycoscience*, *52*(1), pp. 1–5.

Vallon, V. and Komers, R., 2011. Pathophysiology of the diabetic kidney. *Comprehensive Physiology*, *1*(3), p. 1175.

Wilkinson, I. and Cockcroft, J.R., 2007. Cholesterol, lipids and arterial stiffness. *Atherosclerosis, Large Arteries and Cardiovascular Risk*, *44*, pp. 261–277.

Zhao, G., Kan, J., Li, Z. and Chen, Z., 2005. Characterization and immunostimulatory activity of an (1→ 6)-aD-glucan from the root of Ipomoea batatas. *International Immunopharmacology*, *5*(9), pp. 1436–1445.

70 Walnut (*Juglans regia* L.)

Juglans regia L.

Etymology: Latin *juglans* = walnut and *regia* = royal because the Greek called it *karyon basilikon* = royal nut

Family: Juglandaceae

Synonyms: *Juglans duclouxiana* Dode; *Juglans fallax* Dode; *Juglans orientis* Dode; *Juglans sinensis* (C. DC.) Dode

Common names: Walnut; noix (Fr.); nussbaum (Ger.); noz (Port.); грецкий орех (Rus.); nuez (Spa.)

Part used: Seed

Constituents: Fixed oil (unsaturated fatty acids: linoleic acid, linolenic acid, oleic acid) (Greve et al., 1992; Beyhan et al., 2017), tannins (ellagitannins: ellagic acid (400 mg in 4 walnuts), tellimagrandin I) (Shimoda et al., 2009).

Medical history: Walnut trees are mentioned in the Songs of Songs of King Solomon (6:11): "I went down to the grove of nut trees". Dioscorides says of walnuts that they increase bile secretion and are useful for tumors. The Medical School of Salerno (10th century) asserts that walnuts are antidotal but warns against eating too much of them and forbids to eat them with both fish and meat. For Fusch in 16th-century Germany, walnuts are hot to the second degree and dry to the first. Lémery (1716) recommends the oil as carminative and nephritic.

Medicinal uses: High cholesterol (Turkey); gout, arthritis (Iran); sore throat (Pakistan); rheumatism, memory enhancer (India)

Blood pressure: Methanol extract given at 200 mg/kg/day for 15 days to rats poisoned with dexamethasone decreased diastolic blood pressure from 128 to 109 mmHg (similar to Captopril at 25 mg/kg/day), probably because of increased production of NO (Joukar et al., 2017).

Patients with chronic kidney diseases consuming 30 g of walnuts per day for 30 days had a decrease in systolic blood pressure of 4 mmHg (Sanchis et al., 2019). In healthy subjects, intake of walnut decreased diastolic blood pressure and heart rate (Steffen et al., 2021).

Ellagic acid given at 30 mg/kg/day orally for 4 weeks to rats poisoned with L-NAME decreased hypertension and restored plasma nitrate and nitrite (Jordão et al., 2017).

Blood lipids and glucose: Rats on a high-fat diet given a seed coat extract at 200 mg/kg/day were protected against hypertriglyceridemia via the enhancement of fatty acid peroxisomal β-oxidation in the liver (Shimoda et al., 2009).

 DOI: 10.1201/9781003301455-70

In healthy volunteers, intake of walnuts as part of the daily diet increased HDL-cholesterol (Lavedrine et al., 1999). In hyperlipidemic patients, consuming 30 g of walnuts per day for two months increased HDL-cholesterol from 31.8 to 38.1 mg/dL (Tufail et al., 2015). In a subsequent study, patients with chronic kidney diseases consuming 30 g of walnuts per day for 30 days had an increase in HDL-cholesterol of 5.4 mg/dL and a decrease in LDL-cholesterol of 5.4 mg/dL (Sanchis et al., 2019).

Intake of walnut oil at 15 g/day for 3 months by type-2 diabetic patients decreased fasting blood glucose from 158.3 to 137.9 mg/dL (Zibaeenezhad et al., 2016).

Bones and cartilages: Rats fed walnut for 42 days had increased plasma levels of hr-CRP in CFA-induced arthritis and decreased inflammatory cells infiltration, bone erosion, and paw inflammation (Javed et al., 2022).

Brain: Rats fed walnut for 28 days had enhanced cognitive function and cerebral serotoninergic activity (Haider et al., 2011).

Warning: Consumption of walnut in rats causes goiter (Linazasoro et al., 1970).

Comments: (i) Intake of walnut oil has a protective effect on endothelial function (Berryman et al., 2013). In healthy volunteers, the consumption of 15 g of walnuts per day for 4 weeks had no effect on lipid profile or on arterial stiffness (Din et al., 2011), suggesting that at least 30 g of walnut must be taken daily for beneficial effects on the cardiovascular system.

(ii) Intake of walnut oil at 15 g/day for 3 months by type-2 diabetic patients had no effect on blood pressure (Zibaeenezhad et al., 2016), implying that polyphenols in the seed coat account at least in part for blood-pressure-lowering effects.

(iii) It is clear that eating walnut regularly (30 g minimum) is beneficial for the aging cardiovascular system and brain. However, since it contains high amounts of potassium it should be avoided in hyperkaliemic patients.

REFERENCES

Berryman, C.E., Grieger, J.A., West, S.G., Chen, C.Y.O., Blumberg, J.B., Rothblat, G.H., Sankaranarayanan, S. and Kris-Etherton, P.M., 2013. Acute consumption of walnuts and walnut components differentially affect postprandial lipemia, endothelial function, oxidative stress, and cholesterol efflux in humans with mild hypercholesterolemia. *The Journal of Nutrition*, *143*(6), pp. 788–794.

Beyhan, O., Ozcan, A., Ozcan, H., Kafkas, N.E.S.İ.B.E., Kafkas, S., Sutyemez, M. and Ercişli, S., 2017. Fat, fatty acids and tocopherol content of several walnut genotypes. *Notulae Botanicae Horti Agrobotanici Cluj-Napoca*, *45*(2).

Din, J.N., Aftab, S.M., Jubb, A.W., Carnegy, F.H., Lyall, K., Sarma, J., Newby, D.E. and Flapan, A.D., 2011. Effect of moderate walnut consumption on lipid profile, arterial stiffness and platelet activation in humans. *European Journal of Clinical Nutrition*, *65*(2), pp. 234–239.

Fusch, L., 1555. *De Historia Stirpium Commetarii Insignes*. Lugduni Apud Ioan Tornaesium.

Greve, L.C., McGranahan, G., Hasey, J., Snyder, R., Kelly, K., Goldhamer, D. and Labavitch, J.M., 1992. Variation in polyunsaturated fatty acids composition of Persian walnut. *Journal of the American Society for Horticultural Science*, *117*(3), pp. 518–522.

Haider, S., Batool, Z., Tabassum, S., Perveen, T., Saleem, S., Naqvi, F., Javed, H. and Haleem, D.J., 2011. Effects of walnuts (Juglans regia) on learning and memory functions. *Plant Foods for Human Nutrition*, 66, pp. 335–340.

Javed, K., Rakha, A., Butt, M.S., Faisal, M.N., Tariq, U. and Saleem, M., 2022. Evaluating the anti-arthritic potential of walnut (Juglans regia L.) in FCA induced Sprague Dawley rats. *Journal of Food Biochemistry*, 46(10), p. e14327.

Jordão, J.B.R., Porto, H.K.P., Lopes, F.M., Batista, A.C. and Rocha, M.L., 2017. Protective effects of ellagic acid on cardiovascular injuries caused by hypertension in rats. *Planta Medica*, 83(10), pp. 830–836.

Joukar, S., Ebrahimi, S., Khazaei, M., Bashiri, A., Shakibi, M.R., Naderi, V., Shahouzehi, B. and Alasvand, M., 2017. Co-administration of walnut (Juglans regia) prevents systemic hypertension induced by long-term use of dexamethasone: A promising strategy for steroid consumers. *Pharmaceutical Biology*, 55(1), pp. 184–189.

Lavedrine, F., Zmirou, D., Ravel, A., Balducci, F. and Alary, J., 1999. Blood cholesterol and walnut consumption: A cross-sectional survey in France. *Preventive Medicine*, 28(4), pp. 333–339.

Lémery, N., 1716. *Traité universel des drogues simples, mises en ordre alphabétique. Où l'on trouve leurs différens noms . . . et tout ce qu'il y a de particulier dans les animaux, dans les végétaux, et dans les minéraux*. Au dépend de la Companie.

Linazasoro, J.M., Sanchez-Martin, J.A. and Jimenez-Diaz, C.A.R.L.O.S., 1970. Goitrogenic effect of walnut and its action on thyroxine excretion. *Endocrinology*, 86(3), pp. 696–700.

Sanchis, P., Molina, M., Berga, F., Muñoz, E., Fortuny, R., Costa-Bauzá, A., Grases, F. and Buades, J.M., 2019. A pilot randomized crossover trial assessing the safety and short-term effects of walnut consumption by patients with chronic kidney disease. *Nutrients*, 12(1), p. 63.

Shimoda, H., Tanaka, J., Kikuchi, M., Fukuda, T., Ito, H., Hatano, T. and Yoshida, T., 2009. Effect of polyphenol-rich extract from walnut on diet-induced hypertriglyceridemia in mice via enhancement of fatty acid oxidation in the liver. *Journal of Agricultural and Food Chemistry*, 57(5), pp. 1786–1792.

Steffen, L.M., Yi, S.Y., Duprez, D., Zhou, X., Shikany, J.M. and Jacobs Jr, D.R., 2021. Walnut consumption and cardiac phenotypes: The coronary artery risk development in young adults (CARDIA) study. *Nutrition, Metabolism and Cardiovascular Diseases*, 31(1), pp. 95–101.

Tufail, S., Fatima, A., Niaz, K.H.A.L.I.D., Qusoos, A. and Murad, S., 2015. Walnuts increase good cholesterol (HDL-Cholesterol) and prevent coronary artery disease. *Pakistan Journal of Medical & Health Sciences*, 9(4), pp. 1244–1246.

Zibaeenezhad, M., Aghasadeghi, K., Hakimi, H., Yarmohammadi, H. and Nikaein, F., 2016. The effect of walnut oil consumption on blood sugar in patients with diabetes mellitus type 2. *International Journal of Endocrinology and Metabolism*, 14(3).

71 Lettuce (*Lactuca sativa* L.)

Lactuca sativa L.

Etymology: From the Latin *lactuca* = milky sap and *sativus* = cultivated

Synonym: *Lactuca scariola* var. *sativa* (L.) Moris

Family: Asteraceae

Common names: Lettuce; laitue (Fr.); kopfsalat (Ger.); alface (Port.); лат ý к (Rus.); lechuga (Spa.)

Parts used: Leaf, seed

Constituents: Hydroxycinnamic acid derivatives (chlorogenic acid) (Nicolle et al., 2004), flavonol glycosides (isoquercitrin) (Xu et al., 2012), *S*-malic acid 1'-*O*-gentiobioside (Lagemann et al., 2012); sesquiterpenes (Bennett et al., 2002); fibers (Nicolle et al., 2004).

Medical history: The plant was sacred in ancient Egypt and could have been among the bitter herbs mentioned in Exodus (12:8): "and on this night, they shall eat the flesh, roasted over the fire, and unleavened cakes, with bitter herbs they shall eat it". Dioscorides recommends lettuce for stomach and as a galactagogue. It was cold and humid to the third degree in 16th-century Germany (Fusch, 1555), while in Italy, it was recommended for keeping the stomach healthy, for decreasing venereal appetite, as a galactagogue, and for inflammation (Matthioli, 1572). In 18th-century Scotland, physicians used lettuce for liver health, to decrease venereal appetite, as an anti-inflammatory, and for bilious and nephritic pain (Alston, 1770). The latex of lettuce, "lettuce opium" or *lactucarium*, was like opium used to induce sleep and to calm during the 19th century (O'Shaughnessy, 1842).

Medicinal uses: Fever (Iraq); galactagogue, diuretic, gastric ulcers, furuncles (Korea); bronchitis, pertussis (the Philippines)

Blood pressure: The plant produces *S*-malic acid 1'-*O*-gentiobioside, which inhibited ACE *in vitro* with $EC_{50} = 27.8$ µM (Lagemann et al., 2012).

Heart: Gonzalez-Lima et al. (1986) isolated from lettuce latex a substance that prevents cardiac arrythmia and atrial fibrillation in rodents.

Plasma lipids and glucose: The lipid profile of obese mice was improved after taking lettuce for 8 weeks (Han et al., 2018). Seeds of lettuce given at 1 g/kg for 12 weeks to patients receiving atorvastatin (20 mg/day) reduced plasma triglycerides, cholesterol, and LDL-cholesterol and decreased atorvastatin-induced elevation of hepatic enzymes (Moghadam et al., 2020).

Addition of freeze-dried lettuce for 4 weeks to a high-fat diet fed to mice decreased plasma cholesterol from 1.8 to 1.6 g/mL, plasma triglycerides

DOI: 10.1201/9781003301455-71

from 812 to 623 mg/dL, and LDL-cholesterol from 311 to 139 mg/dL (Lee et al., 2009).

Gallbladder: Diet containing 20% of lettuce given for 3 weeks to rats increased the production of fecal bile acids as well as cecal propionic acid (Nicolle et al., 2004).

Brain: Ethanol extract given orally to mice at 200 mg/kg/day for 14 days protected rats against scopolamine-induced dementia (Malik et al., 2018).

REFERENCES

Alston, C., 1770. *Lectures on the Materia Medica: Containing the Natural History of Drugs, their Virtues and Doses: Also Directions for the Study of the Materia Medica; and an Appendix on the Method of Prescribing*. Edward and Charles Dilly.

Bennett, M.H., Mansfield, J.W., Lewis, M.J. and Beale, M.H., 2002. Cloning and expression of sesquiterpene synthase genes from lettuce (Lactuca sativa L.). *Phytochemistry, 60*(3), pp. 255–261.

Fusch, L., 1555. *De Historia Stirpium Commetarii Insignes*. Lugduni Apud Ioan Tornaesium.

Gonzalez-Lima, F., Valedon, A. and Stiehil, W.L., 1986. Depressant pharmacological effects of a component isolated from lettuce, Lactuca sativa. *International Journal of Crude Drug Research, 24*(3), pp. 154–166.

Han, Y., Zhao, C., He, X., Sheng, Y., Ma, T., Sun, Z., Liu, X., Liu, C., Fan, S., Xu, W. and Huang, K., 2018. Purple lettuce (Lactuca sativa L.) attenuates metabolic disorders in diet induced obesity. *Journal of Functional Foods, 45*, pp. 462–470.

Lagemann, A., Dunkel, A. and Hofmann, T., 2012. Activity-guided discovery of (S)-malic acid 1′-O-β-gentiobioside as an angiotensin I-converting enzyme inhibitor in lettuce (Lactuca sativa). *Journal of Agricultural and Food Chemistry, 60*(29), pp. 7211–7217.

Lee, J.H., Felipe, P., Yang, Y.H., Kim, M.Y., Kwon, O.Y., Sok, D.E., Kim, H.C. and Kim, M.R., 2009. Effects of dietary supplementation with red-pigmented leafy lettuce (Lactuca sativa) on lipid profiles and antioxidant status in C57BL/6J mice fed a high-fat high-cholesterol diet. *British Journal of Nutrition, 101*(8), pp. 1246–1254.

Malik, J., Kaur, J. and Choudhary, S., 2018. Standardized extract of Lactuca sativa Linn. and its fractions abrogates scopolamine-induced amnesia in mice: A possible cholinergic and antioxidant mechanism. *Nutritional Neuroscience, 21*(5), pp. 361–372.

Matthioli, P.A., 1572. *Commentaires sur les Six Livres de Pedacius Dioscorides Anazarbeen de la matière medicinale*. A l'Escue de Milan.

Moghadam, M.H., Ghasemi, Z., Sepahi, S., Rahbarian, R., Mozaffari, H.M. and Mohajeri, S.A., 2020. Hypolipidemic effect of Lactuca sativa seed extract, an adjunctive treatment, in patients with hyperlipidemia: A randomized double-blind placebo-controlled pilot trial. *Journal of Herbal Medicine, 23*, p. 100373.

Nicolle, C., Cardinault, N., Gueux, E., Jaffrelo, L., Rock, E., Mazur, A., Amouroux, P. and Rémésy, C., 2004. Health effect of vegetable-based diet: Lettuce consumption improves cholesterol metabolism and antioxidant status in the rat. *Clinical Nutrition, 23*(4), pp. 605–614.

Xu, F., Zou, G.A., Liu, Y.Q. and Aisa, H.A., 2012. Chemical constituents from seeds of Lactuca sativa. *Chemistry of Natural Compounds, 48*(4), pp. 574–576.

72 Bottle Gourd (*Lagenaria siceraria* (Molina) Standl.)

Lagenaria siceraria (Molina) Standl.

Etymology: From the Greek *lagenaria* = bottle and the Hebrew *sekor* = strong fermented drink

Family: Cucurbitaceae

Synonyms: *Cucumis mairei* H. Lév.; *Cucurbita siceraria* Molina; *Cucurbita leucantha* Duchesne; *Lagenaria leucantha* Rusby; *Lagenaria vulgaris* Ser.

Common names: Bottle gourd, calabash; calebasse, courge, gourde des pélerins, gourde massue (Fr.); kalebasse (Ger.); cabaça (Port.); калабас (Rus.); calabaza (Span.)

Part used: Fruit

Constituents: Cucurbitane-type triterpenes (cucurbitacin B) (Glotter et al., 1971).

Medical history: Simeon Seth (10th century) asserts that bottle gourd quenches thirst, induces urination, treats stomach inflammation, decreases libido, and benefits the bladder and lungs. Fusch in the 16th century calls bottle gourd *Cucurbita minor* and defines it as cold and humid to the second degree. The oil of seeds was used for the skin in 18th-century France (Lémery, 1716). The seeds were used in the preparation of the *"quatre semences froides"* or four cold seeds in 19th-century France (Guibour, 1836). Purgative in 19th century India (Drury, 1873).

Medicinal uses: Jaundice (Pakistan); sedative, laxative (India; Bangladesh)

Blood pressure: Powder of fruits given at 500 mg/kg/day for 4 weeks to rats poisoned with L-NAME decreased systolic blood pressure from 162.7 to 130.3 mmHg (Mali et al., 2012). Fruit juice given orally at 200 mL/day for 90 days to dyslipidemic patients decreased systolic blood pressure from 127 to 123 mmHg (Katare et al., 2014).

Blood lipids and glucose: Fruit juice given orally at 200 mL/day for 90 days to dyslipidemic patients decreased fasting glucose from 91.9 to 86.9 mg/dL, cholesterol from 206.2 to 158.8 mg/mL, triglycerides from 128.1 to 104.6 mg/dL, and LDL-cholesterol from 145.3 to 104.3 mg/dL (Katare et al., 2014). In type-2 diabetic patients, fruit juice given orally at 200 mL/day for 90 days decreased plasma fasting glucose, total cholesterol, serum triglycerides, and LDL-cholesterol (Katare et al., 2013).

Heart: Powder of fruits given at 500 mg/kg/day for 4 weeks to rats poisoned with L-NAME rescued the heart against inflammation and necrosis (Mali

DOI: 10.1201/9781003301455-72

et al., 2012). Fruit juice given orally at 400 mg/kg/day to rats for 30 days evoked protective effects against isoproterenol-induced myocardial infarction evidenced by improvement of ECG parameters (Upaganlawar & Balaraman, 2010). Fruit powder given orally at 500 mg/kg/day to rats for 51 days evoked protective effects against isoprenaline-induced myocardial infarction evidenced by halving of plasma creatine kinase-myocardial band, a lower heart rate, and sustained blood pressure, and mitigated blood pressure fall (Mali & Bodhankar, 2010).

Brain: Ethanol extract given orally at 400 mg/kg/day for 6 weeks protected rats against aluminum-chloride-induced amnesia (Prashar et al., 2014).

Warning: Although pharmacological evidence indicates beneficial on the aging cardiovascular system, the plant should not be consumed. The fruits when raw contain a high amount of bitter and poisonous cucurbitacins which induce a few minutes after ingestion vomiting of blood, hypotension, and diarrhea (Ho et al., 2014; Verma & Jaiswal, 2015).

REFERENCES

Fusch, L., 1555. *De Historia Stirpium Commetarii Insignes*. Lugduni Apud Ioan Tornaesium.

Glotter, E., Goldsmith, D., Gross, D., Hanson, J.R., Huneck, S., Johnson, F., Lavie, D., Premuzic, E., Rüdiger, W., Lavie, D. and Glotter, E., 1971. The cucurbitanes, a group of tetracyclic triterpenes. *Fortschritte der Chemie Organischer Naturstoffe/Progress in the Chemistry of Organic Natural Products*, 29, pp. 307–362.

Ho, C.H., Ho, M.G., Ho, S.P. and Ho, H.H., 2014. Bitter bottle gourd (Lagenaria siceraria) toxicity. *The Journal of Emergency Medicine*, 46(6), pp. 772–775.

Katare, C., Saxena, S., Agrawal, S., Joseph, A.Z., Subramani, S.K., Yadav, D., Singh, N., Bisen, P.S. and Prasad, G.B.K.S., 2014. Lipid-lowering and antioxidant functions of bottle gourd (Lagenaria siceraria) extract in human dyslipidemia. *Journal of Evidence-Based Complementary & Alternative Medicine*, 19(2), pp. 112–118.

Katare, C., Saxena, S., Agrawal, S. and Prasad, G.B.K.S., 2013. Alleviation of diabetes induced dyslipidemia by Lagenaria siceraria fruit extract in human type 2 diabetes. *Journal of Herbal Medicine*, 3(1), pp. 1–8.

Lémery, N., 1716. *Traité universel des drogues simples, mises en ordre alphabétique. Où l'on trouve leurs différens noms . . . et tout ce qu'il y a de particulier dans les animaux, dans les végétaux, et dans les minéraux.* Au dépend de la Companie.

Mali, V.R. and Bodhankar, S.L., 2010. Cardioprotective effect of Lagenaria siceraria (LS) fruit powder in isoprenalin-induced cardiotoxicity in rats. *European Journal of Integrative Medicine*, 2(3), pp. 143–149.

Mali, V.R., Mohan, V. and Bodhankar, S.L., 2012. Antihypertensive and cardioprotective effects of the Lagenaria siceraria fruit in N G-nitro-L-arginine methyl ester (L-NAME) induced hypertensive rats. *Pharmaceutical Biology*, 50(11), pp. 1428–1435.

Prashar, Y.A.S.H., Gill, N. and Perween, A.M.B.E.R., 2014. Protective effect of Lagenaria siceraria in reversing Aluminium Chloride induced learning and memory deficits in experimental animal model. *International Journal of Recent Advances in Pharmaceutical Research*, 4, pp. 87–104.

Upaganlawar, A. and Balaraman, R., 2010. Protective effects of Lagenaria siceraria (Molina) fruit juice in isoproterenol induced myocardial infarction. *International Journal of Pharmacology*, 6(5), pp. 645–651.

Verma, A. and Jaiswal, S., 2015. Bottle gourd (Lagenaria siceraria) juice poisoning. *World Journal of Emergency Medicine*, 6(4), p. 308.

73 Banaba (*Lagerstroemia speciosa* (L.) Pers.)

Lagerstroemia speciosa (L.) **Pers.**

Etymology: After the Swedish naturalist Magnus von Lagerström (1696–1759) and from the Latin *speciosa* = splendid

Family: Lythraceae

Synonyms: *Adambea glabra* Lam.; *Lagerstroemia flos-reginae* Retz.; *Lagerstroemia reginae* Roxb.; *Munchausia speciosa* L.

Common names: Banaba, pride of India, queen's crape myrtle; grand goyavier fleur (Fr.); resedá-gigante (Port.); banabá de Filipinas, reina de las flores (Spa.)

Part used: Leaf

DOI: 10.1201/9781003301455-73

Constituents: Ursane-type triterpenes (corosolic acid), tannins (ellagitannins: lagerstroemine, ellagic acid), phenolic acids (gallic acid) (Bai et al., 2008).

Medical history: The Dutch were the first to introduce banaba to the knowledge of Europeans during the 17th century. This tree was the *"flos reginae"* of Rumphius and mentioned in the *Hortus Malabaricus* of Hendrik van Rheede. William Roxburgh gave a description of the plant in the *Plants of the Coast of Corommandel* (1795). Later, Father Blanco (1837), wrote *"este arbol bellissimo"* and provided the native name in the Philippines: "banaba".

Medicinal uses: Laxative (Iraq); Diabetes (Myanmar, Thailand); toothache (Indonesia); diuretic, diarrhea, urinary tract infection, kidney stones, dysmenorrhea, headache, low blood pressure (the Philippines)

Blood pressure: Ethanol extract of leaves given at 400 mg/kg/day for 6 weeks to rats poisoned by L-NAME decreased systolic blood pressure from 171.1 to 126.9 mmHg as well as myocardial ROS and TNF α, increased plasma NO, and improved myocardial cytoarchitecture (Aleissa et al., 2022). In spontaneously hypertensive and obese rats on a high-fat diet, adding 0.07% of corosolic acid to their diet for 14 weeks decreased systolic blood pressure from 209 to 188 mmHg and decreased hr-CRP (Yamaguchi et al., 2006).

Plasma lipids and glucose: Ethanol extract of leaves given orally to type-2 diabetic patients at 48 mg/kg/day for 15 days induced a 30% decrease in glycemia (Judy et al., 2003). Giving 500 mg to obese patients twice a day, before breakfast and before dinner, for 12 weeks caused decreases in systolic blood pressure from 121.5 to 11.3 mmHg, fasting plasma glucose from 5.9 to 5.7 mmol/L, and triglycerides from 2.3 to 1.7 mmol/L and decreased insulin resistance (López-Murillo et al., 2022).

Kidneys: In streptozotocin-induced diabetic rats, oral intake of an extract prevented glucose-induced kidney damage (Aljarba et al., 2021).

Warning: Intake of banaba can induce hypoglycemia when taken with antidiabetic drugs (Posadzki et al., 2013)

Comment: (i) Georg Eberhard Rumphius (1627–1702) was a blind German botanist based in Ambon archipelago from 1654 till his death and the author of a monumental book titled *Herbarium Amboinense* (1741).

(ii) Manuel Blanco (1779–1845) was a Spanish friar and botanist based in Manila and the author of a book titled *Flora de Filipinas. según el sistema de Linneo* (1837).

(iii) William Roxburgh (1751–1815) was a Scottish botanist and physician, in charge of the botanical garden of Madras and author of *Flora Indica; or descriptions of Indian plants* (1820).

(iv) Banaba leaves for teas are commercially available in the Philippines, and capsules of banaba leaf extracts are sold in Germany.

REFERENCES

Aleissa, M.S., Al-Zharani, M., Alneghery, L.M., Hasnain, M.S., Almutairi, B., Ali, D., Alarifi, S. and Alkahtani, S., 2022. Lagerstroemia speciosa ameliorated blood pressure in LNAME induced hypertension in experimental rats through NO/cGMP and oxidative stress modulation. *BioMed Research International, 2022.*

Aljarba, N.H., Hasnain, M.S., AlKahtane, A., Algamdy, H. and Alkahtani, S., 2021. Lagerstroemia speciosa extract ameliorates oxidative stress in rats with diabetic nephropathy by inhibiting AGEs formation. *Journal of King Saud University-Science, 33*(6), p. 101493.

Bai, N., He, K.A.N., Roller, M., Zheng, B., Chen, X., Shao, Z., Peng, T. and Zheng, Q., 2008. Active compounds from Lagerstroemia speciosa, insulin-like glucose uptake-stimulatory/inhibitory and adipocyte differentiation-inhibitory activities in 3T3-L1 cells. *Journal of Agricultural and Food Chemistry, 56*(24), pp. 11668–11674.

Judy, W.V., Hari, S.P., Stogsdill, W.W., Judy, J.S., Naguib, Y.M. and Passwater, R., 2003. Antidiabetic activity of a standardized extract (Glucosol™) from Lagerstroemia speciosa leaves in Type II diabetics: A dose-dependence study. *Journal of Ethnopharmacology, 87*(1), pp. 115–117.

López-Murillo, L.D., González-Ortiz, M., Martínez-Abundis, E., Cortez-Navarrete, M. and Pérez-Rubio, K.G., 2022. Effect of banaba (Lagerstroemia speciosa) on metabolic syndrome, insulin sensitivity, and insulin secretion. *Journal of Medicinal Food, 25*(2), pp. 177–182.

Posadzki, P., Watson, L. and Ernst, E., 2013. Contamination and adulteration of herbal medicinal products (HMPs): An overview of systematic reviews. *European Journal of Clinical Pharmacology, 69*, pp. 295–307.

Yamaguchi, Y., Yamada, K., Yoshikawa, N., Nakamura, K., Haginaka, J. and Kunitomo, M., 2006. Corosolic acid prevents oxidative stress, inflammation and hypertension in SHR/NDmcr-cp rats, a model of metabolic syndrome. *Life Sciences, 79*(26), pp. 2474–2479.

74 Garden Cress (*Lepidium sativum* L.)

Lepidium sativum L.

Etymology: From the Greek *lepis* = a scale and the Latin *sativum* = cultivated

Family: Brassicaceae

Synonym: *Lepidium spinescens* DC.

Common names: Garden cress; cresson de jardin (Fr.); garten kresse (Ger.); agrião do jardim (Port.); кресс сал ат (Rus.); berro de jardín (Spa.)

Parts used: Leaf, seed

Constituents: Glucosinolates (benzyl isothicyanate) (Burow et al., 2007; Yokoyama et al., 2020), ascorbic acid (Sat et al., 2013), fixed oil in the seeds (unsaturated fatty acids: α-linolenic acid, oleic, linoleic acid) (Diwakar et al., 2010).

Medical history: The Medical School of Salerno (10th century) recommends garden cress for hair growth as well as for tooth and gum aches, herpes, and eczema. According to Fusch in 16th-century Germany, cress is dry to the fourth degree. Lémery in 18th-century France advocates distillate of garden cress to treat scurvy, dropsy, rheumatism, kidney stones, and jaundice. In 19th-century France, it was used against scurvy (Moquin-Tandon, 1861).

Medicinal uses: Goiter (Turkey); tonic, aphrodisiac, diuretic (Iraq); bleeding (Pakistan); fractured bones, expulsion of placenta (India)

Blood pressure: Oral administration of an aqueous extract of seeds to hypertensive rats at 20 mg/kg/day for 3 weeks decreased blood pressure (Maghrani et al., 2005).

Plasma lipids and glucose: Rats given cress powder at 6 g/kg in a cholesterol-enriched diet for 2 weeks experienced decreases in plasma cholesterol from 33.4 to 29.3 mg/mL and triglycerides from 43.6 to 30 mg/dL (Althnaian, 2014).

Kidneys: Oral administration of an aqueous extract of seeds to hypertensive rats at 20 mg/kg/day for 3 weeks stimulated the secretion of urine (Maghrani et al., 2005). Aqueous extract of seeds given at the single dose of 100 mg/kg orally increased the urinary secretion of sodium and potassium ions as well as increasing urine volume from 4.7 to 7.1 mL/100 g/h in a manner similar to that of hydrochlorothiazide at 10 mg/kg (Patel et al., 2009).

Bones and cartilages: Addition of seeds to the diet of glucocortisone-induced osteoporotic rats at 7.5% increased femur calcification (Arafa & El-kholey, 2021).

Immune system: Aqueous extract of seeds (1 g boiled in 1 L) given orally to mice at the volume of 1 mL/day for about 20 days increased leukocytes count (Mahassni & Khudauardi, 2017).

DOI: 10.1201/9781003301455-74

Skin and hair: Lyophilized extract of seeds applied topically for 21 days to the skin of rats attenuated testosterone-induced alopecia (Albalawi et al., 2023).

Brain: Aqueous extract of seeds given orally at 20 mg/kg/day to rats for 8 weeks evoked some protection against aluminum-chloride-induced amnesia (Balgoon, 2023).

Warnings: The plant must be avoided by patients with thyroid dysfunction. Feeding rats a diet containing more than 2% seeds caused toxic effects (Adam, 1999).

Comments: (i) With aging, vascular system NADPH oxidase activity increases (Zarzuelo et al., 2013). NADPH oxidase catalyzes the production of superoxide anions, which react with NO to form peroxynitrite and thereby decrease the bioavailability of endothelium-derived NO. This decrease in turn drives age-related endothelial dysfunction and hypertension (Karthik et al., 2011; Zielonka et al., 2016).

(ii) Isothiocyanates are released from glucosinolates by myrosinases upon plant tissue injury. They are electrophiles, meaning that they accept a pair of electrons from nucleophiles and as such activate nuclear factor erythroid-derived 2-like 2 and inhibit the expression of NADPH oxidase (Nakamura & Miyoshi, 2010). Further, isothiocyanates scavenge ROS and thereby increase the bioavailability of NO (Weragoda et al., 2018). Further, being antioxidant, isothiocyanates decrease the formation of LDL in vascular intima and the development of atherosclerosis (Toikka et al., 2000).

(iii) Rats given phenylisothiocyanates orally 5 days per week for 4 weeks experienced a decrease in plasma thyroxine (Speijers et al., 1985). On the luminal side of the apical membrane of thyroid gland cells, iodide is oxidized by thyroid peroxidase, a reaction that requires the presence of hydrogen peroxide generated by NADPH oxidase, explaining at least in part the goitrogenic effects (Kopp, 2012) of isothiocyanates on the thyroid gland.

(iv) Isothiocyanates at high doses are pro-oxidant (Nakamura & Miyoshi, 2010).

(v) It is clear that consuming garden cress salads regularly is beneficial to the aging cardiovascular system.

REFERENCES

Adam, S.E.I., 1999. Effects of various levels of dietary Lepidium sativum L. seeds in rats. *The American Journal of Chinese Medicine*, 27(03n04), pp. 397–405.

Albalawi, M.A., Hafez, A.M., Elhawary, S.S., Sedky, N.K., Hassan, O.F., Bakeer, R.M., El Hadi, S.A., El-Desoky, A.H., Mahgoub, S. and Mokhtar, F.A., 2023. The medicinal activity of lyophilized aqueous seed extract of Lepidium sativum L. in an androgenic alopecia model. *Scientific Reports*, 13(1), p. 7676.

Althnaian, T., 2014. Influence of dietary supplementation of Garden cress (Lepidium sativum L.) on liver histopathology and serum biochemistry in rats fed high cholesterol diet. *Journal of Advanced Veterinary and Animal Research*, 1(4), pp. 216–223.

Arafa, R. and El-kholey, H., 2021. The potential effects of garden cress seeds (Lepidium Sativum L.) on the bone of female rats suffering from osteoporosis مجمع البحوث في مجالات التربية النوعية pp. 1399–1426.

Balgoon, M.J., 2023. Garden Cress (Lepidium sativum) seeds ameliorated aluminum-induced Alzheimer disease in rats through antioxidant, anti-inflammatory, and antiapoptotic effects. *Neuropsychiatric Disease and Treatment*, pp. 865–878.

Burow, M., Bergner, A., Gershenzon, J. and Wittstock, U., 2007. Glucosinolate hydrolysis in Lepidium sativum—identification of the thiocyanate-forming protein. *Plant Molecular Biology*, 63(1), pp. 49–61.

Diwakar, B.T., Dutta, P.K., Lokesh, B.R. and Naidu, K.A., 2010. Physicochemical properties of garden cress (Lepidium sativum L.) seed oil. *Journal of the American Oil Chemists' Society*, 87, pp. 539–548.

Fusch, L., 1555. *De Historia Stirpium Commetarii Insignes*. Lugduni Apud Ioan Tornaesium.

Karthik, D., Viswanathan, P. and Anuradha, C.V., 2011. Administration of rosmarinic acid reduces cardiopathology and blood pressure through inhibition of p22phox NADPH oxidase in fructose-fed hypertensive rats. *Journal of Cardiovascular Pharmacology*, 58(5), pp. 514–521.

Kopp, P., 2012. Thyroid hormone synthesis. *Werner and Ingbar's the Thyroid. A Fundamental and Clinical Text, ed*, 10, pp. 48–74.

Lémery, N., 1716. *Traité universel des drogues simples, mises en ordre alphabétique. Où l'on trouve leurs différens noms . . . et tout ce qu'il y a de particulier dans les animaux, dans les végétaux, et dans les minéraux*. Au dépend de la Companie.

Maghrani, M., Zeggwagh, N.A., Michel, J.B. and Eddouks, M., 2005. Antihypertensive effect of Lepidium sativum L. in spontaneously hypertensive rats. *Journal of Ethnopharmacology*, 100(1–2), pp. 193–197.

Mahassni, S.H. and Khudauardi, E.R., 2017. A pilot study: The effects of an aqueous extract of Lepidium sativum seeds on levels of immune cells and body and organs weights in Mice. *Journal of Ayurvedic and Herbal Medicine*, 3, pp. 27–32.

Moquin-Tandon, A., 1861. *Element de Botanique Médicale*. J.Baillière et Fils.

Nakamura, Y. and Miyoshi, N., 2010. Electrophiles in foods: The current status of isothiocyanates and their chemical biology. *Bioscience, Biotechnology, and Biochemistry*, 74(2), pp. 242–255.

Patel, U., Kulkarni, M., Undale, V. and Bhosale, A., 2009. Evaluation of diuretic activity of aqueous and methanol extracts of Lepidium sativum garden cress (Cruciferae) in rats. *Tropical Journal of Pharmaceutical Research*, 8(3).

Sat, I.G., Yildirim, E., Turan, M. and Demirbas, M., 2013. Antioxidant and nutritional characteristics of garden cress (Lepidium sativum). *Acta Scientiarum Polonorum Hortorum Cultus*, 12(4), pp. 173–179.

Speijers, G.J.A., Danse, L.H.J.C., Van Leeuwen, F.X.R. and Loeber, J.G., 1985. Four-week toxicity study of phenyl isothiocyanate in rats. *Food and Chemical Toxicology*, 23(11), pp. 1015–1017.

Toikka, J.O., Laine, H., Ahotupa, M., Haapanen, A., Viikari, J.S., Hartiala, J.J. and Raitakari, O.T., 2000. Increased arterial intima-media thickness and in vivo LDL oxidation in young men with borderline hypertension. *Hypertension*, 36(6), pp. 929–933.

Weragoda, G.K., Pilkington, R.L., Polyzos, A. and Richard, A.J., 2018. Regioselectivity of aryl radical attack onto isocyanates and isothiocyanates. *Organic & Biomolecular Chemistry*, 16(46), pp. 9011–9020.

Yokoyama, S.I., Kodera, M., Hirai, A., Nakada, M., Ueno, Y. and Osawa, T., 2020. Benzyl isothiocyanate produced by garden cress (Lepidium sativum) prevents accumulation of hepatic lipids. *Journal of Nutritional Science and Vitaminology*, 66(5), pp. 481–487.

Zielonka, J., Zielonka, M., VerPlank, L., Cheng, G., Hardy, M., Ouari, O., Ayhan, M.M., Podsiadły, R., Sikora, A., Lambeth, J.D. and Kalyanaraman, B., 2016. Mitigation of NADPH oxidase 2 activity as a strategy to inhibit peroxynitrite formation. *Journal of Biological Chemistry*, 291(13), pp. 7029–7044.

75 Sponge Gourd (*Luffa aegyptiaca* Mill.)

Luffa aegyptiaca **Mill.**

DOI: 10.1201/9781003301455-75

Etymology: From the Arabic name of the plant, *louff*, and the Latin *aegyptica* = from Egypt

Family: Cucurbitaceae

Synonyms: *Luffa cylindrica* M. Roem.; *Momordica cylindrica* L.; *Momordica luffa* L.

Common names: Dishcloth gourd, Egyptian momordica, loofah, sponge gourd; luffa d'Egypte, patole, pipangalle (Fr.); schwammkürbis (Ger.); cabaça esponja (Port.); губчатая тыква (Rus.); esponja vegetal (Spa.)

Part used: Fruit

Constituents: Triterpenes, steroidal saponins (Xiong et al., 1994; Han et al., 2020), flavonol glycosides, hydroxycinnamic acid derivatives (Du & Wang, 2007).

Medical history: The plant was called *petola* by Rumphius. It is mentioned by Hendrik van Rheede in his *Hortus Malabaricus*. Purgative in 19th-century France (Baillon, 1886). Sponge gourd was used as emetic in 19th-century Sri Lanka.

Medicinal uses: Hypertension (Nepal, the Philippines); conjunctivitis (Sikkim); asthma (Korea)

Blood pressure: Total saponin extract given at 200 mg/kg/day for 28 days to progesterone-induced obese rats decreased blood pressure from 135.5 to 125 mmHg (D'silva et al., 2021).

Blood lipids and glucose: Intake of fruits decreased plasma lipids (Kanthal et al., 2010). A single oral 200 mg/kg dose of aqueous extract of fruits given to alloxan-induced diabetic rats decreased plasma glucose by 32% (Glibenclamide at 10 mg/kg: 57.1%) after 10 hours (Swain et al., 2013). Total saponin extract from fruits given at 200 mg/kg/day for 28 days to progesterone-induced obese rats decreased glycemia from 120 to 94.1 mg/dL as well as cholesterol and triglycerides (D'silva et al., 2021).

Warning: Aqueous and alcoholic extracts of the fruits were poisonous to mice at an oral dose above 2g/kg (Al-Snafi, 2019).

REFERENCES

Al-Snafi, A.E., 2019. Constituents and pharmacology of Luffa cylindrica-A review. *IOSR Journal of Pharmacy*, 9(9), pp. 68–79.

D'silva, W.W., Biradar, P.R. and Patil, A., 2021. Luffa cylindrica: A promising herbal treatment in progesterone induced obesity in mice. *Journal of Diabetes & Metabolic Disorders*, 20, pp. 329–340.

Du, Q. and Wang, K., 2007. Preparative separation of phenolic constituents in the fruits of Luffa cylindrica (L.) roem using slow rotary countercurrent chromatography. *Journal of Liquid Chromatography & Related Technologies*, 30(13), pp. 1915–1922.

Han, Y., Zhang, X., Qi, R., Li, X., Gao, Y., Zou, Z., Cai, R. and Qi, Y., 2020. Lucyoside B, a triterpenoid saponin from Luffa cylindrica, inhibits the production of inflammatory mediators via both nuclear factor-κB and activator protein-1 pathways in activated macrophages. *Journal of Functional Foods*, 69, p. 103941.

Kanthal, L.K., De, S., Aneela, S., Choudhury, N.S.K., Padhy, I.P. and Mridha, D., 2010. Pharmacological study on hypolipidemic effect of luffa-aegyptiaca MILL fruit. *Advances in Pharmacology and Toxicology*, *11*(1), p. 35.

Swain, T., Sahoo, R.K. and Kar, D.M., 2013. Phytomedicinal potential of Luffa cylindrica (L.) Reom extracts. JOURNAL OF PURE AND APPLIED MICROBIOLOGY, March 2013. Vol. 7(1), p1-7

Xiong, S.L., Fang, Z.P. and Zeng, X.Y., 1994. Chemical constituents of Luffa cylindrica (L.) Roem. *Zhongguo Zhong yao za zhi= Zhongguo Zhongyao Zazhi= China Journal of Chinese Materia Medica*, *19*(4), pp. 233–234.

76 Common White Horehound (*Marrubium vulgare* L.)

Marrubium vulgare L.

Etymology: After an ancient Roman city, *Marruvium*, and Latin *vulgare* = common

Family: Lamiaceae

Synonym: *Marrubium hamatum* Kunth

Common names: Common white horehound; bon riblet, grand bonhome, marrube blanc (Fr.); andorn (Ger.); marrohlo (Port.); шандра обыкновенная (Rus.); camarruego (Spa.)

Part used: Aerial part

Constituents: Diterpenes (marrubenol, marrubiin) (El Bardai et al., 2003), essential oil (β-bisabolene) (Hamdaoui et al., 2013), 6-octadecynoic acid (Ohtera et al., 2013).

Medical history: Dioscorides defines common white horehound as dry to the third degree and hot to the second degree and recommends it for cough, pneumonia, cachexia, ptysis (tuberculosis?), female fertility, and jaundice; Galen recommends common white horehound for deobstructing the liver and the pancreas and for treating pneumonia (Galen). In 16th-century Italy, Matthioli (1572) prescribed it for painful liver and jaundice. While in England, it was used for obstruction of blood vessels overflowing of the gall bladder, old cough, phlegm, and asthma (Parkinson, 1640) and in Scotland for pneumonia, asthma, lung affections (Alston, 1770). In 18th-century France, common white horehound was used for obstructions of the liver and pancreas, asthma and hemoptysis (Lémery, 1716). The plant was brewed in 19th-century England into an ale, and in the rest of the Western world, it was used for chronic pulmonary complaints as well as for uterine and hepatic infections (Pereira, 1843).

Medicinal uses: For cough, dyspepsia (1–2 g in 250 mL of boiling water, 3 times daily) (European Union); hypertension, diabetes (Algeria); cold, cough, diuretic, carminative (Iraq); abdominal pain, carminative (Turkey); lung troubles, cough (Pakistan)

Blood pressure: Aqueous extract given to hypertensive rats caused a decrease in systolic blood pressure via a mechanism independent from NO (El Bardai et al., 2001, 2004). Marrubenol and marrubiin relaxed rings of rat aorta

DOI: 10.1201/9781003301455-76

exposed to potassium chloride by blocking L-type calcium ions channels (El Bardai et al., 2003).

Heart: Methanol extract of aerial parts given orally for 2 days at 40 mg twice a day protected rats against isoproterenol-induced cardiac injuries as evidenced by improved mean arterial blood pressure from 66 to 108 mmHg (normal 107 mmHg), increased left ventricular systolic pressure, corrected ventricular dysfunction, and amelioration of cardiac cytoarchitecture (Yousefi et al., 2013).

Plasma lipids and glucose: Infusion of 1 g leaves given 3 times a day before meals to type-2 diabetic patients for 21 days decreased plasma glucose level by 0.6% and cholesterol and triglycerides by 4.1 and 5.7%, respectively (Herrera-Arellano et al., 2004). Methanol extract given to streptozotocin-induced diabetic rats for 28 days at the daily dose of 500 mg/kg decreased glycaemia, total cholesterol, LDL-cholesterol, and triglycerides and increased insulinaemia (Elberry et al., 2015). A single oral dose of ethanol extract decreased glycemia in healthy rats (Vergara-Galicia et al., 2012).

Immune system: Marrubin given at 100 mg/kg protected rodent against ovalbumin-induced allergic edema (Stulzer et al., 2006).

Warning: Aqueous extract oral LD_{50} = 12 g/kg in mice (Rhallaba et al., 2014).

Comments: (i) The leaves of this herb are covered with whitish hairs, giving the leaves some resemblance to lung tissues. Because of this, the plant was used for lung affection. This is the theory of signatures used by our peers. It may sound a bit unscientific today, but looking at a walnut, one could see a brain, while looking at tamarind petals, one could see blood capillaries.

(ii) L-type calcium ions channels are involved in insulin secretion (Velasco et al., 2016), and marrubiol and marrubin might modulate insulin secretion. 6-Octadecynoic acid induced triglyceride accumulation by adipocytes *in vitro* via the stimulation of peroxisome proliferator-activated receptor γ (PPAR-γ) (Ohtera et al., 2013). PPAR-γ agonists are used to treat insulin resistance (Olefsky et al., 2000).

(iii) Infusions of common white horehound could potentially by beneficial for the aging cardiovascular system. More experiments are needed.

REFERENCES

Alston, C., 1770. *Lectures on the Materia Medica: Containing the Natural History of Drugs, their Virtues and Doses: Also Directions for the Study of the Materia Medica; and an Appendix on the Method of Prescribing*. Edward and Charles Dilly.

El Bardai, S., Lyoussi, B., Wibo, M. and Morel, N., 2001. Pharmacological evidence of hypotensive activity of Marrubium vulgare and Foeniculum vulgare in spontaneously hypertensive rat. *Clinical and Experimental Hypertension (New York, NY: 1993)*, 23(4), pp. 329–343.

El Bardai, S., Lyoussi, B., Wibo, M. and Morel, N., 2004. Comparative study of the antihypertensive activity of Marrubium vulgare and of the dihydropyridine calcium antagonist amlodipine in spontaneously hypertensive rat. *Clinical and Experimental Hypertension*, 26(6), pp. 465–474.

El Bardai, S., Morel, N., Wibo, M., Fabre, N., Llabres, G., Lyoussi, B. and Quetin-Leclercq, J., 2003. The vasorelaxant activity of marrubenol and marrubiin from Marrubium vulgare. *Planta Medica*, *69*(01), pp. 75–77.

Elberry, A.A., Harraz, F.M., Ghareib, S.A., Gabr, S.A., Nagy, A.A. and Abdel-Sattar, E., 2015. Methanolic extract of Marrubium vulgare ameliorates hyperglycemia and dyslipidemia in streptozotocin-induced diabetic rats. *International Journal of Diabetes Mellitus*, *3*(1), pp. 37–44.

Hamdaoui, B., Wannes, W.A., Marrakchi, M., Brahim, N.B. and Marzouk, B., 2013. Essential oil composition of two Tunisian horehound species: Marrubium vulgare L. and Marrubium aschersonii Magnus. *Journal of Essential Oil Bearing Plants*, *16*(5), pp. 608–612.

Herrera-Arellano, A., Aguilar-Santamaria, L., Garcia-Hernandez, B., Nicasio-Torres, P. and Tortoriello, J., 2004. Clinical trial of Cecropia obtusifolia and Marrubium vulgare leaf extracts on blood glucose and serum lipids in type 2 diabetics. *Phytomedicine*, *11*(7–8), pp. 561–566.

Lémery, N., 1716. *Traité universel des drogues simples, mises en ordre alphabétique. Où l'on trouve leurs différens noms . . . et tout ce qu'il y a de particulier dans les animaux, dans les végétaux, et dans les minéraux.* Au dépend de la Companie.

Matthioli, P.A., 1572. *Commentaires sur les Six Livres de Pedacius Dioscorides Anazarbeen de la matière medicinale.* A l'Escue de Milan.

Ohtera, A., Miyamae, Y., Nakai, N., Kawachi, A., Kawada, K., Han, J., Isoda, H., Neffati, M., Akita, T., Maejima, K. and Masuda, S., 2013. Identification of 6-octadecynoic acid from a methanol extract of Marrubium vulgare L. as a peroxisome proliferator-activated receptor γ agonist. *Biochemical and Biophysical Research Communications*, *440*(2), pp. 204–209.

Parkinson, J., 1640. *Theatrum Botanicum: The Theater of Plants: Or, An Herball of Large Extent: Containing Therein a More Ample and Exact History and Declaration of the Physicall Herbs and Plants . . . Distributed Into Sundry Classes Or Tribes, for the More Easie Knowledge of the Many Herbes of One Nature and Property.* Tho. Cotes.

Pereira, J., 1843. *The Elements of Materia Medica and Therapeutics.* Lea and Blanchard.

Olefsky, J.M., 2000. Treatment of insulin resistance with peroxisome proliferator–activated receptor γ agonists. *The Journal of Clinical Investigation*, *106*(4), pp. 467–472.

Rhallaba, A., Chakira, S., Elbadaouia, K. and lmolek Alaouia, T., 2014. Evaluation of acute and subacute toxicity of aqueous extracts from Marrubium vulgare L in rodents. *Journal of Pharmacy Research*, *8*(5), pp. 684–688.

Stulzer, H.K., Tagliari, M.P., Zampirolo, J.A., Cechinel-Filho, V. and Schlemper, V., 2006. Antioedematogenic effect of marrubiin obtained from Marrubium vulgare. *Journal of Ethnopharmacology*, *108*(3), pp. 379–384.

Velasco, M., Díaz-García, C.M., Larqué, C. and Hiriart, M., 2016. Modulation of ionic channels and insulin secretion by drugs and hormones in pancreatic beta cells. *Molecular Pharmacology*, *90*(3), pp. 341–357.

Vergara-Galicia, J., Aguirre-Crespo, F., Tun-Suarez, A., Aguirre-Crespo, A., Estrada-Carrillo, M., Jaimes-Huerta, I., Flo-res-Flores, A., Estrada-Soto, S. and Ortiz-Andrade, R., 2012. Acute hypoglycemic effect of ethanolic extracts from Marrubium vulgare. *Phytopharmacology*, *3*(1), pp. 54–60.

Yousefi, K., Soraya, H., Fathiazad, F., Khorrami, A., Hamedeyazdan, S., Maleki-Dizaji, N. and Garjani, A., 2013. Cardioprotective effect of methanolic extract of *Marrubium vulgare* L. on isoproterenol-induced acute myocardial infarction in rats. *Indian Journal of Experimental Biology*, *51*, pp. 653–660.

77 Lemon Balm (*Melissa officinalis* L.)

Melissa officinalis L.

Etymology: From the Latin *meli* = honey bee and *officinalis* = of medicinal value

Family: Lamiaceae

Synonym: *Melissa bicornis* Klokov

Common names: Lemon balm, melissa; mélisse officinale (Fr.); melissenblätter (Ger.); melissa (Port.); лимонный бальзам (Rus.); melisa (Spa)

Part used: Leaf

Constituents: Hydroxycinnamic acid derivatives (rosmarinic acid) (Chen et al., 2017), essential oil (citronellal, geranial, neral) (Shabby et al., 1995; Hăncianu et al., 2008).

Medical history: Pliny the Elder recommends lemon balm for inflammations. The plant was considered by Middle Ages Arab physicians including Avicenna to be hot and dry to the second degree and useful for heart heath, depression, and "purging the melancholic vapors" (?). Serapion the Younger indicates that lemon balm rejoices the heart and helps digestion, invigorates, and prevents heart palpitations. In 12th-century Byzantium, lemon balm was taken for insomnia (Simeon Seth). It was an antiseptic in the 16th century in Italy (Matthioli, 1572). In the 18th-century Scotland, the plant was used to cheer the spirit and for palpitations, fainting, loss of memory, apoplexy, palsy, epilepsy, mania, and bareness (Alston, 1770). In the same era in France, physicians used lemon balm for hysteria, paralysis, and fevers (Lémery, 1716). In 19th-century Europe and North America, it was used for fever (Pereira, 1843).

Medicinal uses: Stress, insomnia (1.5–4.5 g of leaves in 150 mL of boiling water 1–3 times daily) (European Union); asthma, cardiovascular diseases, nephritis, forgetfulness, diabetes (Turkey)

Blood pressure: Intake of capsules containing 400 mg of lemon balm 3 times per day for 4 weeks resulted in a decrease in diastolic and systolic pressure (Shekarriz et al., 2021). Type-2 diabetic patients taking 500 mg capsules twice daily for 3 months had a decrease of both diastolic and systolic blood pressure (Nayebi et al., 2019). Intake of 3 g/day of lemon balm for 2 months by patients with stable angina pectoris increased plasma NO and decreased systolic and diastolic blood pressure (Javid et al., 2018).

Heart: Rats given drinking water containing aqueous extract at 50 mg/mL experienced prolongation of QTc and JT intervals (Joukar & Asadipour,

2015). Intake of aqueous extract at 100 mg/kg/day for 7 days decreased the rat heart rate from 377 to 264 beat/min (Joukar et al., 2016). Intake of 3 g/day of lemon balm for 2 months by patients with stable angina pectoris increased the ejection fraction (Javid et al., 2018).

Plasma lipids and glucose: In type-2 diabetic patients, taking 500 mg capsules twice daily for 3 months decreased plasma triglycerides (Nayebi et al., 2019). Type-2 diabetic patients taking 700 mg of lemon balm hydroalcoholic extract twice daily for 12 weeks had decreases in plasma glucose, triglycerides, LDL-cholesterol, hs-CRP, and systolic blood pressure (Asadi et al., 2019).

Brain: c daily for 16 weeks led to improved cognition (Akhondzadeh et al., 2003).

Warnings: (i) The oral LD_{50} of essential oil in mice is 2.5 g /kg (Stojanović et al., 2019). Patients with thyroid dysfunction should not use this plant (Parimal et al., 2020).

(ii) Use cautiously in patients taking hormonal agents and sedatives (Posadzki et al., 2013).

Comment: It is clear that lemon balm in the form of an infusion, as per the European Union's recommendation, is beneficial for elderly people.

REFERENCES

Akhondzadeh, S., Noroozian, M., Mohammadi, M., Ohadinia, S., Jamshidi, A.H. and Khani, M., 2003. Melissa officinalis extract in the treatment of patients with mild to moderate Alzheimer's disease: A double blind, randomised, placebo controlled trial. *Journal of Neurology, Neurosurgery & Psychiatry, 74*(7), pp. 863–866.

Alston, C., 1770. *Lectures on the Materia Medica: Containing the Natural History of Drugs, their Virtues and Doses: Also Directions for the Study of the Materia Medica; and an Appendix on the Method of Prescribing*. Edward and Charles Dilly.

Asadi, A., Shidfar, F., Safari, M., Hosseini, A.F., Fallah Huseini, H., Heidari, I. and Rajab, A., 2019. Efficacy of Melissa officinalis L. (lemon balm) extract on glycemic control and cardiovascular risk factors in individuals with type 2 diabetes: A randomized, double-blind, clinical trial. *Phytotherapy Research, 33*(3), pp. 651–659.

Chen, S.G., Leu, Y.L., Cheng, M.L., Ting, S.C., Liu, C.C., Wang, S.D., Yang, C.H., Hung, C.Y., Sakurai, H., Chen, K.H. and Ho, H.Y., 2017. Anti-enterovirus 71 activities of Melissa officinalis extract and its biologically active constituent rosmarinic acid. *Scientific Reports, 7*(1), pp. 1–16.

Hăncianu, M., Aprotosoaie, A.C., Gille, E., Poiată, A., Tuchiluş, C., Spac, A. and Stănescu, U., 2008. Chemical composition and in vitro antimicrobial activity of essential oil of Melissa officinalis L. from Romania. *Revista medico-chirurgicala a Societatii de Medici si Naturalisti din Iasi, 112*(3), pp. 843–847.

Javid, A.Z., Haybar, H., Dehghan, P., Haghighizadeh, M.H., Mohaghegh, S.M., Ravanbakhsh, M., Mohammadzadeh, A. and Bahrololumi, S.S., 2018. The effects of Melissa officinalis on echocardiography, exercise test, serum biomarkers, and blood pressure in patients with chronic stable angina. *Journal of Herbal Medicine, 11*, pp. 24–29.

Joukar, S. and Asadipour, H., 2015. Evaluation of Melissa officinalis (Lemon Balm) Effects on Heart Electrical System. *Research in Cardiovascular Medicine, 4*(2), pp. e27013–e27013.

Joukar, S., Asadipour, H., Sheibani, M., Najafipour, H. and Dabiri, S., 2016. The effects of Melissa officinalis (lemon balm) pretreatment on the resistance of the heart to myocardial injury. *Pharmaceutical Biology*, *54*(6), pp. 1005–1013.

Lémery, N., 1716. *Traité universel des drogues simples, mises en ordre alphabétique. Où l'on trouve leurs différens noms . . . et tout ce qu'il y a de particulier dans les animaux, dans les végétaux, et dans les minéraux*. Au dépend de la Companie.

Matthioli, P.A., 1572. *Commentaires sur les Six Livres de Pedacius Dioscorides Anazarbeen de la matière medicinale*. A l'Escue de Milan.

Nayebi, N., Esteghamati, A., Meysamie, A., Khalili, N., Kamalinejad, M., Emtiazy, M. and Hashempur, M.H., 2019. The effects of a Melissa officinalis L. based product on metabolic parameters in patients with type 2 diabetes mellitus: A randomized double-blinded controlled clinical trial. *Journal of Complementary and Integrative Medicine*, *16*(3).

Parimal, K.P., Mayuresh, H.R., Jagdish, B.R. and Satish, M.S., 2020. Herbal anti-thyroid drugs: An overview. *Research Journal of Pharmacy and Technology*, *13*(10), pp. 5045–5051.

Pereira, J., 1843. *The Elements of Materia Medica and Therapeutics*. Lea and Blanchard.

Posadzki, P., Watson, L. and Ernst, E., 2013. Contamination and adulteration of herbal medicinal products (HMPs): An overview of systematic reviews. *European Journal of Clinical Pharmacology*, *69*, pp. 295–307.

Shabby, A.S., El-Gengaihi, S. and Khattab, M., 1995. Oil of Melissa officinalis L., as affected by storage and herb drying. *Journal of Essential Oil Research*, *7*(6), pp. 667–669.

Shekarriz, Z., Shorofi, S.A., Nabati, M., Shabankhani, B. and Yousefi, S.S., 2021. Effect of Melissa officinalis on systolic and diastolic blood pressures in essential hypertension: A double-blind crossover clinical trial. *Phytotherapy Research*, *35*(12), pp. 6883–6892.

Stojanović, N.M., Randjelović, P.J., Mladenović, M.Z., Ilić, I.R., Petrović, V., Stojiljković, N., Ilić, S. and Radulović, N.S., 2019. Toxic essential oils, part VI: Acute oral toxicity of lemon balm (Melissa officinalis L.) essential oil in BALB/c mice. *Food and Chemical Toxicology*, *133*, p. 110794.

78 Bitter Gourd (*Momordica charantia* L.)

Momordica charantia L.

Etymology: From the Latin *mordeo* = I bite

Family: Cucurbitaceae

Synonyms: *Cucumis argyi* H. Lév.; *Momordica chinensis* Spreng.; *Momordica indica* L.; *Momordica muricata* Willd.; *Momordica sinensis* Spreng.; *Sicyos fauriei* H. Lév.

Common names: Balsam apple, bitter gourd; courge amère (Fr.); bitterer kürbis (Ger.); cabaco amargo (Port.); горькая тыква (Rus.); calabaza amarga (Spa.)

Part used: Fruit

Constituents: Cucurbitan-type triterpene saponins (charantin) (Murakami et al., 2001; Wang et al., 2014).

Medical history: Bitter gourd was unknown to the physicians of ancient Rome and Middle Ages Europe. The 17th-century Dutch botanist Rumphius, who lived in Indonesia, calls it as *amara indica*. It was used in curries on the Malabar coast during the 18th century (Aublet, 1775), as also noted by William Roxburgh. Rheede in the 17th century describes the plant in detail in his *Hortus Malabaricus* under the name of "*pandi-pavel*" and notes its use by natives for skin infections including leprosy.

Medicinal uses: Diabetes (Turkey, India, Thailand; the Philippines; Malaysia, Indonesia), fever (Myanmar, the Philippines); high blood pressure (India); cholera (Mauritius); intestinal worms (India, Myanmar, the Philippines); anemia (the Philippines)

Blood pressure: Aqueous extract of fruits given orally to type-1 diabetic rats at 1.5 g/kg/day for 28 days decreased systolic blood pressure from about 140 to 130 mmHg and diastolic blood pressure from about 100 to 80 mmHg. At the aortic level, this regimen decreased MDA and increased NO, increased endothelial NO expression, reduced the thickness of the tunica media, and protected the thoracic aorta cytoarchitecture (Abas et al., 2015).

Plasma lipids and glucose: Aqueous extract given orally to alloxan-induced diabetic mice at 250 mg/kg for 21 days decreased glycemia from 222.5 to 112.3 mg/dL (Sharma et al., 2014). Aqueous extract of fruits given orally to to type-1 diabetic rats at 1.5 g/kg/day for 28 days normalized plasma cholesterol and triglycerides (Abas et al., 2015).

The glucose tolerance of type-2 diabetic patients consuming 2 g of bitter gourd per day for 12 weeks increased.

DOI: 10.1201/9781003301455-78

In streptozotocin-induced type-2 diabetic mice, oral administration of a saponin extract at 40 mg/kg/day for 5 weeks normalized glycemia, decreased hepatic NF-kB, and induced the phosphorylation of hepatic AMPK (Wang et al., 2019).

Kidneys: Aqueous extract given orally to alloxan-induced diabetic mice at 250 mg/kg for 21 days caused decreases in plasma urea, uric acid, and creatinine and protected kidneys against proximal and distal convoluted tubules, glomeruli, and Bowman's capsules cytoarchitecture alterations (Sharma et al., 2014).

Bones and cartilages: Aqueous extract given orally for 12 days to mice attenuated CFA-induced arthritis (Kola et al., 2018).

Skin and air: Extract given orally to mice at the daily dose of 50 mg/kg/day 3 times a week for 3 years improved skin moisture retention, reduced the formation of wrinkles, and decreased dermal matrix metalloprotease-1 and hyaluronidase activity (Hiramoto et al., 2020).

Brain: Butanol extract given orally to mice at 200 mg/kg/day for 14 days evoked some protection against amyloid protein β-induced dementia with decreased lipid peroxidation and NO in the brain (Sin et al., 2021).

Warnings: Eating too much bitter gourd gives very unpleasant tiredness and is even dangerous because of hypoglycemia. In mice, the oral LD_{50} of aqueous extract is above 2 g/kg (Kola et al., 2018).

Comments: (i) The activation of AMP-activated protein kinase inhibits NADPH oxidase (Rodríguez et al., 2020).

Increasing systolic blood pressure activates NADPH oxidase, which releases ROS (Santos et al., 2014).

TNF α activates NADPH oxidase, which releases superoxide ions that react with NO, resulting in endothelial dysfunction and the production of hydrogen peroxide and as a consequence thickening of the media (Zalba et al., 2001).

(ii) Jean Baptiste Christophore Fusée Aublet (1720–1778) was a French pharmacist based in French Guiana and the author of a book titled *Histoire des plantes de la Guiane Françoise* (1775).

(iii) Butanol extracts saponins.

(iv) NO is released by astrocytes surrounding amyloid β plaques in Alzheimer's disease (Wallace et al., 1997).

(v) Intake of boiled bitter gourd at a dietary dose with dishes once or twice a week could potentially be beneficial for aging.

REFERENCES

Abas, R., Othman, F. and Thent, Z.C., 2015. Effect of Momordica charantia fruit extract on vascular complication in type 1 diabetic rats. *EXCLI Journal, 14*, p. 179.

Aublet, F., 1775. *Histoire des plantes de la Guiane françoise*. Didot.

Cortez-Navarrete, M., Martinez-Abundis, E., Perez-Rubio, K.G., Gonzalez-Ortiz, M. and Méndez-del Villar, M., 2018. Momordica charantia administration improves insulin secretion in type 2 diabetes mellitus. *Journal of Medicinal Food, 21*(7), pp. 672–677.

Hiramoto, K., Orita, K., Yamate, Y. and Kobayashi, H., 2020. Role of Momordica charantia in preventing the natural aging process of skin and sexual organs in mice. *Dermatologic Therapy*, *33*(6), p. e14243.

Kola, V., Mondal, P., Thimmaraju, M.K., Mondal, S. and Rao, N.V., 2018. Antiarthritic potential of aqueous and ethanolic fruit extracts of "Momordica charantia" using different screening models. *Pharmacognosy Research*, *10*(3).

Murakami, T., Emoto, A., Matsuda, H. and Yoshikawa, M. (2001). Medicinal foodstuffs. XXI. Structures of new cucurbitane-type triterpene glycosides, goyaglycosides-a,-b,-c,-d,-e,-f,-g, and-h, and new oleanane-type triterpene saponins, goyasaponins I, II, and III, from the fresh fruit of Japanese Momordica charantia L. *Chemical and Pharmaceutical Bulletin*, *49*(1), pp. 54–63.

Rodríguez, C., Contreras, C., Sáenz-Medina, J., Muñoz, M., Corbacho, C., Carballido, J., García-Sacristán, A., Hernandez, M., López, M., Rivera, L. and Prieto, D., 2020. Activation of the AMP-related kinase (AMPK) induces renal vasodilatation and downregulates Nox-derived reactive oxygen species (ROS) generation. *Redox Biology*, *34*, p. 101575.

Santos, C.X., Nabeebaccus, A.A., Shah, A.M., Camargo, L.L., Filho, S.V. and Lopes, L.R., 2014. Endoplasmic reticulum stress and Nox-mediated reactive oxygen species signaling in the peripheral vasculature: Potential role in hypertension. *Antioxidants & Redox Signaling*, *20*(1), pp. 121–134.

Sharma, B., Siddiqui, M., Ram, G., Yadav, R.K., Kumari, A., Sharma, G. and Jasuja, N.D., 2014. Rejuvenating of kidney tissues on alloxan induced diabetic mice under the effect of Momordica charantia. *Advances in Pharmaceutics*, *2014*.

Sin, S.M., Kim, J.H., Cho, E.J. and Kim, H.Y., 2021. Cognitive improvement effects of momordica charantia in amyloid beta-induced Alzheimer's disease mouse model. *Journal of Applied Biological Chemistry*, *64*(3), pp. 299–307.

Wallace, M.N., Geddes, J.G., Farquhar, D.A. and Masson, M.R., 1997. Nitric oxide synthase in reactive astrocytes adjacent to β-amyloid plaques. *Experimental Neurology*, *144*(2), pp. 266–272.

Wang, H.Y., Kan, W.C., Cheng, T.J., Yu, S.H., Chang, L.H. and Chuu, J.J., 2014. Differential anti-diabetic effects and mechanism of action of charantin-rich extract of Taiwanese Momordica charantia between type 1 and type 2 diabetic mice. *Food and Chemical Toxicology*, *69*, pp. 347–356.

Wang, Q., Wu, X., Shi, F. and Liu, Y., 2019. Comparison of antidiabetic effects of saponins and polysaccharides from Momordica charantia L. in STZ-induced type 2 diabetic mice. *Biomedicine & Pharmacotherapy*, *109*, pp. 744–750.

Zalba, G., José, G.S., Moreno, M.U., Fortuño, M.A., Fortuño, A., Beaumont, F.J. and Díez, J., 2001. Oxidative stress in arterial hypertension: Role of NAD (P) H oxidase. *Hypertension*, *38*(6), pp. 1395–1399.

79 Moringa (*Moringa oleifera* Lam.)

Moringa oleifera Lam.

Etymology: From the Tamil name of the plant, *murungai*, and the Latin *oleifera* = bearing oil

Family: Moringaceae

Synonym: *Moringa pterygosperma* Gaertn.

Common names: Drumstick tree, horse-radish tree, moringa; meerrettichbaum (Ger.)

Parts used: Leaf, seed

Constituents: Benzoacetonitrile glycosides (niazirin and niazirinin), thiocarbamate glycosides (niazimin A and B), isothiocyanates (El Haddad et al., 2019).

Medical history: Moringa was unknown to Roman and Middle Ages physicians. Ancient Sanskrit medical texts recommend taking moringa to induce urination (Dymock, 1884). It was also known to Avicenna. Moringa was used for "melancholic humors", for snake bites, and for leprosy in 16th-century India (Acosta, 1578). In 19th-century Java, it was used for dropsy, while in India, it was a remedy for headache, hysteria, rheumatism, sexual impotence, and fatigue (Ainslie, 1813; Mohideen Sheriff Khan Bahadur, 1891).

Medicinal uses: Aphrodisiac (Pakistan); body aches, colds, fever, rheumatism, hypertension (Bangladesh); heart stimulant (Myanmar); aphrodisiac, increase sperm, heart problems, flatulence, diuretic (India); rheumatism (Nepal); blood pressure, diabetes, increase breast milk, wounds (the Philippines)

Blood pressure: Isothiocyanates and niaziminin A and B decreased blood pressure in rodents (Faizi et al., 1994). Aqueous extract of leaves given orally at 60 mg/kg/day for 3 weeks attenuated blood pressure and tachycardia induced by L-NAME poisoning via anti-ROS effects and increased production of NO (Aekthammarat et al., 2019, 2020). Consumption of cooked leaves by individual ingesting 7 g of salt/day prevented increased blood pressure (Chan Sun et al., 2020).

Heart: Seeds given at 750 mg/day for 8 weeks to hypertensive rats decreased left ventricular anterior wall thickness, interseptal thickness on diastole, and left ventricular fibrosis by activating peroxisome proliferator-activated receptor-α and δ (Randriamboavonjy et al., 2016).

Plasma glucose and lipids: Aqueous extract of leaves given at 1 mg/g for 30 days to rats on a high-fat diet decreased serum cholesterol from 115 to 103.2 mg/dL

DOI: 10.1201/9781003301455-79

(normal: 90 mg/dL) (Ghasi et al., 2000). Rabbits fed a diet containing 10% leaves for 8 weeks had decreased plasma cholesterol (Anthony & Ashawe, 2014). A single 300 mg/kg oral dose of aqueous extract of leaves decreased glycemia within 6 hours by about 33%, and this effect was similar to that of tolbutamide at 200 mg/kg in healthy rats (Edoga et al., 2013). Methanol extract of seeds given orally at 100 mg/kg/day for 10 weeks to rats poisoned with a high-fat diet decreased plasma cholesterol (Ajayi et al., 2020).

Kidneys: A single oral dose of 50 mg/kg of alcoholic extract of leaves increased the urinary excretion of sodium ions in rats (Tahkur et al., 2016).

Bones and cartilages: Intake of 1 g/day of leaves by post-menopausal women for 12 weeks increased bone density (Brown et al., 2016).

Immune system: Methanol extract of leaves given at 1 g/day for 14 days to rodents prevented leukocytes count fall induced by cyclophosphamide (Nfambi et al., 2015).

Skin and hair: Extract of leaves (5%) given daily and externally to rats for 30 days increased the number of hair follicles in anagen phase (Builders et al., 2014).

Brain: Aqueous extract of leaves given to rats orally at 250 mg/kg/day protected mice against colchicine-induced dementia (Ganguly et al., 2010).

Comments: (i) The Portuguese physician Cristóbal Acosta (1525–1594) wrote a book titled *Tractado de las drogas y medicinas de las Indias orientales* (1578).

(ii) Mohideen Sheriff Khan Bahadur (?–1891) was an Indian surgeon in Madras and the author of a book titled *Materia medica of Madras* (1891).

(iii) In the destitute villages of the Philippines, moringa is eaten daily, perhaps explaining the longevity of some of its inhabitants who are deprived of food and formal medicines.

(iv) It is clear that consuming moringa regularly is beneficial for aging.

(v) Seeds of moringa have the ability to clean water (Bichi, 2013). There is the gut feeling that moringa seeds could interfere with the intestinal absorption of lipids.

(vi) Aqueous extract given at 1 g/kg to rats did not evoke toxic effects (Adedapo et al., 2009).

REFERENCES

Acosta, C., 1578. *Tractado de las drogas, y medicinas de las Indias*. por Martin de Victoria.

Adedapo, A.A., Mogbojuri, O.M. and Emikpe, B.O., 2009. Safety evaluations of the aqueous extract of the leaves of Moringa oleifera in rats. *Journal of Medicinal Plants Research*, 3(8), pp. 586–591.

Aekthammarat, D., Pannangpetch, P. and Tangsucharit, P., 2019. Moringa oleifera leaf extract lowers high blood pressure by alleviating vascular dysfunction and decreasing oxidative stress in L-NAME hypertensive rats. *Phytomedicine*, 54, pp. 9–16.

Aekthammarat, D., Tangsucharit, P., Pannangpetch, P., Sriwantana, T. and Sibmooh, N., 2020. Moringa oleifera leaf extract enhances endothelial nitric oxide production leading to relaxation of resistance artery and lowering of arterial blood pressure. *Biomedicine & Pharmacotherapy*, 130, p. 110605.

Ainslie, W., 1813. *Materia Indica of Hindoostan*. Government Press.

Ajayi, T.O., Moody, J.O., Odumuwagun, O.J. and Olugbuyiro, J.A., 2020. Lipid altering potential of Moringa oleifera lam seed extract and isolated constituents in wistar rats. *African Journal of Biomedical Research*, 23(1), pp. 77–85.

Anthony, V.P. and Ashawe, D., 2014. The effect of Moringa oleifera leaf meal (molm) on the hematological parameters and the cholesterol level of rabbits. *American Journal of Biological, Chemical and Pharmaceutical Sciences*, 2(3), pp. 1–6.

Bichi, M.H., 2013. A review of the applications of Moringa oleifera seeds extract in water treatment. *Civil and Environmental Research*, 3(8), pp. 1–10.

Brown, J., Merritt, E., Mowa, C.N. and McAnulty, S., 2016. *Effect of Moringa Oleifera on Bone Density in Post Menopausal Women* (Master thesis, Appalachian State University).

Builders, P.F., Iwu, I.W., Mbah, C.C., Builders, M.I. and Audu, M.M., 2014. Moringa oleifera ethosomes a potential hair growth activator: Effect on rats. *J Pharm Biomed Sci*, 4(7):611–618.

Chan Sun, M., Ruhomally, Z.B., Boojhawon, R. and Neergheen-Bhujun, V.S., 2020. Consumption of Moringa oleifera Lam leaves lowers postprandial blood pressure. *Journal of the American College of Nutrition*, 39(1), pp. 54–62.

Dymock, W., 1884. *The Vegetable Materia Medica of Western India*. Education Society Press.

Edoga, C.O., Njoku, O.O., Amadi, E.N. and Okeke, J.J., 2013. Blood sugar lowering effect of Moringa oleifera Lam in albino rats. *International Journal of Scientific & Technology*, 3(1), pp. 88–90.

El-Haddad, A.E., El-Deeb, E.M., Koheil, M.A., El-Khalik, S.M.A. and El-Hefnawy, H.M., 2019. Nitrogenous phytoconstituents of genus Moringa: Spectrophotometrical and pharmacological characteristics. *Medicinal Chemistry Research*, 28, pp. 1591–1600.

Faizi, S., Siddiqui, B.S., Saleem, R., Siddiqui, S., Aftab, K. and Gilani, A.U.H., 1994. Isolation and structure elucidation of new nitrile and mustard oil glycosides from Moringa oleifera and their effect on blood pressure. *Journal of Natural Products*, 57(9), pp. 1256–1261.

Ganguly, R., Hazra, R., Ray, K. and Guha, D., 2010. Effect of Moringa oleifera in experimental model of Alzheimer's disease: Role of antioxidants. *Annals of Neurosciences*, 12(3), pp. 33–36.

Ghasi, S., Nwobodo, E. and Ofili, J.O., 2000. Hypocholesterolemic effects of crude extract of leaf of Moringa oleifera Lam in high-fat diet fed Wistar rats. *Journal of Ethnopharmacology*, 69(1), pp. 21–25.

Nfambi, J., Bbosa, G.S., Sembajwe, L.F., Gakunga, J. and Kasolo, J.N., 2015. Immunomodulatory activity of methanolic leaf extract of Moringa oleifera in Wistar albino rats. *Journal of Basic and Clinical Physiology and Pharmacology*, 26(6), pp. 603–611.

Randriamboavonjy, J.I., Loirand, G., Vaillant, N., Lauzier, B., Derbré, S., Michalet, S., Pacaud, P. and Tesse, A., 2016. Cardiac protective effects of Moringa oleifera seeds in spontaneous hypertensive rats. *American Journal of Hypertension*, 29(7), pp. 873–881.

Sheriff, M., 1891. *Materia Medica of Madras*. Government Press.

Tahkur, R.S., Soren, G., Pathapati, R.M. and Buchineni, M., 2016. Diuretic activity of moringa oleifera leaves extract in swiss albino rats. *The Pharma Innovation*, 5(3, Part A), p. 8.

80 Common Watercress (*Nasturtium officinale* W.T. Aiton)

Nasturtium officinale W.T. Aiton

Etymology: From the Latin *nasus tordus* = induces cough and *officinale* = found in pharmacies

Family: Brassicaceae

Synonyms: *Rorippa nasturtium-aquaticum* (L.) Hayek; *Sisymbrium nasturtium-aquaticum* L.

Common names: Common watercress; cresson alenois, cresson des jardins, cresson de fontaine (Fr.); brunnenkresse (Ger.); agrião comum (Port.); кресс водяной (Rus.); berro (Spa.)

Part used: Leaf

Constituents: Glucosinolates, isothiocyanates (Theunis et al., 2022).

Medical history: Common watercress was known to Dioscorides, who prescribed it to excite venereal appetite and induce menses. Romans used to cook watercress in soup to remove phlegm from the lungs. The plant was called *nasturtium* by Pliny the Elder, and Galen recommended it for sciatica. The Medical School of Salerno (10th century) advocates common watercress for hair growth: *"crines retinere fluentes"*. According to Matthioli (1572), those who eat common watercress regularly have a sharpened intelligence. Used for salad in 18th-century France, and Lémery (1716) asserts that it purifies the blood and improves breathing. In 18-th century Scotland, physicians considered commo watercress as a hot and dry remedy to the second degree for scurvy, kidney stones, dropsy, and obstruction of the pancreas and the liver. It was used as a diuretic in 19th-century France (Guibourt, 1836)

Medicinal uses: Abdominal pain (Iran); cough (Pakistan)

Blood pressure: Common watercress added as 2.5% of a diet given to spontaneously hypertensive rats for 11 weeks decreased systolic blood pressure (Hayashi et al., 2014).

Blood lipids and glucose: Intake of common watercress in animals decreases plasma lipids and glucose. Oral administration of a hydroalcoholic extract at 500 mg/kg per day for 10 days to hypercholesterolemic rats decreased plasma cholesterol, triglycerides, and LDL-cholesterol by 34.2, 30.1, and 52.9%, respectively, while raising HDL-cholesterol by 27% (Bahramikia & Yazdanparast, 2008). Aqueous extract given orally at 75 mg/kg/day

DOI: 10.1201/9781003301455-80

for 4 weeks to streptozotocin-induced diabetic rats decreased glycemia (Shahrokhi et al., 2009). Watercress added to 2.5% of a diet given to spontaneously hypertensive rats for 11 weeks decreased plasma cholesterol, LDL-cholesterol, and triglycerides (Hayashi et al., 2014). Hydroalcoholic extract of aerial parts given orally to streptozotocin-induced diabetic rats at 200 mg/kg/day for 4 weeks decreased plasma glucose from about 500 to 400 mg/dL (normal: about 100 mg/dL) and cholesterol from about 800 to 400 mg/dL (normal: about 75 mg/dL), as well as triglycerides (Mousa-Al-Reza Hadjzadeh et al., 2015). Common watercress added to 2.5% of diet given to spontaneously hypertensive rats for 11 weeks decreased systolic blood pressure, plasma cholesterol, LDL-cholesterol, and triglycerides (Hayashi et al., 2014).

Kidneys: Hydroalcoholic extract given at 750 mg/kg/day for 30 days attenuated the formation of kidney stones induced by ethylene glycol poisoning in rodents (Mehrabi et al., 2016).

Bones and cartilages: Ethanol extract given orally to mice at 500 mg/kg/day for 6 days attenuated arthritis induced by the subplantar injection of formaldehyde as efficiently as indomethacin given orally at 10 mg/kg/day for 6 days (Mostafazadeh et al., 2022).

Brain: Ethanol extract given to mice orally at 80 mg/kg/day for 14 days attenuated dementia induced by dexamethasone (Ruanglertboon et al., 2019).

Warnings: Eating watercress from rivers and ponds poses the risk of *Fasciola hepatica* infection (Carrada-Bravo, T., 2003). The plant must be avoided by patients with thyroid problems (Theunis et al., 2022). The oral LD_{50} of hydroalcoholic extract is above 5 g/kg in rats (Sadeghi et al., 2014).

Comments: (i) Sterol regulatory element-binding proteins are transcription factors (Jeon & Osborne, 2012) which, once activated, command the synthesis of enzymes in charge of cholesterol and fatty acid synthesis (Horton et al., 2002). Allyl isothiocyanate inhibits the activation of sterol regulatory element-binding proteins (Miyata et al., 2016).

(ii) It is clear that regular intake of watercress at a dietary dose is beneficial for aging. In France, recipes of delicious watercress soups are available.

REFERENCES

Alston, C., 1770. *Lectures on the Materia Medica: Containing the Natural History of Drugs, their Virtues and Doses: Also Directions for the Study of the Materia Medica; and an Appendix on the Method of Prescribing*. Edward and Charles Dilly.

Bahramikia, S. and Yazdanparast, R., 2008. Effect of hydroalcoholic extracts of Nasturtium officinale leaves on lipid profile in high-fat diet rats. *Journal of Ethnopharmacology*, *115*(1), pp. 116–121.

Carrada-Bravo, T., 2003. Fascioliasis: Diagnosis, epidemiology and treatment. *Revista de gastroenterología de Mexíco*, *68*(2), pp. 135–142.

Guibourt, N.J.B.G., 1836. *Histoire abrégée des drogues simples*. Méquignon-Marvis Père et fils.

Hayashi, A., Takahashi, R. and Kimoto, K., 2014. Effects of watercress (Nasturtium officinale) intake on blood pressure and lipid metabolism in spontaneously hypertensive rats. *Nippon Eiyo Shokuryo Gakkaishi*, *67*(4), pp. 185–191.

Horton, J.D., Goldstein, J.L. and Brown, M.S., 2002. SREBPs: Activators of the complete program of cholesterol and fatty acid synthesis in the liver. *The Journal of Clinical Investigation*, *109*(9), pp. 1125–1131.

Jeon, T.I. and Osborne, T.F., 2012. SREBPs: Metabolic integrators in physiology and metabolism. *Trends in Endocrinology & Metabolism*, *23*(2), pp. 65–72.

Lémery, N., 1716. *Traité universel des drogues simples, mises en ordre alphabétique. Où l'on trouve leurs différens noms . . . et tout ce qu'il y a de particulier dans les animaux, dans les végétaux, et dans les minéraux.* Au dépend de la Companie.

Matthioli, P.A., 1572. *Commentaires sur les Six Livres de Pedacius Dioscorides Anazarbeen de la matière medicinale.* A l'Escue de Milan.

Mehrabi, S., Askarpour, E., Mehrabi, F. and Jannesar, R., 2016. Effects of hydrophilic extract of Nasturtium officinale on prevention of ethylene glycol induced renal stone in male Wistar rats. *Journal of Nephropathology*, *5*(4), p. 123.

Miyata, S., Inoue, J., Shimizu, M. and Sato, R., 2016. Allyl isothiocyanate suppresses the proteolytic activation of sterol regulatory element-binding proteins and de novo fatty acid and cholesterol synthesis. *Bioscience, Biotechnology, and Biochemistry*, *80*(5), pp. 1006–1011.

Mostafazadeh, M., Sadeghi, H., Sadeghi, H., Zarezade, V., Hadinia, A. and Kokhdan, E.P., 2022. Further evidence to support acute and chronic anti-inflammatory effects of Nasturtium officinale. *Research in Pharmaceutical Sciences*, *17*(3), p. 305.

Mousa-Al-Reza Hadjzadeh, Z.R., Moradi, R. and Ghorbani, A., 2015. Effects of hydroalcoholic extract of watercress (Nasturtium officinale) leaves on serum glucose and lipid levels in diabetic rats. *Indian Journal of Physiology and Pharmacology*, *59*(2), pp. 223–230.

Ruanglertboon, W., Kumarnsit, E., Dej-Adisai, S., Vongvatcharanon, U. and Udomuksorn, W., 2019. The neuroprotective effect of Nasturtium officinale on learning ability and density of parvalbumin neurons in the hippocampus of neurodegenerative-induced mice model. *Sains Malaysiana*, *48*(10), pp. 2191–2199.

Sadeghi, H., Mostafazadeh, M., Sadeghi, H., Naderian, M., Barmak, M.J., Talebianpoor, M.S. and Mehraban, F., 2014. In vivo anti-inflammatory properties of aerial parts of Nasturtium officinale. *Pharmaceutical Biology*, *52*(2), pp. 169–174.

Shahrokhi, N., Hadad, M.K., Keshavarzi, Z. and Shabani, M., 2009. Effects of aqueous extract of water cress on glucose and lipid plasma in streptozotocin induced diabetic rats. *Pakistan Journal of Physiology*, *5*(2), pp. 6–10.

Theunis, M., Naessens, T., Peeters, L., Brits, M., Foubert, K. and Pieters, L., 2022. Optimization and validation of analytical RP-HPLC methods for the quantification of glucosinolates and isothiocyanates in Nasturtium officinale R. Br and Brassica oleracea. *LWT*, *165*, p. 113668.

81 Nigella (*Nigella sativa* L.)

Nigella sativa **L.**

Etymology: From the French word *nigelle* and the Latin *sativus* = cultivated

Family: Ranunculaceae

Common names: Black cumin, nigella; cumin noir, nielle, nigelle, poivrette (Fr.); nigela (Port.); нигелла (Rus.)

Part used: Seed

Constituents: Fixed oil (thymoquinone, unsaturated fatty acids: oleic acid, linoleic acid) (Burits & Bucar, 2000; Atta, 2003).

Medical history: Ancient Sanskrit medical texts recommend nigella after childbirth (Dutt, 1877). Dioscorides calls it *melanthium* and advocates its use as diuretic, for difficulty breathing, for expelling intestinal worms, and for toothache but warns that it is lethal at high doses. Galen, Simeon Seth, and Avicenna use it for heart failure (orthopnea). Simeon Seth uses nigella for kidney stones, while Avicenna prescribes it for fever and depression. According to Sunni narrations, the Prophet of Islam said, "This black cumin is healing for all diseases except death" (Sahih al-Bukhari 5687, Book 70, Hadith 10). For Fusch (1555), nigella is hot and dry to the third degree. French physicians in the 18th century use nigella for flatulence, to induce menstruation, to increase production of milk, and for intestinal worms (Lémery, 1716). In 19th-century India, Unani physicians used nigella as a diuretic and carminative (Ainslie, 1813).

Medicinal uses: Anthelmintic, colds, rheumatism, stomachache (Turkey); diuretic (Iraq); carminative, galactagogue (Myanmar)

Blood pressure: Intake of 2.5 mL of oil twice a day for 8 weeks decreased systolic and diastolic blood pressure in healthy volunteers (Fallah Huseini et al., 2013). In patients with mild hypertension, aqueous extract given at 200 mg twice a day for 8 weeks caused decreases in diastolic and systolic blood pressure by about 2 mmHg (Dehkordi & Kamkhah, 2008).

In type-2 diabetic patients, intake of 2g/day of nigella for a year decreased systolic blood pressure from 134.3 to 129.3 mmHg and diastolic blood pressure from 79.8 to 76.6 mmHg (Badar et al., 2017).

Thymoquinone mixed at the concentration of 10 mg/L in the drinking water of rats poisoned with L-NAME for 4 weeks normalized systolic blood pressure from about 180 to 120 mmHg (Khattab & Nagi, 2007).

Plasma lipids and glucose: Aqueous extract given at 200 mg twice a day for 8 weeks to mild hypertensive patients decreased plasma cholesterol from

191.5 to 167.1 mg/dL, LDL-cholesterol from 121.5 to 101.6 mg/dL, and tri-glycerides from 124.4 to 97.4 mg/dL (Dehkordi & Kamkhah, 2008). Seeds given at 1 g twice a day for 12 weeks to type-2 diabetic patients decreased fasting blood glucose from 204 to 169 mg/dL (Bamosa et al., 2010). In type-2 diabetic patients, intake of 2g/day for a year decreased plasma choles-terol from 199.2 to 180.6 mg/dL, LDL-cholesterol from 120.7 to 114.2 mg/dL, and triglycerides from 189.7 to 169.6 mg/dL (Badar et al., 2017).

Thymoquinone given orally at 10 mg/kg/day for 8 weeks to rabbits poi-soned with a high-cholesterol diet decreased plasma cholesterol by 26% and aortic MDA by 73% and decreased the development of aortic atheroma (Ragheb et al., 2008).

Kidneys: In rats, oil given orally at 0.3 mL/day for 7 days attenuated kidney damages induced by bilateral renal artery occlusion (Bayrak et al., 2008). Thymoquinone mixed at the concentration of 10 mg/L in the drinking water of rats poisoned with L-NAME for 4 weeks decreased plasma creatinine to normal levels (Khattab & Nagi, 2007).

Capsules of seeds given at 500 mg twice a day for 10 weeks induced com-plete stone excretion in kidney stone patients (Ardakani Movaghati et al., 2019).

Heart rate: In type-2 diabetic patients, intake of 2g/day for a year decreased heart rate from 88.1 to 82.6 bpm (Badar et al., 2017).

Bones and cartilages: In ovariectomized rats, oral administration of nigella at 800 mg/kg/day for 12 weeks increased plasma ions calcium and cortical and trabecular bone thickness and decreased plasma alkaline phosphatase, collagen terminal telopeptide, MDA, TNF α, and interleukin-6 (Seif, 2014).

Patients with rheumatoid arthritis taking oil at 500 mg twice a day for a month had decreased swelling of joints and morning stiffness (Gheita & Kenawy, 2012).

Brain: Thymoquinone given orally and daily at 40 mg/kg/day for 14 days improved the cognitive function of rats with aluminum chloride/D-galactose-induced Alzheimer's disease. This regimen decreased cerebral interleukin-1β as well as TNF α and amyloid proteins β deposition (Abulfadl et al., 2018).

Ethanol extract given orally at 400 mg/kg for 21 days protected rats against chlorpromazine-induced Parkinson's disease (Sandhu & Rana, 2013).

Warnings: Consumption must be avoided in pregnancy (Keshri et al., 1995). Patients on warfarin must not take nigella because thymoquinone inhibits cytochrome P450 2C9 (Albassam et al., 2018).

Comment: Thymoquinone activates AMPK and enhances SIRT1 expression (Yang et al., 2016).

REFERENCES

Abulfadl, Y.S., El-Maraghy, N.N., Ahmed, A.E., Nofal, S., Abdel-Mottaleb, Y. and Badary, O.A., 2018. Thymoquinone alleviates the experimentally induced Alzheimer's dis-ease inflammation by modulation of TLRs signaling. *Human & Experimental Toxicol-ogy*, *37*(10), pp. 1092–1104.

Ainslie, W., 1813. *Materia Indica of Hindoostan*. Government Press.

Albassam, A.A., Ahad, A., Alsultan, A. and Al-Jenoobi, F.I., 2018. Inhibition of cytochrome P450 enzymes by thymoquinone in human liver microsomes. *Saudi Pharmaceutical Journal*, 26(5), pp. 673–677.

Ardakani Movaghati, M.R., Yousefi, M., Saghebi, S.A., Sadeghi Vazin, M., Iraji, A. and Mosavat, S.H., 2019. Efficacy of black seed (Nigella sativa L.) on kidney stone dissolution: A randomized, double-blind, placebo-controlled, clinical trial. *Phytotherapy Research*, 33(5), pp. 1404–1412.

Atta, M.B., 2003. Some characteristics of nigella (Nigella sativa L.) seed cultivated in Egypt and its lipid profile. *Food Chemistry*, 83(1), pp. 63–68.

Badar, A., Kaatabi, H., Bamosa, A., Al-Elq, A., Abou-Hozaifa, B., Lebda, F., Alkhadra, A. and Al-Almaie, S., 2017. Effect of Nigella sativa supplementation over a one-year period on lipid levels, blood pressure and heart rate in type-2 diabetic patients receiving oral hypoglycemic agents: Nonrandomized clinical trial. *Annals of Saudi Medicine*, 37(1), pp. 56–63.

Bamosa, A.O., Kaatabi, H., Lebdaa, F.M., Elq, A.M. and Al-Sultanb, A., 2010. Effect of Nigella sativa seeds on the glycemic control of patients with type 2 diabetes mellitus. *Indian Journal of Physiology and Pharmacology*, 54(4), pp. 344–354.

Bayrak, O., Bavbek, N., Karatas, O.F., Bayrak, R., Catal, F., Cimentepe, E., Akbas, A., Yildirim, E., Unal, D. and Akcay, A., 2008. Nigella sativa protects against ischaemia/reperfusion injury in rat kidneys. *Nephrology Dialysis Transplantation*, 23(7), pp. 2206–2212.

Burits, M. and Bucar, F., 2000. Antioxidant activity of Nigella sativa essential oil. *Phytotherapy Research*, 14(5), pp. 323–328.

Dehkordi, F.R. and Kamkhah, A.F., 2008. Antihypertensive effect of Nigella sativa seed extract in patients with mild hypertension. *Fundamental & Clinical Pharmacology*, 22(4), pp. 447–452.

Dutt, U.C., 1877. *The Materia Medica of the Hindus*. Thacker, Spink, & Co.

Fallah Huseini, H., Amini, M., Mohtashami, R., Ghamarchehre, M.E., Sadeqhi, Z., Kianbakht, S. and Fallah Huseini, A., 2013. Blood pressure lowering effect of Nigella sativa L. seed oil in healthy volunteers: A randomized, double-blind, placebo-controlled clinical trial. *Phytotherapy Research*, 27(12), pp. 1849–1853.

Fusch, L., 1555. *De Historia Stirpium Commetarii Insignes*. Lugduni Apud Ioan Tornaesium.

Gheita, T.A. and Kenawy, S.A., 2012. Effectiveness of Nigella sativa oil in the management of rheumatoid arthritis patients: A placebo controlled study. *Phytotherapy Research*, 26(8), pp. 1246–1248.

Keshri, G., Singh, M.M., Lakshmi, V. and Kamboj, V.P., 1995. Post-coital contraceptive efficacy of the seeds of Nigella sativa in rats. *Indian Journal of Physiology and Pharmacology*, 39, pp. 59–59.

Khattab, M.M. and Nagi, M.N., 2007. Thymoquinone supplementation attenuates hypertension and renal damage in nitric oxide deficient hypertensive rats. *Phytotherapy Research: An International Journal Devoted to Pharmacological and Toxicological Evaluation of Natural Product Derivatives*, 21(5), pp. 410–414.

Lémery, N., 1716. *Traité universel des drogues simples, mises en ordre alphabétique. Où l'on trouve leurs différens noms . . . et tout ce qu'il y a de particulier dans les animaux, dans les végétaux, et dans les minéraux*. Au dépend de la Companie.

Ragheb, A., Elbarbry, F., Prasad, K., Mohamed, A., Ahmed, M.S. and Shoker, A., 2008. Attenuation of the development of hypercholesterolemic atherosclerosis by thymoquinone. *International Journal of Angiology*, 17(04), pp. 186–192.

Sandhu, K.S. and Rana, A.C., 2013. Evaluation of anti parkinson's activity of Nigella sativa (kalonji) seeds in chlorpromazine induced experimental animal model. *Mortality*, 22(5), p. 23.

Seif, A.A., 2014. Nigella Sativa reverses osteoporosis in ovariectomized rats. *BMC Complementary and Alternative Medicine*, 14, pp. 1–8.

Yang, Y., Bai, T., Yao, Y.L., Zhang, D.Q., Wu, Y.L., Lian, L.H. and Nan, J.X., 2016. Upregulation of SIRT1-AMPK by thymoquinone in hepatic stellate cells ameliorates liver injury. *Toxicology Letters*, 262, pp. 80–91.

82 Sweet Basil (*Ocimum basilicum* L.)

Ocimum basilicum L.

Etymology: From the Greek *ozo* = I smell and *basilikon* = royal

Family: Lamiaceae

Synonym: *Ocimum thyrsiflorum* L.

Common names: Sweet basil; basilic (Fr.); basilikum (Ger.); manjericão (Port.); albahaca (Spa.)

Part used: Leaf

Constituents: Essential oil (methyl chavicol, linalool, methyl eugenol) (Sajjadi, 2006; Telci et al., 2006; Chalchat & Özcan, 2008); rosmarinic acid (Jayasinghe et al., 2003).

Medical history: Dioscorides recommends sweet basil for flatulence, stomachache, inflammation, and as diuretic. Simeon Seth in 12th-century Byzantium attests that the smell of basil is good for the head and heart and "changes sorrow into gaity". In the 16th-century Germany, Fusch (1555) defines sweet basil as hot to the second degree and humid to the fourth degree, while Matthioli (1572) in Italy uses the juice of sweet basil to wash the eyes.

Medicinal uses: Infusion made of 2–4 g of leaves in 150 mL of water is taken for flatulence in Europe (Czygan, 2004). Carminative (Turkey; Pakistan, Iran; India; Bangladesh, the Philippines); diuretic (Turkey); fever (Pakistan; Thailand; Papua New Guinea); jaundice, headaches (Indonesia); urinary problems (the Philippines); weakness, headache (Papua New Guinea)

Blood pressure: Aqueous extract given at 200 mg/kg/day to rats with hypertension (induced by left renal artery occlusion) reduced systolic blood pressure from 201 to 177 mmHg and the diastolic blood pressure from 100 to 85 mmHg with a decrease in plasma levels of endothelin (Umar et al., 2010).

Heart: In isolated hearts of frogs, an extract evoked positive ionotropic and negative chronotropic actions mediated through ß-adrenergic receptors (Muralidharan & Dhananjayan, 2004). In rats poisoned with fructose-enriched diet, rosmarinic acid at 10 mg/kg reduced myocardial damage and blood pressure via the inhibition of NADPH oxidase (Karthik et al., 2011). Rosmarinic acid given orally at 50 mg/kg/day to type-2 diabetic rats for 28 days lowered plasma LDL-cholesterol and decreased oxidative stress in cardiac tissue (Zych et al., 2019). In hypertensive rats, rosmarinic acid reduced systolic blood pressure and inhibited ACE activity (Ferreira et al., 2018).

DOI: 10.1201/9781003301455-82

Plasma lipids and cholesterol: In rats fed a high-cholesterol diet, extract of basil reduced plasma triglycerides and cholesterol (Amrani et al., 2006; Gökçe et al., 2021). Rosmarinic acid given orally at 50 mg/kg/day for 28 days lowered plasma LDL-cholesterol in type 2 diabetic rats (Zych et al., 2019).

Kidneys: Rosmarinic acid (3 mg/kg) increased the urinary excretion of sodium ions (Moser et al., 2020).

Bones and cartilages: Aqueous extract given at 400 mg/kg/day for 8 weeks protected rats against dexamethasone-induced decreases in femoral bone mineral density (Hozayen et al., 2016). Methanol extract given orally at the dose of 400 mg/kg/day for 10 days attenuated arthritis induced by CFA-induced arthritis (Phadtare et al., 2013).

Skin and hair: Ethanol extract applied externally at night on the cheeks of volunteers daily for 12 weeks increased skin moisture and decreased wrinkles (Rasul & Akhtar, 2011).

Brain: Aqueous extract given orally for 14 days improved the cognition of mice poisoned with glutamate (Ayuob et al., 2018).

Warning: Giving 500 mg/kg of an hydroalcoholic extract in rodents depleted platelets, hematocrit, and red blood cells (Rasekh et al., 2012). Because of methyl chavicol, basil must not be given during pregnancy and lactation or taken for more than a week (Czygan, 2004).

Comments: (i) Aging mice given rosmarinic acid orally at 200 mg/kg/day for 30 days experienced normalization in the expression of hepatic and kidney SOD, catalase, and glutathione peroxidase (GPx) (Zhang et al., 2015). Superoxide dismutase, catalase, and glutathione peroxidase are enzymes that protect cellular mitochondria, lipids, proteins, and DNA against the superoxide anion, hydroxyl radicals, and hydrogen peroxide. During the process of aging, the expression of these enzymes decreases.

(ii) Rosmarinic acid given to mice protected mitochondria in hepatocytes (Zhang et al., 2015).

REFERENCES

Amrani, S., Harnafi, H., El Houda Bouanani, N., Aziz, M., Serghini Caid, H., Manfredini, S., Besco, E., Napolitano, M. and Bravo, E., 2006. Hypolipidaemic activity of aqueous ocimum basilicum extract in acute hyperlipidaemia induced by triton WR-1339 in rats and its antioxidant property. *Phytotherapy Research: An International Journal Devoted to Pharmacological and Toxicological Evaluation of Natural Product Derivatives*, 20(12), pp. 1040–1045.

Ayuob, N.N., El Wahab, M.G.A., Ali, S.S. and Abdel-Tawab, H.S., 2018. Ocimum basilicum improve chronic stress-induced neurodegenerative changes in mice hippocampus. *Metabolic Brain Disease*, *33*, pp. 795–804.

Chalchat, J.C. and Özcan, M.M., 2008. Comparative essential oil composition of flowers, leavesand stems of basil (Ocimum basilicum L.) used as herb. *Food Chemistry*, *110*(2), pp. 501–503.

Czygan, F.C., 2004. *Herbal Drugs and Phytopharmaceuticals: A Handbook for Practice on a Scientific Basis*. CRC Press.

Ferreira, L.G., Evora, P.R.B., Capellini, V.K., Albuquerque, A.A., Carvalho, M.T.M., da Silva Gomes, R.A., Parolini, M.T. and Celotto, A.C., 2018. Effect of rosmarinic acid on the arterial blood pressure in normotensive and hypertensive rats: Role of ACE. *Phytomedicine*, *38*, pp. 158–165.

Fusch, L., 1555. *De Historia Stirpium Commetarii Insignes*. Lugduni Apud Ioan Tornaesium.

Gökçe, Y., Kanmaz, H., Er, B., Sahin, K. and Hayaloglu, A.A., 2021. Influence of purple basil (Ocimum basilicum L.) extract and essential oil on hyperlipidemia and oxidative stress in rats fed high-cholesterol diet. *Food Bioscience*, *43*, p. 101228.

Hozayen, W.G., El-Desouky, M.A., Soliman, H.A., Ahmed, R.R. and Khaliefa, A.K., 2016. Antiosteoporotic effect of Petroselinum crispum, Ocimum basilicum and Cichorium intybus L. in glucocorticoid-induced osteoporosis in rats. *BMC Complementary and Alternative Medicine*, *16*, pp. 1–11.

Jayasinghe, C., Gotoh, N., Aoki, T. and Wada, S., 2003. Phenolics composition and antioxidant activity of sweet basil (Ocimum basilicum L.). *Journal of Agricultural and Food Chemistry*, *51*(15), pp. 4442–4449.

Karthik, D., Viswanathan, P. and Anuradha, C.V., 2011. Administration of rosmarinic acid reduces cardiopathology and blood pressure through inhibition of p22phox NADPH oxidase in fructose-fed hypertensive rats. *Journal of Cardiovascular Pharmacology*, *58*(5), pp. 514–521.

Matthioli, P.A., 1572. *Commentaires sur les Six Livres de Pedacius Dioscorides Anazarbeen de la matière medicinale*. A l'Escue de Milan.

Moser, J.C., Cechinel-Zanchett, C.C., Mariano, L.N.B., Boeing, T., da Silva, L.M. and de Souza, P., 2020. Diuretic, natriuretic and Ca2+-sparing effects induced by rosmarinic and caffeic acids in rats. *Revista Brasileira de Farmacognosia*, *30*(4), pp. 588–592.

Muralidharan, A. and Dhananjayan, R., 2004. Cardiac stimulant activity of Ocimum basilicum Linn. extracts. *Indian Journal of Pharmacology*, *36*(3), p. 163.

Phadtare, S., Pandit, R., Shinde, V. and Mahadik, K., 2013. Comparative phytochemical and pharmacological evaluations of two varieties of Ocimum basilicum for antiarthritic activity. *Journal of Pharmacognosy and Phytochemistry*, *2*(2), pp. 158–167.

Rasekh, H.R., Hosseinzadeh, L., Mehri, S., Kamli-Nejad, M., Aslani, M. and Tanbakoosazan, F., 2012. Safety assessment of Ocimum basilicum hydroalcoholic extract in wistar rats: Acute and subchronic toxicity studies. *Iranian Journal of Basic Medical Sciences*, *15*(1), p. 645.

Rasul, A. and Akhtar, N., 2011. Formulation and in vivo evaluation for anti-aging effects of an emulsion containing basil extract using non-invasive biophysical techniques. *DARU: Journal of Faculty of Pharmacy, Tehran University of Medical Sciences*, *19*(5), p. 344.

Sajjadi, S.E., 2006. Analysis of the essential oils of two cultivated basil (Ocimum basilicum L.) from Iran. *DARU Journal of Pharmaceutical Sciences*, *14*(3), pp. 128–130.

Telci, I., Bayram, E., Yılmaz, G. and Avcı, B., 2006. Variability in essential oil composition of Turkish basils (Ocimum basilicum L.). *Biochemical Systematics and Ecology*, *34*(6), pp. 489–497.

Umar, A., Imam, G., Yimin, W., Kerim, P., Tohti, I., Berké, B. and Moore, N., 2010. Antihypertensive effects of Ocimum basilicum L. (OBL) on blood pressure in renovascular hypertensive rats. *Hypertension Research*, *33*(7), pp. 727–730.

Zhang, Y., Chen, X., Yang, L., Zu, Y. and Lu, Q., 2015. Effects of rosmarinic acid on liver and kidney antioxidant enzymes, lipid peroxidation and tissue ultrastructure in aging mice. *Food & Function*, *6*(3), pp. 927–931.

Zych, M., Wojnar, W., Borymski, S., Szałabska, K., Bramora, P. and Kaczmarczyk-Sedlak, I., 2019. Effect of rosmarinic acid and sinapic acid on oxidative stress parameters in the cardiac tissue and serum of type 2 diabetic female rats. *Antioxidants*, *8*(12), p. 579.

83 Olive (*Olea europaea* L.)

Olea europaea **L.**

Etymology: From the Greek *elaia* = olive tree and the Latin *europea* = from Europe

Family: Oleaceae

Common names: Olive; olijf (Ger.); azeitonas (Port.); оливка (Rus.); aceituna (Spa.)

Part used: Fruit

Constituents: Fixed oil (unsaturated fatty acids: oleic acid, linoleic acid; phytosterols) (Ozkan et al., 2017), phenolics (Ocakoglu et al., 2009), seco-iridoid glycosides (oleuropein) (Omar, 2010).

Medical history: Olives have been used since the dawn of time in the Middle East ("A land of honey and olive trees"; Deuteronomy 8,8). Dioscorides recommends olive oil for teeth and gum problems. In the 23rd chapter of the Coran, verse 20 is written "And a tree issuing from Mount Sinai which bears oil and seasoning for all to eat". Olive oil was used for dysentery and colic in 18th-century France (Lémery, 1716). In 19th-century Europe, it was used as a mild laxative (Guibourt, 1836).

Medicinal uses: Wounds, sprain (Turkey)

Blood pressure: Elderly hypertensive patients taking 60 g of oil daily for 4 weeks experienced a normalization of systolic and diastolic blood pressure (Perona et al., 2004). Intake of 25 mL/day of olive oil by healthy volunteers for 9 weeks caused a slight decrease in blood pressure (Bondia-Pons et al., 2007).

Plasma lipids and glucose: Elderly patients taking 60 g of oil daily for 4 weeks had decreases in total and LDL-cholesterol (Perona et al., 2004). Intake of 25 mL/day by healthy volunteers for 9 weeks caused a slight decrease in plasma cholesterol of about 0.1 mmol/L (Bondia-Pons et al., 2007). In healthy young volunteers, consumption of 50 mL of olive oil decreased glycemia and increased HDL-cholesterol (Oliveras-López et al., 2012).

Heart: Consumption of 50 g/day of olive oil daily for about 4 years resulted in decreased risk of atrial fibrillation (Martínez-González et al., 2014). Intake of about 20 to 30 g/day of olive oil decreases the risks of stroke (Donat-Vargas et al., 2022).

DOI: 10.1201/9781003301455-83

Bones and cartilages: Phenolic extract of oil given to mice for 12 days at 200 mg/kg/day decreased joint edema, inflammatory cells migration, cartilage degradation, and bone erosion (Rosillo et al., 2014).

Immune system: Volunteers adding olive oil in their diet for 8 weeks showed decreased expression of intercellular adhesion molecule 1 (ICAM 1) (Yaqoob et al., 1998)

Comments: (i) The formation of arterial atheroma starts with the adhesion of circulating monocytes to ICAM 1, expressed by activated arterial endothelium insults due to the accumulation of foam cells in arterial walls (Sanmarco et al., 2018).

(ii) Extra virgin olive oil is beneficial at a normal dietary dose for the elderly.

REFERENCES

Bondia-Pons, I., Schröder, H., Covas, M.I., Castellote, A.I., Kaikkonen, J., Poulsen, H.E., Gaddi, A.V., Machowetz, A., Kiesewetter, H. and López-Sabater, M.C., 2007. Moderate consumption of olive oil by healthy European men reduces systolic blood pressure in non-Mediterranean participants. *The Journal of Nutrition*, *137*(1), pp. 84–87.

Donat-Vargas, C., Sandoval-Insausti, H., Peñalvo, J.L., Iribas, M.C.M., Amiano, P., Bes-Rastrollo, M., Molina-Montes, E., Moreno-Franco, B., Agudo, A., Mayo, C.L. and Laclaustra, M., 2022. Olive oil consumption is associated with a lower risk of cardiovascular disease and stroke. *Clinical Nutrition*, *41*(1), pp. 122–130.

Guibourt, N.J.B.G., 1836. *Histoire abrégée des drogues simples*. Méquignon-Marvis Père et fils.

Lémery, N., 1716. *Traité universel des drogues simples, mises en ordre alphabétique. Où l'on trouve leurs différens noms . . . et tout ce qu'il y a de particulier dans les animaux, dans les végétaux, et dans les minéraux*. Au dépend de la Companie.

Martínez-González, M.Á., Toledo, E., Arós, F., Fiol, M., Corella, D., Salas-Salvadó, J., Ros, E., Covas, M.I., Fernández-Crehuet, J., Lapetra, J. and Muñoz, M.A., 2014. Extravirgin olive oil consumption reduces risk of atrial fibrillation: The PREDIMED (Prevención con Dieta Mediterránea) trial. *Circulation*, *130*(1), pp. 18–26.

Ocakoglu, D., Tokatli, F., Ozen, B. and Korel, F., 2009. Distribution of simple phenols, phenolic acids and flavonoids in Turkish monovarietal extra virgin olive oils for two harvest years. *Food Chemistry*, *113*(2), pp. 401–410.

Oliveras-López, M.J., Innocenti, M., Martín Bermudo, F., López-García de la Serrana, H. and Mulinacci, N., 2012. Effect of extra virgin olive oil on glycaemia in healthy young subjects. *European Journal of Lipid Science and Technology*, *114*(9), pp. 999–1006.

Omar, S.H., 2010. Oleuropein in olive and its pharmacological effects. *Scientia Pharmaceutica*, *78*(2), pp. 133–154.

Ozkan, A., Aboul-Enein, H.Y., Kulak, M. and Bindak, R., 2017. Comparative study on fatty acid composition of olive (Olea europaea L.), with emphasis on phytosterol contents. *Biomedical Chromatography*, *31*(8), p. e3933.

Perona, J.S., Cañizares, J., Montero, E., Sánchez-Domínguez, J.M., Catalá, A. and Ruiz-Gutiérrez, V., 2004. Virgin olive oil reduces blood pressure in hypertensive elderly subjects. *Clinical Nutrition*, *23*(5), pp. 1113–1121.

Rosillo, M.Á., Alcaraz, M.J., Sánchez-Hidalgo, M., Fernández-Bolaños, J.G., Alarcón-de-la-Lastra, C. and Ferrándiz, M.L., 2014. Anti-inflammatory and joint protective effects of extra-virgin olive-oil polyphenol extract in experimental arthritis. *The Journal of Nutritional Biochemistry*, *25*(12), pp. 1275–1281.

Sanmarco, L.M., Eberhardt, N., Ponce, N.E., Cano, R.C., Bonacci, G. and Aoki, M.P., 2018. New insights into the immunobiology of mononuclear phagocytic cells and their relevance to the pathogenesis of cardiovascular diseases. *Frontiers in Immunology*, 8, p. 1921.

Yaqoob, P., Knapper, J.A., Webb, D.H., Williams, C.M., Newsholme, E.A. and Calder, P.C., 1998. Effect of olive oil on immune function in middle-aged men. *The American Journal of Clinical Nutrition*, 67(1), pp. 129–135.

84 Wild Marjoram (*Origanum vulgare* L.)

Origanum vulgare L.

Etymology: From the Greek *oros* = mountain and *ganos* = pride and the Latin *vulgare* = common

Synonyms: *Origanum normale* D. Don; *Origanum creticum* Lour.

Common names: Wild marjoram, oregano; origan (Fr.); orégano (Port.; Spa.); орегано (Rus.)

Part used: Leaf

Constituents: Essential oil (carvacrol, thymol) (Vokou et al., 1993), hydroxy-cinnamic acid derivative (rosmarinic acid) (Silva et al., 2022).

Medical history: For ancient Roman physicians, including Galen, oregano was hot and dry to the third degree and used for bronchitis. In 16th-century Italy, it was used for jaundice (Matthioli, 1572). Oregano was used for fatigue, asthma, and indigestion in 18th-century Scotland (Alston, 1770). In 19th-century, in Europe and North America, it was used for cough and asthma (Pereira, 1843).

Medicinal uses: Stomachache (Turkey, the Philippines, India); headache (Turkey, India); asthma (Turkey, Pakistan, India); hypertension, epilepsy, toothache, cold (Turkey); fever, inflammation, bronchitis, rheumatism (India); tonic (Pakistan, India); carminative, antispasmodic, urinary problems (Pakistan); asthma, cough, cold (the Philippines)

Blood pressure: Ethanol extract given at 20 mg/kg/day for 14 days to rats poisoned with ethylene glycol decreased systolic blood pressure from 181 to 134 mmHg and diastolic blood pressure from 123 to 91.4 mmHg and improved aortic cytoarchitecture (ElSawy & Mosa, 2021).

Heart: In rats, injecting carvacrol at 100 µg/kg intraperitoneally blocked cardiac L-type calcium ions channel, thereby compelling bradycardia, and decreasing blood pressure (Aydin et al., 2007).

Plasma lipids and glucose: Aqueous extract given orally to diabetic rats at 20 mg/kg decreased glycemia (Lemhadri et al., 2004). Feeding rats with 100 mg/kg/day for 30 days of an alcohol extract of oregano decreased LDL-cholesterol from 58.7 to 29.1 mg/mL, increased HDL-cholesterol from 19.4 to 56.1 mg/mL, and decreased plasma triglycerides from 76 to 42 mg/mL (Foroozandeh et al., 2016).

Kidneys: Ethanol extract given at 20 mg/kg/day for 14 days to rats poisoned with ethylene glycol prevented the renal deposition of calcium oxalate stones and decreased plasma creatinine and urea (ElSawy & Mosa, 2021).

Brain: Aqueous extract given orally at 350 mg/kg/day for 14 days improved the cognitive function of rats (Ghaderi et al., 2020).

DOI: 10.1201/9781003301455-84

Comments: (i) Essential oil given to rats at 200 mg/kg/day for 90 days did not evoke side effects (Llana-Ruiz-Cabello et al., 2017).

(ii) Increased plasma volume due to elevated glycemia owed to osmosis and increased vascular resistance account for hypertension in diabetes mellitus (Ohishi, 2018).

(iii) Hyperinsulinemia in obese patients stimulates sodium reabsorption at the renal level (Modan et al., 1985; Nosadini et al., 1993).

REFERENCES

Alston, C., 1770. *Lectures on the Materia Medica: Containing the Natural History of Drugs, their Virtues and Doses: Also Directions for the Study of the Materia Medica; and an Appendix on the Method of Prescribing.* Edward and Charles Dilly.

Aydin, Y., Kutlay, Ö., Ari, S., Duman, S., Uzuner, K. and Aydin, S., 2007. Hypotensive effects of carvacrol on the blood pressure of normotensive rats. *Planta Medica*, *73*(13), pp. 1365–1371.

ElSawy, N.A. and Mosa, O.F., 2021. The antiurolithic activity of Origanum vulgare on rats treated with ethylene glycol and ammonium chloride: Possible pharmaco-biochemical and ultrastructure effects. *Current Urology*, *15*(2), p. 119.

Foroozandeh, M., Bigdeli, M. and Rahnema, M., 2016. The effect of hydro alcoholic extract of Origanum vulgare on weight and serum lipid profile in male Wistar rats. *Journal of Jahrom University of Medical Sciences*, *14*, pp. 51–57.

Ghaderi, A., Karimi, S.A., Talaei, F., Shahidi, S., Faraji, N. and Komaki, A., 2020. The effects of aqueous extract of Origanum vulgare on learning and memory in male rats. *Journal of Herbmed Pharmacology*, *9*(3), pp. 239–244.

Lemhadri, A., Zeggwagh, N.A., Maghrani, M., Jouad, H. and Eddouks, M., 2004. Anti-hyper-glycaemic activity of the aqueous extract of Origanum vulgare growing wild in Tafilalet region. *Journal of Ethnopharmacology*, *92*(2–3), pp. 251–256.

Llana-Ruiz-Cabello, M., Maisanaba, S., Puerto, M., Pichardo, S., Jos, A., Moyano, R. and Camean, A.M., 2017. A subchronic 90-day oral toxicity study of Origanum vulgare essential oil in rats. *Food and Chemical Toxicology*, *101*, pp. 36–47.

Matthioli, P.A., 1572. *Commentaires sur les Six Livres de Pedacius Dioscorides Anazarbeen de la matière medicinale.* A l'Escue de Milan.

Modan, M., Halkin, H., Almog, S., Lusky, A., Eshkol, A., Shefi, M., Shitrit, A. and Fuchs, Z., 1985. Hyperinsulinemia. A link between hypertension obesity and glucose intoler-ance. *The Journal of Clinical Investigation*, *75*(3), pp. 809–817.

Nosadini, R., Sambataro, M., Thomaseth, K., Pacini, G., Cipollina, M.R., Brocco, E., Solini, A., Carraro, A., Velussi, M., Frigato, F. and Crepaldi, G., 1993. Role of hyperglycemia and insulin resistance in determining sodium retention in non-insulin-dependent diabe-tes. *Kidney International*, *44*(1), pp. 139–146.

Ohishi, M., 2018. Hypertension with diabetes mellitus: Physiology and pathology. *Hyperten-sion Research*, *41*(6), pp. 389–393.

Pereira, J., 1843. *The Elements of Materia Medica and Therapeutics.* Lea and Blanchard.

Silva, M.L.E., Lucarini, R., Dos Santos, F.F., Martins, C.H., Pauletti, P.M., Januario, A.H., Santos, M.F.C. and Cunha, W.R., 2022. Hypoglycemic effect of rosmarinic acid-rich infusion (RosCE) from Origanum vulgare in alloxan-induced diabetic rats. *Natural Product Research*, *36*(17), pp. 4519–4525.

Vokou, D., Kokkini, S. and Bessiere, J.M., 1993. Geographic variation of Greek oregano (Orig-anum vulgare ssp. hirtum) essential oils. *Biochemical Systematics and Ecology*, *21*(2), pp. 287–295.

85 Java Tea (*Orthosiphon aristatus* (Blume) Miq.)

Orthosiphon aristatus (**Blume**) **Miq.**

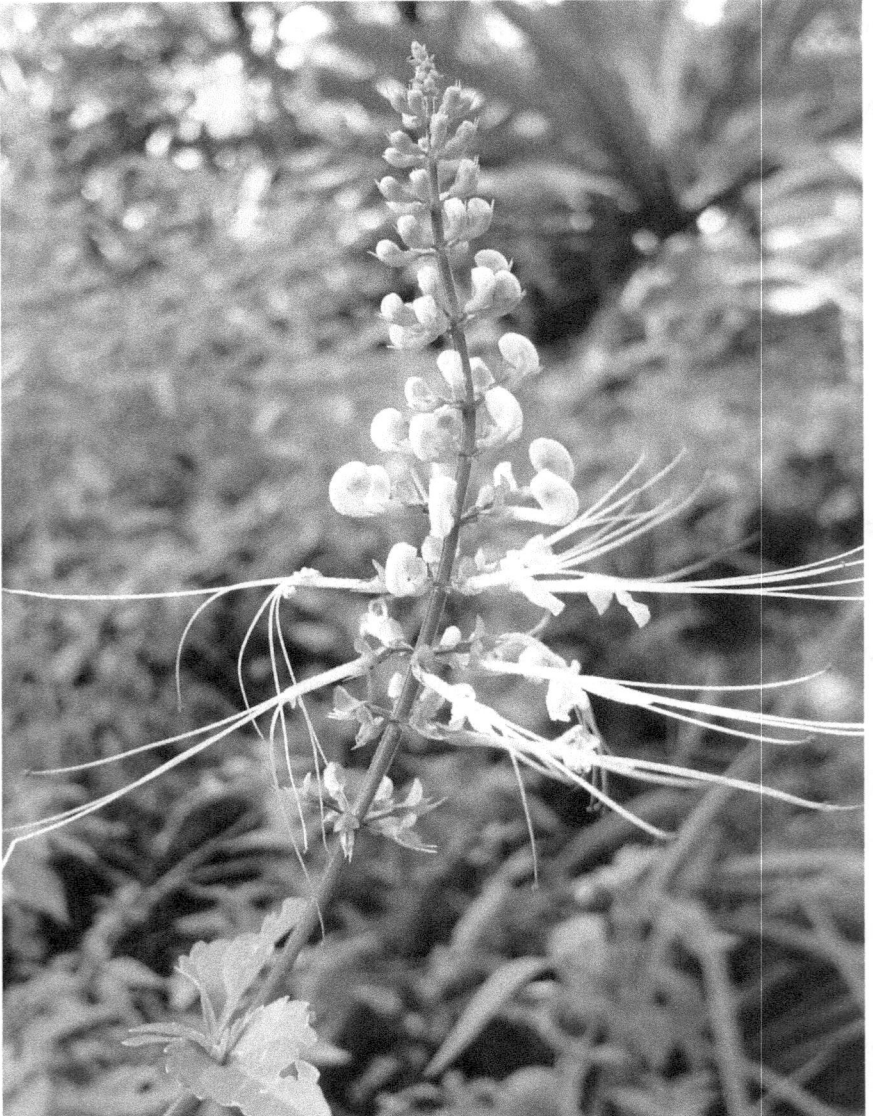

DOI: 10.1201/9781003301455-85

Etymology: From the Greek *orthos* = straight and *syphon* = tube and the Latin *aristatus* = aristate

Family: Lamiaceae

Synonyms: *Clerodendrum spicatum* Thunb.; *Clerodendranthus stamineus* (Bentham) Kudo; *Ocimum aristatum* Bl.; *Orthosiphon stamineus* Benth.

Common names: Cat's whiskers, Java tea; orthosiphon aristé (Fr.); orthosiphonblätter (Ger.); ortossifão (Port.); ortosifón (Spa.)

Part used: Leaf

Constituents: Chromenes (methylripariochromene A, orthochromene A, acetovanillochromene) (Matsubara et al., 1999), flavonols (sinensetin = 5,6,7,3′,4′-pentamethoxyflavone) (Liu et al., 2022).

Medical history: In Indonesia, Java tea has been used as a diuretic and for kidney stones since the dawn of time. The Dutch brought awareness of the diuretic virtues of this plant to Europeans. Karl Ludwig von Blume describes Java tea in his *Bijdragen tot de flora van Nederlandsch Indië*, as does van der Burg in his *De Geneesheer in Nederlandsch-Indië* (1885) (Pols, 2009). Java tea was used as a diuretic and for gout, pyelonephritis, kidney stones, and chronic cystitis in 19th-century Western world medical practice.

Medicinal uses: Diuretic (2–3 g in 150 mL of boiling water) (European Union); kidney disorders, bladder diseases, urinary problems (Myanmar); gallstone, jaundice (Thailand); hypertension (Malaysia); diuretic (Indonesia); diabetes, hypertension, kidney diseases (the Philippines)

Blood pressure: Methanol extract given orally to rats at 250 mg/kg once a day for 2 weeks decreased systolic blood pressure from 149 to 108 mmHg in a manner similar to Irbesartan at 20 mg/kg/day (Azizan et al., 2012).

Methylripariochromene A given subcutaneously to stroke-prone spontaneously hypertensive rats at a single dose of 100 mg decreased systolic blood pressure by about 25 mmHg after 24 hours via a mechanism involving, at least in part, calcium ions channel blockade by orthochromene A_and acetovanillochromene (Matsubara et al., 1999; Ohashi et al., 2000).

Heart: Methylripariochromene A given subcutaneously at a single dose of 100 mg/kg to stroke-prone spontaneously hypertensive rats decreased heart rate by about 80 bpm after 3 hours (Matsubara et al., 1999).

Plasma glucose and lipids: Aqueous extract given orally at 1 g/kg/day for 14 days to streptozotocin-induced diabetic rats decreased glycemia from 392.7 to 341.2 mg/dL (Glibenclamide at 5 mg/kg: 383.6 to 326.8 mg/dL) and plasma triglycerides down to 64 mg/dL (diabetic untreated: 110.8 mg/dL). The extract at the concentration of 100 μg/mL increased the secretion of insulin induced by glucose in isolated rat pancreas (Sriplang et al., 2007).

Kidneys: Methylripariochromene A given subcutaneously at a single dose of 100 mg/kg to stroke-prone spontaneously hypertensive rats increased urine secretion from 9.6 to 28.7 mL/kg/3h (Matsubara et al., 1999).

Bones and cartilages: Sinensetin given to rats at 20 mg/kg/day delayed the progression of osteoarthritis with protection of cartilage. *In vitro*, sinensetin

prevented the secretion of interleukin-6 and TNF α and decreased the production of matrix metalloproteinase-9 and -13 by chondrocytes exposed to
interleukin-1β (Liu et al., 2022).
Brain: Ethanol extract given orally to rats at 200 mg/kg/day for 10 days attenuated dementia induced by the intracerebroventricular injection of streptozotocin (Retinasamy et al., 2020).

Warnings: The oral LD_{50} in rat for aqueous extract is above 5g/kg
(Pariyani et al., 2015). Intake of an extract at 900 mg (mixed with orlistat
and green tea) for more than 2 months caused hypoglycemia (plasma glucose < 3.89 mmol/L) and hyperinsulinemia (Kim & Cho, 1999). This exem-
plifies that when herbal products are not manufactured and dispensed under
professional pharmaceutical control, populations are at risk. It is surpris-
ing to see on the internet a myriad of herbal preparations available with the
potential to poison people. For instance, flavonoid concentrate capsules are
everywhere, but the manufacturers seem to have forgotten that pure concentrated flavonoids are pro-oxidants and therefore carcinogenic (Procházková
et al., 2011). Why is the discipline of medicinal plants not taught to standards
that would protect populations against poisoning and adulteration in 2023?
Comment: In the synovial fluid, activated fibroblast-like synoviocytes secrete
interleukin-1β and TNF α, which in turn activate chondrocytes that yield
matrix metalloproteinase-1 and -13. These metalloproteinases in turn catalyze the degradation of collagen in cartilage (Abd Razak et al., 2021). In
rheumatic arthritis, crystals of uric acid activate fibroblast-like synoviocytes (Chen et al., 2011).

REFERENCES

Abd Razak, H.R.B., Chew, D., Kazezian, Z. and Bull, A.M., 2021. Autologous protein solution: A promising solution for osteoarthritis? *EFORT Open Reviews*, 6(9), p. 716.
Azizan, N.A., Ahmad Mohamed, K., Ahmad, M.Z. and Asmawi, Z., 2012. The in vivo antihypertensive effects of standardized methanol extracts of Orthosiphon stamineus on
spontaneous hypertensive rats: A preliminary study. *African Journal of Pharmacy and
Pharmacology*, 6(6), pp. 376–379.
Chen, D.P., Wong, C.K., Tam, L.S., Li, E.K. and Lam, C.W., 2011. Activation of human fibroblast-like synoviocytes by uric acid crystals in rheumatoid arthritis. *Cellular & Molecu-
lar Immunology*, 8(6), pp. 469–478.
Kim, J.H. and Cho, H.C., 1999. A case of hypoglycemia caused by orthosiphon in a non-
diabetic man. *The Journal of Korean Diabetes*, 11(1), pp. 86–89.
Liu, Z., Liu, R., Wang, R., Dai, J., Chen, H., Wang, J. and Li, X., 2022. Sinensetin attenuates IL-1β-induced cartilage damage and ameliorates osteoarthritis by regulating SER-
PINA3. *Food & Function*, 13(19), pp. 9973–9987.
Matsubara, T., Bohgaki, T., Watarai, M., Suzuki, H., Ohashi, K. and Shibuya, H., 1999. Antihypertensive actions of methylripariochromene A from Orthosiphon aristatus, an Indonesian traditional medicinal plant. *Biological and Pharmaceutical Bulletin*, 22(10),
pp. 1083–1088.
Ohashi, K., Bohgaki, T. and Shibuya, H., 2000. Antihypertensive substance in the leaves of
kumis kucing (Orthosiphon aristatus) in Java Island. *Yakugaku zasshi: Journal of the
Pharmaceutical Society of Japan*, 120(5), pp. 474–482.

Pariyani, R., Safinar Ismail, I., Azam, A.A., Abas, F., Shaari, K. and Sulaiman, M.R., 2015. Phytochemical screening and acute oral toxicity study of Java tea leaf extracts. *BioMed Research International*, 2015.

Pols, H., 2009. European physicians and botanists, indigenous herbal medicine in the Dutch East Indies, and colonial networks of mediation. *East Asian Science, Technology and Society: An International Journal*, *3*(2–3), pp. 173–208.

Procházková, D., Boušová, I. and Wilhelmová, N., 2011. Antioxidant and prooxidant properties of flavonoids. *Fitoterapia*, *82*(4), pp. 513–523.

Retinasamy, T., Shaikh, M.F., Kumari, Y., Zainal Abidin, S.A. and Othman, I., 2020. Orthosiphon stamineus standardized extract reverses streptozotocin-induced Alzheimer's disease-like condition in a rat model. *Biomedicines*, *8*(5), p. 104.

Sriplang, K., Adisakwattana, S., Rungsipipat, A. and Yibchok-Anun, S., 2007. Effects of Orthosiphon stamineus aqueous extract on plasma glucose concentration and lipid profile in normal and streptozotocin-induced diabetic rats. *Journal of Ethnopharmacology*, *109*(3), pp. 510–514.

86 Parsley (*Petroselinum crispum* (Mill.) Fuss.)

***Petroselinum crispum* (Mill.) Fuss.**

Etymology: From the Greek *petro* = rock and *selino* = parsley and the Latin *crispum* = curly

Synonyms: *Apium crispum* Mill.; *Apium petroselinum* L.

Common names: Parsley; persil (Fr.); petersilie (Ger.); salsinha (Port.); петрушка (Rus.); perejil (Spa.)

Parts used: Leaf, seed

Constituents: Essential oil (apiole, myristicin), flavonol glycosides, furanocoumarins (oxypeucedanin, bergapten) (Gbolade and Lockwood, 1999; Marthe, 2020).

Medical history: Parsley was known to Dioscorides as a diuretic and emmenagogue and as beneficial for inflammation of the digestive system. In 16th-century Germany, Fusch (1555) defines parsley as hot and dry to the third degree, while in Italy, Matthioli (1572) uses it for inflamed eyes, as a diuretic, for swollen breasts, and as an emetic antidote.

Medicinal uses: Urinary tract complain, 2–4 g/day (European Union); kidney stones (Morocco); abdominal ache, shortness of breath (Turkey)

Blood pressure: A decoction of seeds (20 g in 100 mL of boiled water, for 5 minutes) given to rats evoked a decrease in blood pressure from about 100 to 70 mmHg (Campos et al., 2009). Aqueous extract of aerial part given orally at 160 mg/kg protected rats against L-NAME-induced hypertension via blockage of voltage operated and receptor operated calcium ions channels (Ajebli & Eddouks, M., 2019).

Kidneys: A decoction of seeds (20 g in 100 mL of boiled water, for 5 minutes) given to rats increased the excretion of urine flow from 3.9 to 5.1 µL/min as well as increasing the concentration of sodium and potassium ions in the urine (Campos et al., 2009). Parsley given at 7 g/kg/day for 7 days to rats protected liver and kidneys against uricemia (Rahmat et al., 2019).

Plasma lipids and glucose: A decoction of leaves (100 g in 1 L of water and boiled for 30 min) given orally at 2 g/kg per day for 28 days to streptozotocin-induced rats decreased glycemia from 172.5 to 102.1 mg/dL (normoglycemic control: 71.5 mg/dL), prevented weight loss, and at the aortic level, decreased lipid peroxidation and increased glutathione (Sener et al., 2003).

Geese given parsley at 240 g/day for 3 months experienced decreases in total cholesterol and triglycerides and in LDL-cholesterol from 83.1 to 32.7 mg/dL, and increased HDL cholesterol (Al-Daraji et al., 2012).

 DOI: 10.1201/9781003301455-86

Uric acid: Aqueous extract given orally to hyperuricemic mice at 7 g/kg/day decreased uricemia (Soliman et al., 2020)

Bones and cartilages: Aqueous extract of leaves given orally at 400 mg/kg/day for 8 weeks to rats prevented loss of bone calcification induced by dexamethasone (Hozayen et al., 2016).

Brain: Aqueous extract given to rats orally at 2 g/kg/day for 14 days prevented scopolamine-induced dementia.

Warnings: Leaf ethanol extract was hepatotoxic and nephrotoxic at 1g/kg orally (Awe & Banjoko, 2013). The plant must be avoided in pregnancy (Ajmera et al., 2019), and it interacts with warfarin (Kurtaran et al., 2021).

Comments: (i) Depletion of glutathione and increase in ROS are part of the physiology aging (Ajebli & Eddouks, M., 2019). Volunteers consuming 20 g of parsley per day for 2 weeks had increased erythrocyte glutathione reductase and SOD activity (Nielsen et al., 1999).

(ii) Aging, high alcohol, red meat consumption, and genetic predisposition result in hyperuricemia, defined as a plasma concentration of uric acid above 7 mg/dL for men or 5.7 mg/dL for women. When uric acid reaches high levels, crystals of monosodium urate precipitate in the articulations to cause gout, with severe pain in the joints along with swelling and redness. Hyperuricemia is associated with risk of kidney stones and coronary heart diseases (Sakhaee et al., 2012).

REFERENCES

Ajebli, M. and Eddouks, M., 2019. Antihypertensive activity of Petroselinum crispum through inhibition of vascular calcium channels in rats. *Journal of Ethnopharmacology*, *242*, p. 112039.

Ajmera, P., Kalani, S. and Sharma, L., 2019. Parsley-benefits & side effects on health. *International Journal of Physiology, Nutrition and Physical Education*, *4*, pp. 1236–1242.

Al-Daraji, H.J., Al-Mashadani, H.A., Mirza, H.A., Al-Hassani, A.S. and Al-Hayani, W.K., 2012. The effect of utilization of parsley (Petroselinum crispum) in local Iraqi geese diets on blood biochemistry. *Journal of American Science*, *8*(8), pp. 427–432.

Awe, E.O. and Banjoko, S.O., 2013. Biochemical and haematological assessment of toxic effects of the leaf ethanol extract of Petroselinum crispum (Mill) Nyman ex AW Hill (Parsley) in rats. *BMC Complementary and Alternative Medicine*, *13*(1), pp. 1–6.

Bradley, P.R., 1992. *British Herbal Compendium. Volume 1. A Handbook of Scientific Information on Widely Used Plant Drugs. Companion to Volume 1 of the British Herbal Pharmacopoeia*. British Herbal Medicine Association.

Campos, K.E.D., Balbi, A.P.C. and Alves, M.J.Q.D.F., 2009. Diuretic and hipotensive activity of aqueous extract of parsley seeds (Petroselinum sativum Hoffm.) in rats. *Revista Brasileira de Farmacognosia*, *19*, pp. 41–45.

Fusch, L., 1555. *De Historia Stirpium Commetarii Insignes*. Lugduni Apud Ioan Tornaesium.

Gbolade, A.A. and Lockwood, G.B., 1999. Petroselinum crispum (Mill.) Nyman (parsley): In vitro culture, production and metabolism of volatile constituents. *Medicinal and Aromatic Plants*, *XI*, pp. 324–336.

Hozayen, W.G., El-Desouky, M.A., Soliman, H.A., Ahmed, R.R. and Khaliefa, A.K., 2016. Antiosteoporotic effect of Petroselinum crispum, Ocimum basilicum and Cichorium intybus L. in glucocorticoid-induced osteoporosis in rats. *BMC Complementary and Alternative Medicine*, *16*, pp. 1–11.

Kurtaran, M., Koc, N.S., Aksun, M.S., Yildirim, T., Yilmaz, Ş.R. and Erdem, Y., 2021. Petroselinum crispum, a commonly consumed food, affects sirolimus level in a renal transplant recipient: A case report. *Therapeutic Advances in Drug Safety*, *12*, p. 20420986211009358.

Marthe, F., 2020. Petroselinum crispum (Mill.) Nyman (Parsley). In *Medicinal, Aromatic and Stimulant Plants* (pp. 435–466). Springer.

Matthioli, P.A., 1572. *Commentaires sur les Six Livres de Pedacius Dioscorides Anazarbeen de la matière medicinale*. A l'Escue de Milan.

Nielsen, S.E., Young, J.F., Daneshvar, B., Lauridsen, S.T., Knuthsen, P., Sandström, B. and Dragsted, L.O., 1999. Effect of parsley (Petroselinum crispum) intake on urinary apigenin excretion, blood antioxidant enzymes and biomarkers for oxidative stress in human subjects. *British Journal of Nutrition*, *81*(6), pp. 447–455.

Rahmat, A., Ahmad, N.S.S. and Ramli, N.S., 2019. Parsley (Petroselinum crispum) supplementation attenuates serum uric acid level and improves liver and kidney structures in oxonate-induced hyperuricemic rats. *Oriental Pharmacy and Experimental Medicine*, *19*(4), pp. 393–401.

Sakhaee, K., Maalouf, N.M. and Sinnott, B., 2012. Kidney stones 2012: Pathogenesis, diagnosis, and management. *The Journal of Clinical Endocrinology & Metabolism*, *97*(6), pp. 1847–1860.

Sener, G., Saçan, Ö., Yanardag, R. and Ayanoglu-Dülger, G., 2003. Effects of parsley (Petroselinum crispum) on the aorta and heart of STZ induced diabetic rats. *Plant Foods for Human Nutrition*, *58*(3), pp. 1–7.

Soliman, M.M., Nassan, M.A., Aldhahrani, A., Althobaiti, F. and Mohamed, W.A., 2020. Molecular and histopathological study on the ameliorative impacts of Petroselinum crispum and Apium graveolens against experimental hyperuricemia. *Scientific Reports*, *10*(1), p. 9512.

87 Avocado (*Persea americana* Mill.)

Persea americana Mill.

Etymology: From the Greek *persea* = a fruit tree of ancient Egypt and the Latin *americana* = from America

Synonym: *Laurus persea* L.

Common names: Avocado; avocat (Fr.); abacate (Port.); авокадо (Rus.); palta (Spa.)

Part used: Fruit

Constituent: Fixed oil (unsaturated fatty acids: oleic acid, linoleic acid) (Carvalho et al., 2015).

Medical history: Mesoamerican Indians have consumed avocado for millennia (since about 15,000 BC). Spanish physicians in the 15th century, Spanish physicians were the first to bring awareness on avocado to Europeans. Francisco Hernandez de Toledo in his *Nova plantarum, animalium, et mineralium Mexicanorum* (1651) describes avocado fruits in these terms: "*extra nigro, intra virescenti, pinguis qua butyrum aemulatur naturae et sapore viridium nucim*", which can be translated as "black outside, greenish inside, so rich that butter rivals the nature and taste of a green nut". He reports that that natives call avocado "*Ahvaca quahulitl*" and use it as an aphrodisiac ("*venerem . . . excitantis*") (p. 89).

Avocado was introduced by Spanish explorers to Indonesia by 1750 and the Philippines in 1890 (María Elena Galindo-Tovar et al., 2007).

Medicinal uses: Venereal diseases (Gabon); gallstones (Turkey); hypertension, diabetes, diarrhea, bronchitis, inflammation (India); diabetes, diarrhea, stomachache (the Philippines);

Blood pressure: The oil expressed from the fruit pulp given to hypertensive rats at 3.6 g/kg or 45 days caused decreases in systolic and diastolic blood pressure (Márquez-Ramírez et al., 2021).

Plasma lipids and glucose: Avocado pulp given for 7 weeks at 2 g/kg to rats poisoned with fructose and a cholesterol-enriched diet decreased total plasma cholesterol from 95.6 to 58.2 mmol/L, triglycerides from 1.4 to 1 mmol/L, HDL-cholesterol from 89.5 to 52.8 mmol/L, and atherogenic index from 16.8 to 10.5 (Pahua-Ramos et al., 2014).

Obese subjects given 68 g of avocado with had a decrease in plasma insulin (Park et al., 2018).

DOI: 10.1201/9781003301455-87

Kidneys: The oil expressed from the pulp given to hypertensive rats at 3.6 g/kg
for 45 days mitigated renal oxidative insults and ameliorated renal endothelium-dependent vasodilation (Márquez-Ramírez et al., 2021).

Skin: Aqueous extract of peels at 10% of an ointment applied twice a day for
4 weeks to rats increased skin hydration, collagen content, and skin elasticity (Lister et al., 2021).

Comments: (i) Linoleic acid and other unsaturated fatty acids, because of their
double bonds, are able to scavenge ROS and therefore increase the bioavailability of NO in vascular endothelium (Márquez-Ramírez et al., 2021).
In volunteers, levels of linoleic acid in plasma triglycerides were inversely
associated with blood pressure (Grimsgaard et al., 1999). In vascular smooth
muscle cells, linoleic acid induces relaxation by, at least in part, activation of
sodium/potassium-ATPase pumps (Pomposiello et al., 1998).

(ii) A diet enriched with oleic and linoleic acids decreases plasma cholesterol
(Chan et al., 1991) and hs-CRP (Yoneyama et al., 2007).

(iii) It is clear that intake of avocado is beneficial for aging. However, avocado
contains high amounts of potassium and must be avoided in hyperkaliemic
patients.

(iv) Delicious avocado drinks are available in Indonesia, while the Philippines
is home to excellent local varieties.

REFERENCES

Carvalho, C.P., Bernal, E.J., Velásquez, M.A. and Cartagena, V.J.R., 2015. Fatty acid content
of avocados (Persea americana Mill. cv. Hass) in relation to orchard altitude and fruit
maturity stage. *Agronomía Colombiana, 33*(2), pp. 220–227.

Chan, J.K., Bruce, V.M. and McDonald, B.E., 1991. Dietary α-linolenic acid is as effective as
oleic acid and linoleic acid in lowering blood cholesterol in normolipidemic men. *The
American Journal of Clinical Nutrition, 53*(5), pp. 1230–1234.

Galindo-Tovar, M.E., Arzate-Fernández, A.M., Ogata-Aguilar, N. and Landero-Torres, I.,
2007. The avocado (Persea americana, Lauraceae) crop in Mesoamerica: 10,000 years
of history. *Harvard Papers in Botany, 12*(2), pp. 325–334.

Grimsgaard, S., Bønaa, K.H., Jacobsen, B.K. and Bjerve, K.S., 1999. Plasma saturated and linoleic fatty acids are independently associated with blood pressure. *Hypertension, 34*(3),
pp. 478–483.

Lister, I.N., Amiruddin, H.L., Fachrial, E. and Girsang, E., 2021. Anti-aging effectiveness of
avocado peel extract ointment (Persea Americana Mill.) against hydration, collagen,
and elasticity levels in Wistar rat. *Journal of Pharmaceutical Research International*,
pp. 173–184.

Márquez-Ramírez, C.A., Olmos-Orizaba, B.E., García-Berumen, C.I., Calderón-Cortés, E.,
Montoya-Pérez, R., Saavedra-Molina, A., Rodríguez-Orozco, A.R. and Cortés-Rojo, C.,
2021. Avocado oil prevents kidney injury and normalizes renal vasodilation after adrenergic stimulation in hypertensive rats: Probable role of improvement in mitochondrial
dysfunction and oxidative stress. *Life, 11*(11), p. 1122.

Pahua-Ramos, M.E., Garduño-Siciliano, L., Dorantes-Alvarez, L., Chamorro-Cevallos, G.,
Herrera-Martínez, J., Osorio-Esquivel, O. and Ortiz-Moreno, A., 2014. Reduced-calorie
avocado paste attenuates metabolic factors associated with a hypercholesterolemic-high
fructose diet in rats. *Plant Foods for Human Nutrition, 69*(1), pp. 18–24.

Park, E., Edirisinghe, I. and Burton-Freeman, B., 2018. Avocado fruit on postprandial markers of cardio-metabolic risk: A randomized controlled dose response trial in overweight and obese men and women. *Nutrients, 10*(9), p. 1287.

Pomposiello, S.I., Alva, M., Wilde, D.W. and Carretero, O.A., 1998. Linoleic acid induces relaxation and hyperpolarization of the pig coronary artery. *Hypertension, 31*(2), pp. 615–620.

Yoneyama, S., Miura, K., Sasaki, S., Yoshita, K., Morikawa, Y., Ishizaki, M., Kido, T., Naruse, Y. and Nakagawa, H., 2007. Dietary intake of fatty acids and serum C-reactive protein in Japanese. *Journal of Epidemiology, 17*(3), pp. 86–92.

88 Vietnamese Coriander (*Polygonum odoratum* Lour.)

Polygonum odoratum Lour.

Etymology: From the Greek *poly* = many and *gono* = knees and the Latin *odoratum* = fragrant

Family: Polygonaceae

Synonym: *Persicaria odorata* (Lour.) Soják

Common names: Vietnamese coriander; laksa; coriandre Vietnamienne (Fr.); coentro Vietnamita (Port.); Vietnamesischer koriander (Ger.); вьетнамский кориандр (Rus.); cilantro Vietnamita (Spa.)

Part used: Leaf

Constituents: Essential oil (dodecanal) (Hunter et al., 1997), homoisoflavanones (Zhou et al., 2015), flavone glycosides (Ganbaatar et al., 2015).

Medical history: João de Loureiro (1717–1791) was a Portuguese botanist and Jesuit missionary who spent more than 30 years of his life in Vietnam. He provides for the first time in his *Flora Cochinchinensis* this description of Vietnamese coriander: "*planta ista addat optimum condimentum, et piscibus, praecipue affis*", which can be translated as "this plant gives an excellent condiment for meat and fish, especially fish" (p. 243).

Medicinal use: Joint pain (Vietnam)

Plasma glucose and lipids: In rats with partial removal of pancreas and fed a high-fat diet, intake of Vietnamese coriander at 300g/kg/day for 8 weeks increased intake of glucose by peripheral tissues (Choi & Park, 2002).

Comment: In obese patients, chronic intake of sugar and fats translates with time into insulin resistance, meaning that more insulin is needed for peripheral tissues to absorb glucose because insulin receptors become less sensitive to insulin. Thus, the pancreas needs to secrete more insulin until the exhaustion of pancreatic β-cells, leading to type-2 diabetes (Choi & Park, 2002). Intake of saturated fatty acids, such as the palmitic acid in palm oil and animal fats, contribute to the malfunction of insulin receptors.

DOI: 10.1201/9781003301455-88

REFERENCES

Choi, S.B. and Park, S., 2002. The effects of water extract of Polygonatum Odoratum (Mill) Druce on insulin resistance in 90% pancreatectomized rats. *Journal of Food Science*, *67*(6), pp. 2375–2379.

Ganbaatar, C., Gruner, M., Mishig, D., Duger, R., Schmidt, A.W. and Knölker, H.J., 2015. Flavonoid glycosides from the aerial parts of Polygonatum odoratum (Mill.) Druce growing in Mongolia. *The Open Natural Products Journal*, *8*(1).

Hunter, M.V., Brophy, J.J., Ralph, B.J. and Bienvenu, F.E., 1997. Composition of Polygonum odoratum Lour. from southern Australia. *Journal of Essential Oil Research*, *9*(5), pp. 603–604.

Zhou, X., Yuping, Z., Zhao, H., Liang, J., Zhang, Y. and Shi, S., 2015. Antioxidant homoisoflavonoids from Polygonatum odoratum. *Food Chemistry*, *186*, pp. 63–68.

89 Sour Cherry (*Prunus cerasus* L.)

Prunus cerasus L.

Etymology: From the Greek *proynos* = a plum tree and *Cerasus* = a town from which the plant was brought to ancient Rome

Family: Rosaceae

Synonyms: *Cerasus caproniana* (L.) DC.; *Cerasus cerasus* (L.) Eaton & Wright; *Cerasus vulgaris* Mill.

Common names: Pie cherry, sour cherry, tart cherry; aigriottes, cerisier aigre, griottes (Fr.); sauerkirsche (Ger.); cerezo ácido (Spa.)

Part used: Fruit

Constituents: Anthocyanins (cyanidin-3-glucosylrutinoside, cyanidin-3-rutinoside) (Chandra et al., 1992; Damar & Ekşi, 2012), condensed tannins (Çevik et al., 2013).

Medical history: Roman physicians, including Dioscorides, considered sour cherries good for the stomach ("*ventri utilia sunt*"; Dioscorides). The Medical School of Salerno (10th century) recommends sour cherries for good health ("*cerasa si comedas, faciunt tibi randia dona*"), for removing kidney stones, and for "creat[ing] fresh blood". Sour cherry was defined as cold and humid by Simeon Seth in 12th-century Byzantium. In 16th-century Italy, it was used for fever (Matthioli, 1572), while in France, it was given to increase the appetite of the sick and was recommended during pregnancy (Daléchamps, 1615). Lémery (1716) in 18th-century France recommends sour cherries for cooling, for a healthy brain, and for epilepsy, while in Scotland, Alston (1770) advocates sour cherries to quench thirst, to induce defecation, and to treat scurvy.

Medicinal use: Hypertension (Turkey); kidney stones (Iran)

Blood pressure: Sour cherry juice given to hypertensive volunteers at the single oral dose of 60 mL (corresponding to the consumption of 180 fruits) caused after 2 hours a decrease in systolic blood pressure of 7 mmHg (Keane et al., 2016).

Plasma glucose and lipids: Methanol extract given orally at 200 mg/kg/day for 60 days to alloxan-induced diabetic rats decreased glycemia from 13.6 to 8.1 mmol/L and decreased plasma triglycerides (Xia & Xiao, 2019).

Uric acid: Sour cherries at a single oral dose of 60 mL ingested by healthy volunteers decreased plasma uric acid from about 500 to 300 µM/L after 8 hours (Bell et al., 2014).

DOI: 10.1201/9781003301455-89

Immune system: Intake of ethyl acetate extract by mice at 50 mg/kg increased leukocyte count and circulating phagocytosis (Abid et al., 2012).

Brain: Sour cherries juice given to mice for 8 weeks attenuated galactosamine-induced dementia (Cui & Li, 2009).

Commentary: It is clear that consuming fresh sour cherries regularly is beneficial for aging.

REFERENCES

Abid, S., Khajuria, A., Parvaiz, Q., Sidiq, T., Bhatia, A., Singh, S., Ahmad, S., Randhawa, M.K., Satti, N.K. and Dutt, P., 2012. Immunomodulatory studies of a bioactive fraction from the fruit of Prunus cerasus in BALB/c mice. *International Immunopharmacology*, *12*(4), pp. 626–634.

Alston, C., 1770. *Lectures on the Materia Medica: Containing the Natural History of Drugs, their Virtues and Doses: Also Directions for the Study of the Materia Medica; and an Appendix on the Method of Prescribing*. Edward and Charles Dilly.

Bell, P.G., Gaze, D.C., Davison, G.W., George, T.W., Scotter, M.J. and Howatson, G., 2014. Montmorency tart cherry (Prunus cerasus L.) concentrate lowers uric acid, independent of plasma cyanidin-3-O-glucosiderutinoside. *Journal of Functional Foods*, *11*, pp. 82–90.

Çevik, N.A.Z.A.N., Kizilkaya, B.A.Y.R.A.M. and Türker, G.Ü.L.E.N., 2013. The condensed tannin content of fresh fruits cultivated in ida mountains, çanakkale, turkey. *New Knowledge Journal of Science*, *2*, pp. 49–51.

Chandra, A., Nair, M.G. and Iezzoni, A., 1992. Evaluation and characterization of the anthocyanin pigments in tart cherries (Prunus cerasus L.). *Journal of Agricultural and Food Chemistry*, *40*(6), pp. 967–969.

Cui, X. and Li, X., 2009. Effect of Prunus cerasus L. juices on learning and memory in senile mice induced by D-galactose. *Journal of Beihua University*, *10*(6), pp. 501–503.

Damar, İ. and Ekşi, A., 2012. Antioxidant capacity and anthocyanin profile of sour cherry (Prunus cerasus L.) juice. *Food Chemistry*, *135*(4), pp. 2910–2914.

Daléchamps, 1615. *De l' histoire generale des plantes simples*. Chez Heritier Guillaume Rouille.

Keane, K.M., George, T.W., Constantinou, C.L., Brown, M.A., Clifford, T. and Howatson, G., 2016. Effects of Montmorency tart cherry (Prunus Cerasus L.) consumption on vascular function in men with early hypertension. *The American Journal of Clinical Nutrition*, *103*(6), pp. 1531–1539.

Lémery, N., 1716. *Traité universel des drogues simples, mises en ordre alphabétique. Où l'on trouve leurs différens noms . . . et tout ce qu'il y a de particulier dans les animaux, dans les végétaux, et dans les minéraux*. Au dépend de la Companie.

Matthioli, P.A., 1572. *Commentaires sur les Six Livres de Pedacius Dioscorides Anazarbeen de la matière medicinale*. A l'Escue de Milan.

Xiao, G. and Xiao, X., 2019. Antidiabetic effect of hydro-methanol extract of Prunus cerasus L fruits and identification of its bioactive compounds. *Tropical Journal of Pharmaceutical Research*, *18*(3), pp. 597–602.

90 Pomegranate (*Punica granatum* L.)

Punica granatum L.

Etymology: From the Latin *punica* = from Carthage and *granatum* = from Granada in Spain

Family: Lythraceae

Common names: Pomegranate; grenade (Fr.); granatapfel (Ger.); romã (Port.); granata (Spa.)

Part used: Fruit

Constituents: Tannins (ellagitannins: punicalagin, ellagic acid), phenolic acids (gallic acid) (Feng et al., 2017), anthocyanins (Miguel et al., 2004), ascorbic acid (Al-Maiman & Ahmad, 2002), citric acid (Poyrazoğlu et al., 2002).

Medical history: In ancient Sanskrit medical texts, pomegranate is called *dadima*. In the Songs of Songs (8:2) of King David, one can read: "I would give you to drink some spiced wine, of the juice of my pomegranate". Dioscorides calls it *malum punicum*. Arab physicians of the Middle Ages including Avicenna use pomegranate as a diuretic and for stomach health and dysentery. In second chapter of the Coran, one can read, "And We produce gardens of grapevines and olives and pomegranates, similar yet varied". In 16th-century France, pomegranate was used for hemoptysis, nausea, bilious affections, and intestinal worms (Daléchamps, 1615). In 19th-century Europe and North America, pomegranate was given to allay thirst and to refresh (Pereira, 1843).

Medicinal uses: Diarrhea, diabetes, shortness of breath (Turkey); diarrhea (Afghanistan); dysentery, heart palpitation, cardiotonic, blood pressure, tonic, cooling (Pakistan); quenching thirst, laxative, jaundice, stomachache, tonic, dysentery. strengthening the heart, astringent, urinary tract infection (India); bronchitis (Nepal); intestinal diseases, dysentery (Bangladesh); dyspepsia, diarrhea (China); dyspepsia (Korea); hepatitis, carminative (the Philippines)

Blood pressure: Elderly patients with artery stenosis taking concentrated juice daily for a year had thinning of the left and right common carotid arteries intima and media. This regimen decreased arterial stenosis, as evidenced by decreased internal carotid peak systolic velocity of both left and right carotid arteries by 21% as well as end diastolic velocity by 44%. Systolic blood pressure was reduced by 12% (Aviram et al., 2004).

DOI: 10.1201/9781003301455-90

Pomegranate juice given at 50 mL/day for 2 weeks to hypertensive patients caused a decrease in systolic blood pressure from 155 to 147 mmHg as well as decreased plasma ACE activity (Aviram & Dornfeld, 2001). Intake of capsules of an extract of fruits daily for 8 weeks decreased systolic blood pressure (Stockton et al., 2017). Fresh juice taken between meals at 150 mL/day for 2 weeks decreased systolic blood pressure from 130.9 to 124.5 mmHg and diastolic blood pressure from 80 to 76 mmHg and decreased plasma vascular endothelial adhesion molecule 1 (Asgary et al., 2014).

Heart: Rats given ellagic acid at 15 mg/kg/day orally for 10 days were protected against isoproterenol-induced myocardial infarction (Kannan & Quine, 2013).

Juice taken daily by patients with ischemic coronary heart disease for 3 months decreased stress-induced ischemia (Sumner et al., 2005).

Kidneys: Rats orally given the daily dose of 3 mL/kg pomegranate juice for 21 days had decreased levels of plasma creatinine from 0.5 to 0.3 mg/%, and renal MDA decreased by 33% (Moneim et al., 2011).

Bones and cartilages: Butanol extract of peels given orally at 75 mg/kg/day for 28 days to rats with CFA-induced arthritis caused a reduction in joint diameter from 3.2 to 0.5 mm (Gautam et al., 2018).

Skin and hair: Intake of 75 mg of punicalagin by volunteers orally and daily for 4 weeks decreased wrinkle formation by about 6% (Chakkalakal et al., 2022).

Brain: Ethanol extract of seeds given orally to mice at 500 mg/kg/day for 21 days attenuated scopolamine-induced dementia (Kumar et al., 2009).

Warning: Tannin intake at a high dose (2 g/kg) results in liver injuries (Oler et al., 1976).

Comments: (i) Intake of pomegranate on an empty stomach might have better effects on blood pressure as tannins do not react with dietary proteins. Fresh juice could be more effective than bottled ones.

(ii) The formation of arterial atheroma starts with the adhesion of circulating monocytes to vascular cell adhesion molecule-1 (VCAM-1) expressed by activated arterial endothelium (Dancksy et al., 2001;hayse et al., 2020). Since intake of pomegranate juice decreases the expression of VCAM-1, it could prevent atherosclerosis.

(iii) The glycemia and LDL-cholesterol of elderly volunteers taking concentrated juice daily for a year were not modified (Aviram et al., 2004). Similarly, fresh juice taken between meals at 150 mL/day for 2 weeks had no influence on plasma cholesterol, triglycerides, or glucose (Asgary et al., 2014).

(iv) It is clear that the intake of freshly prepared pomegranate juice is beneficial for aging.

REFERENCES

Al-Maiman, S.A. and Ahmad, D., 2002. Changes in physical and chemical properties during pomegranate (Punica granatum L.) fruit maturation. *Food Chemistry*, *76*(4), pp. 437–441.

Asgary, S., Sahebkar, A., Afshani, M.R., Keshvari, M., Haghjooyjavanmard, S. and Rafieian-Kopaei, M., 2014. Clinical evaluation of blood pressure lowering, endothelial function improving, hypolipidemic and anti-inflammatory effects of pomegranate juice in hypertensive subjects. *Phytotherapy Research*, *28*(2), pp. 193–199.

Aviram, M. and Dornfeld, L., 2001. Pomegranate juice consumption inhibits serum angiotensin converting enzyme activity and reduces systolic blood pressure. *Atherosclerosis*, *158*(1), pp. 195–198.

Aviram, M., Rosenblat, M., Gaitini, D., Nitecki, S., Hoffman, A., Dornfeld, L., Volkova, N., Presser, D., Attias, J., Liker, H. and Hayek, T., 2004. Pomegranate juice consumption for 3 years by patients with carotid artery stenosis reduces common carotid intima-media thickness, blood pressure and LDL oxidation. *Clinical Nutrition*, *23*(3), pp. 423–433.

Chakkalakal, M., Nadora, D., Gahoonia, N., Dumont, A., Burney, W., Pan, A., Chambers, C.J. and Sivamani, R.K., 2022. Prospective randomized double-blind placebo-controlled study of oral pomegranate extract on skin wrinkles, biophysical features, and the gut-skin axis. *Journal of Clinical Medicine*, *11*(22), p. 6724.

Daléchamps, 1615. *De l' histoire generale des plantes simples*. Chez Heritier Guillaume Rouille.

Dansky, H.M., Barlow, C.B., Lominska, C., Sikes, J.L., Kao, C., Weinsaft, J., Cybulsky, M.I. and Smith, J.D., 2001. Adhesion of monocytes to arterial endothelium and initiation of atherosclerosis are critically dependent on vascular cell adhesion molecule-1 gene dosage. *Arteriosclerosis, Thrombosis, and Vascular Biology*, *21*(10), pp. 1662–1667.

Feng, L., Yin, Y., Fang, Y. and Yang, X., 2017. Quantitative determination of punicalagin and related substances in different parts of pomegranate. *Food Analytical Methods*, *10*(11), pp. 3600–3606.

Gautam, R.K., Sharma, S., Sharma, K. and Gupta, G., 2018. Evaluation of antiarthritic activity of butanol fraction of Punica granatum Linn. Rind extract against Freund's complete adjuvant-induced arthritis in rats. *Journal of Environmental Pathology, Toxicology and Oncology*, *37*(1).

Kannan, M.M. and Quine, S.D., 2013. Ellagic acid inhibits cardiac arrhythmias, hypertrophy and hyperlipidaemia during myocardial infarction in rats. *Metabolism*, *62*(1), pp. 52–61.

Kumar, S., Maheshwari, K. and Singh, V., 2009. Protective effects of Punica granatum seeds extract against aging and scopolamine induced cognitive impairments in mice. *African Journal of Traditional, Complementary and Alternative Medicines*, *6*(1).

Miguel, G., Fontes, C., Antunes, D., Neves, A. and Martins, D., 2004. Anthocyanin concentration of "Assaria" pomegranate fruits during different cold storage conditions. *Journal of Biomedicine and Biotechnology*, *2004*(5), p. 338.

Moneim, A.E.A., Dkhil, M.A. and Al-Quraishy, S., 2011. Studies on the effect of pomegranate (Punica granatum) juice and peel on liver and kidney in adult male rats. *Journal of Medicinal Plants Research*, *5*(20), pp. 5083–5088.

Oler, A., Neal, M.W. and Mitchell, E.K., 1976. Tannic acid: Acute hepatotoxicity following administration by feeding tube. *Food and Cosmetics Toxicology*, *14*(6), pp. 565–569.

Poyrazoğlu, E., Gökmen, V. and Artık, N., 2002. Organic acids and phenolic compounds in pomegranates (Punica granatum L.) grown in Turkey. *Journal of Food Composition and Analysis*, *15*(5), pp. 567–575.

Pereira, J., 1843. *The Elements of Materia Medica and Therapeutics*. Lea and Blanchard.

Stockton, A., Farhat, G., McDougall, G.J. and Al-Dujaili, E.S., 2017. Effect of pomegranate extract on blood pressure and anthropometry in adults: A double-blind placebo-controlled randomised clinical trial. *Journal of Nutritional Science*, 6, p. e39.

Sumner, M.D., Elliott-Eller, M., Weidner, G., Daubenmier, J.J., Chew, M.H., Marlin, R., Raisin, C.J. and Ornish, D., 2005. Effects of pomegranate juice consumption on myocardial perfusion in patients with coronary heart disease. *The American Journal of Cardiology*, 96(6), pp. 810–814.

Thayse, K., Kindt, N., Laurent, S. and Carlier, S., 2020. VCAM-1 target in non-invasive imaging for the detection of atherosclerotic plaques. *Biology*, 9(11), p. 368.

91 Pear (*Pyrus communis* L.)

Pyrus communis L.

Etymology: From the Latin *pyrus* = pear tree and *communis* = common

Family: Rosaceae

Common names: Pear; poire (Fr.); birne (Ger.); pera (Port.; Spa.); грушевое (Rus.)

Part used: Fruit

Constituents: Tannins (condensed: proanthocyanidins), hydroxycinnamic acid derivatives, flavone glycosides (Fischer et al., 2007).

Medical history: Pears were recommended by Dioscorides for stomach health, for bleeding, and as an astringent. Good for the stomach (Galen). The Medical School of Salerno (10th century) warns against the dangers of eating raw pears and advises ingesting pear compote instead. Pears were used as famine food in 16th-century Italy (Matthioli, 1572). In 18th-century France, pears were used by physicians as an astringent and for stomach health.

Medicinal uses: Anxiety, fever (Pakistan); diuretic (India)

Blood pressure: In obese patients, consumption of 2 pears (about 178 g) per day for 12 weeks reduced systolic blood pressure by about 5 mmHg (Johnson et al., 2016).

Plasma glucose and lipids: Ethyl acetate extract of fruits given at 200 mg/kg/day for 11 days to dexamethasone-induced diabetic rats decreased plasma glucose from 381.3 to 126.6 mg/dL, cholesterol from 217.4 to 153 mg/dL, and triglycerides from 248.7 to 144.1 mg/dL (Velmurugan & Bhargava, 2013).

Gallbladder: Bile acid binds to pear dry matter at 0.4 μmol/100 mg (Kahlon & Smith, 2007).

Skin and hair: Hydroalcoholic extract in 5% formulation administered to healthy volunteers topically for 3 months improved skin elasticity and moisture (Imam et al., 2021).

REFERENCES

Fischer, T.C., Gosch, C., Pfeiffer, J., Halbwirth, H., Halle, C., Stich, K. and Forkmann, G., 2007. Flavonoid genes of pear (Pyrus communis). *Trees, 21*, pp. 521–529.

Imam, S., Shaheen, N., Abidi, S., Siddique, H., Iffat, W., Azhar, I. and Mahmood, Z.A., 2021. Moisturizing efficacy of a cream formulation containing fruit extracts for combating skin aging. *Latin American Journal of Pharmacy, 40*(9), pp. 2099–2104.

DOI: 10.1201/9781003301455-91

Johnson, S.A., Navaei, N., Pourafshar, S., Akhavan, N.S., Elam, M.L., Foley, E., Clark, E.A., Payton, M.E. and Arjmandi, B.H., 2016. Fresh pear (Pyrus communis) consumption may improve blood pressure in middle-aged men and women with metabolic syndrome. *The FASEB Journal, 30*, pp. 1175.12.

Kahlon, T.S. and Smith, G.E., 2007. In vitro binding of bile acids by bananas, peaches, pineapple, grapes, pears, apricots and nectarines. *Food Chemistry, 101*(3), pp. 1046–1051.

Matthioli, P.A., 1572. *Commentaires sur les Six Livres de Pedacius Dioscorides Anazarbeen de la matière medicinale.* A l'Escue de Milan.

Velmurugan, C. and Bhargava, A.N.U.R.A.G., 2013. Anti-diabetic and hypolipidemic activity of fruits of Pyrus communis L. in hyperglycemic rats. *Asian Journal of Pharmaceutical and Clinical Research, 6*(5), pp. 108–111.

92 Radish (*Raphanus sativus* L.)

Raphanus sativus L.

Etymology: From the Latin *raphanus* = radish and *sativus* = cultivated

Family: Brassicaceae

Common names: Radish; radis (Fr.); rettisch (Ger.); rabanete (Port.); рéдька (Rus.); rábano (Spa.)

Part used: Root

Constituents: Glucosinolates (Barillari et al., 2005), fibers, potassium (about 300 mg/g) (Goyeneche et al., 2015).

Medical history: Dioscorides recommends radish as a diuretic and for cough, chest discomfort, and dropsy, while Pliny the Elder asserts that it expel urinary stones. It was considered hot to the third degree and dry to the second by European physicians in the 16th century (Fusch, 1555).

Medicinal uses: Diuretic, urinary stones (Iraq); jaundice, urinary problems, (Pakistan); gallstones (Korea)

Plasma glucose and lipids: Juice of radish given for 6 days to mice poisoned by a diet enriched with 2% cholesterol and 0.5% cholic acid decreased cholesterol from 82 to 79.8 mg/dL and HDL-cholesterol from 67 to 62.8 mg/dL and removed gallstones (Castro-Torres et al., 2012). The juice given to rats at 150 mL/kg for 9 days decreased total cholesterol from 6.2 to 5.2 mmol/L (Kocsis et al., 2002).

Gall bladder: Juice of radish given for 6 days to mice poisoned by a diet enriched with 2% cholesterol and 0.5% cholic acid removed gall stones (Castro-Torres et al., 2012). Fibers of radish given to rats at 5% of diet for 7 weeks decreased plasma cholesterol while increasing fecal bile acid content (Sannoumaru et al., 1996).

Bones and cartilages: Juice given to rats at the volume of 3 mL/day for 7 days attenuated formaldehyde-induced edema (Kamble et al., 2015).

Comments: (i) Bile is a necessary secretion for the emulsification of dietary fats and the removal of excess cholesterol from the body. It consists of cholesterol, phospholipids, and predominantly bile salts. With aging and diet enriched with cholesterol there is the risk of cholesterol precipitation and the formation of stones in the gall bladder (Marschall & Einarsson, 2007).

(ii) Glucosinolates must not be ingested by patients with thyroid dysfunction.

DOI: 10.1201/9781003301455-92

(iii) It is clear that the intake of fresh radish is beneficial for aging. However, radish contains high amounts of potassium and must be avoided in hyperkaliemic patients.

REFERENCES

Barillari, J., Cervellati, R., Paolini, M., Tatibouët, A., Rollin, P. and Iori, R., 2005. Isolation of 4-methylthio-3-butenyl glucosinolate from Raphanus sativus sprouts (Kaiware Daikon) and its redox properties. *Journal of Agricultural and Food Chemistry*, *53*(26), pp. 9890–9896.

Castro-Torres, I.G., Naranjo-Rodríguez, E.B., Domínguez-Ortíz, M.Á., Gallegos-Estudillo, J. and Saavedra-Vélez, M.V., 2012. Antilithiasic and hypolipidaemic effects of Raphanus sativus L. var. niger on mice fed with a lithogenic diet. *Journal of Biomedicine and Biotechnology*, *2012*.

Fusch, L., 1555. *De Historia Stirpium Commetarii Insignes*. Lugduni Apud Ioan Tornaesium.

Goyeneche, R., Roura, S., Ponce, A., Vega-Gálvez, A., Quispe-Fuentes, I., Uribe, E. and Di Scala, K., 2015. Chemical characterization and antioxidant capacity of red radish (Raphanus sativus L.) leaves and roots. *Journal of Functional Foods*, *16*, pp. 256–264.

Kamble, S., Ahmed, M.Z., Ramabhimaiaha, S. and Patil, P., 2015. Anti-inflammatory activity of Raphanus sativus L in acute and chronic experimental models in albino rats. *Biomedical and Pharmacology Journal*, *6*(2), pp. 173–177.

Kocsis, I., Lugasi, A., Hagymási, K., Kéry, A., Fehér, J., Szoke, É. and Blázovics, A., 2002. Beneficial properties of black radish root (Raphanus sativus L. var. niger) squeezed juice in hyperlipidemic rats: Biochemical and chemiluminescence measurements. *Acta Alimentaria*, *31*(2), pp. 185–190.

Marschall, H.U. and Einarsson, C., 2007. Gallstone disease. *Journal of Internal Medicine*, *261*(6), pp. 529–542.

Sannoumaru, Y., Shimizu, J., Nakamura, K., Hayakawa, T., Takita, T. and Innami, S., 1996. Effects of semi-purified dietary fibers isolated from Lagenaria siceraria, Raphanus sativus and Lentinus edodes on fecal steroid excretions in rats. *Journal of Nutritional Science and Vitaminology*, *42*(2), pp. 97–110.

93 Blackcurrant (*Ribes nigrum* L.)

Ribes nigrum L.

Etymology: From the Latin *ribes* = currant and *nigrum* = black

Family: Grossulariaceae

Synonyms: *Grossularia nigra* (L.) Rupr.; *Ribes olidum* Moench; *Ribes pauciflorum* Turcz. ex Ledeb.

Common names: Blackcurrant; cassis (Fr.); schwarze johannisbeere (Ger.); groselha preta. (Port.); черная смородина (Rus.); grosella nigra (Spa.)

Part used: Fruit

Constituents: Anthocyanins (about 250 mg/100 g), ascorbic acid (about 130–200 mg/100 mL) (Vagiri et al., 2013), citric acid (Zheng et al., 2009).

Medical history: Dioscorides used blackcurrant as an astringent and for fever. In 16th-century Germany, physicians considered blackcurrant cold and dry to the first degree (Fusch,1555). In 17th-century England, physicians used blackcurrant for cooling and to promote appetite and quench thirst (Parkinson, 1640). The French physician and chemist Etienne-François Geoffroy (1672–1731), in his *Tractatus de materia medica* (1742), advises intake of blackcurrant for fever. It was used to make liquors in 19th-century France (Guibourt, 1836) and for cooling, sore throat, and fevers in 19th-century North America.

Medicinal uses: Diuretic (Turkey)

Blood pressure: An extract given to elderly patients at 600 mg/day decreased systolic blood pressure from 136 to 130 mmHg and diastolic blood pressure from 84 to 78 mmHg (Cook et al., 2020). An extract taken at 300 mg twice a day for 7 days by elderly volunteers reduced carotid–femoral pulse-wave velocity and blood pressure (Okamoto et al., 2020).

Plasma lipids and glucose: Powder of blackcurrant added at 3% of diet for 3 months decreased plasma cholesterol in rats from 213.4 to 140.5 mg/dL, LDL-cholesterol from 43 to 31.7 mg/dL, and triglycerides from 269.8 to 151.4 mg/dL (Nanashima et al., 2020).

Adipose tissues: Powder of blackcurrant added as 3% of the diet of rats without ovaries for 3 months decreased visceral fat from about 35 to 20 g (Nanashima et al., 2020).

Kidneys: Blackcurrant juice given at 330 mL per day increased urinary pH and citric acid excretion and prevented uric acid kidney stone formation (Kessler et al., 2002).

DOI: 10.1201/9781003301455-93

Brain: Anthocyanins given orally as part of the diet to Alzheimer's mouse models for about 10 months evoked some protection again the formation of amyloid protein β plaques (Vepsäläinen et al., 2013).

Comments: (i) With menopause, decreased estrogen causes weight gain and increased plasma lipids (Nanashima et al., 2020).

(ii) With aging and dehydration, decreased urinary pH uric acid forms stones in the kidney that can be removed by increased intake of water and by increasing urinary pH and citric acid (Gutman, 1968).

(iii) Citric acid attenuates the formation of kidney stones by forming complexes with calcium ions (Gul & Monga, 2014).

REFERENCES

Cook, M.D., Sandu, BSc, A.K. and Joyce, PhD, J.P., 2020. Effect of New Zealand blackcurrant on blood pressure, cognitive function and functional performance in older adults. *Journal of Nutrition in Gerontology and Geriatrics*, *39*(2), pp. 99–113.

Fusch, L., 1555. *De Historia Stirpium Commetarii Insignes*. Lugduni Apud Ioan Tornaesium.

Guibourt, N.J.B.G., 1836. *Histoire abrégée des drogues simples*. Méquignon-Marvis Père et fils.

Gul, Z. and Monga, M., 2014. Medical and dietary therapy for kidney stone prevention. *Korean Journal of Urology*, *55*(12), pp. 775–779.

Gutman, A.B., 1968. Uric acid nephrolithiasis. *The American Journal of Medicine*, *45*(5), pp. 756–779.

Kessler, T., Jansen, B. and Hesse, A., 2002. Effect of blackcurrant-, cranberry-and plum juice consumption on risk factors associated with kidney stone formation. *European Journal of Clinical Nutrition*, *56*(10), pp. 1020–1023.

Nanashima, N., Horie, K., Yamanouchi, K., Tomisawa, T., Kitajima, M., Oey, I. and Maeda, H., 2020. Blackcurrant (Ribes nigrum) extract prevents dyslipidemia and hepatic steatosis in ovariectomized rats. *Nutrients*, *12*(5), p. 1541.

Okamoto, T., Hashimoto, Y., Kobayashi, R., Nakazato, K. and Willems, M.E.T., 2020. Effects of blackcurrant extract on arterial functions in older adults: A randomized, double-blind, placebo-controlled, crossover trial. *Clinical and Experimental Hypertension*, *42*(7), pp. 640–647.

Vagiri, M., Ekholm, A., Öberg, E., Johansson, E., Andersson, S.C. and Rumpunen, K., 2013. Phenols and ascorbic acid in black currants (Ribes nigrum L.): Variation due to genotype, location, and year. *Journal of Agricultural and Food Chemistry*, *61*(39), pp. 9298–9306.

Vepsäläinen, S., Koivisto, H., Pekkarinen, E., Mäkinen, P., Dobson, G., McDougall, G.J., Stewart, D., Haapasalo, A., Karjalainen, R.O., Tanila, H. and Hiltunen, M., 2013. Anthocyanin-enriched bilberry and blackcurrant extracts modulate amyloid precursor protein processing and alleviate behavioral abnormalities in the APP/PS1 mouse model of Alzheimer's disease. *The Journal of Nutritional Biochemistry*, *24*(1), pp. 360–370.

Zheng, J., Yang, B., Tuomasjukka, S., Ou, S. and Kallio, H., 2009. Effects of latitude and weather conditions on contents of sugars, fruit acids, and ascorbic acid in black currant (Ribes nigrum L.) juice. *Journal of Agricultural and Food Chemistry*, *57*(7), pp. 2977–2987.

94 Rosemary (*Rosmarinus officinalis* L.)

Rosmarinus officinalis L.

Etymology: From the Latin *ros* dew, *marinus* = sea, and *officinalis* = of medicinal value

Family: Lamiaceae

Common names: Rosemary; romarin (Fr.); rosmarin (Ger.); alecrim (Port.); romero (Spa.)

Part used: Leaf

Constituents: Essential oil, diterpenes (carnosic acid, carnosol), Hydroxycinnamic acid derivatives (rosmarinic acid) (Del Bano et al., 2003).

Medical history: Dioscorides recommends rosemary for jaundice and to strengthen memory ("*cerebrum corroborat*"), and for Galen, it is a cure-all. Known to Arab physicians (Ibn Baytar). Hot and dry (Fusch, 1555). Used as a spice of choice for meat from remote times in Southern Europe. It was a tonic and used for vertigo, asthma, heart palpitations, and inflammation in the 18th-century lectures of Alston (1770). In 19th-century Europe, it was used to grow hair and to facilitate digestion and for anxiety (Guibourt, 1836; Pereira, 1843).

Medicinal uses: Carminative (1–2 g in in 150 mL of boiling water, to be taken 2–3 times daily) (European Union). Dyspepsia, antihypertensive (Turkey); flatulence, palpitations (the Philippines)

Blood glucose and lipids: Oral administration of rosemary leaf extract (0.5 mL prepared from 50 g of rosemary soaked in 150 mL hot water) to streptozotocin-induced diabetic rats caused significant declines in the blood levels of triglycerides, total cholesterol, and LDL-cholesterol but increased HDL-cholesterol (Al-Jamal & Alqadi, 2011).

Heart: Prophylactic oral intake of carnosic acid at 50 mg/kg evoked antioxidative protection against isoproterenol-induced myocardial stress via the translocation of nuclear factor erythroid 2-related factor 2 and the expression of heme-oxygenase-1 (Sahu et al., 2014).

Bones and cartilages: Rosemary added as 5% of a low-calcium diet for 8 weeks increased femoral bone mineral density in rats (Elbahnasawy et al., 2019). Aqueous extract given to rats attenuated CFA-induced arthritis (de Almeida Gonçalves et al., 2018).

Brain: Aqueous extract given orally to rats at the daily dose of 200 mg/kg attenuated scopolamine-induced dementia (Ozarowski et al., 2013).

DOI: 10.1201/9781003301455-94

Warnings: In patients with low blood pressure, intake of essential oil (1 mL every 8 hours for 44 weeks) increased blood pressure (Fernández et al., 2014). Not to be used when taking anticoagulants (Posadzki et al., 2013).

Comments: (i) When cells endure oxidative insults, nuclear factor erythroid-derived 2-like 2 translocates into the nucleus to induce the expression of genes coding for antioxidant enzymes such as heme-oxygenase 1 (Francis-queti-Ferron et al., 2019).Carnosol induces the translocation of nuclear factor erythroid-derived 2-like 2 and the expression of heme oxidase-1 (Wang et al., 2016).

(ii) It is clear that adding rosemary to dishes at a normal dietary dose or taking light rosemary infusions is beneficial to cognitive functions in aging.

REFERENCES

Al-Jamal, A.R. and Alqadi, T., 2011. Effects of rosemary (Rosmarinus officinalis) on lipid profile of diabetic rats. *Jordan Journal of Biological Sciences*, 4(4), pp. 199–204.

Alston, C., 1770. *Lectures on the Materia Medica: Containing the Natural History of Drugs, their Virtues and Doses: Also Directions for the Study of the Materia Medica; and an Appendix on the Method of Prescribing*. Edward and Charles Dilly.

de Almeida Gonçalves, G., de Sa-Nakanishi, A.B., Comar, J.F., Bracht, L., Dias, M.I., Barros, L., Peralta, R.M., Ferreira, I.C. and Bracht, A., 2018. Water soluble compounds of Rosmarinus officinalis L. improve the oxidative and inflammatory states of rats with adjuvant-induced arthritis. *Food & Function*, 9(4), pp. 2328–2340.

Del Bano, M.J., Lorente, J., Castillo, J., Benavente-García, O., Del Rio, J.A., Ortuño, A., Quirin, K.W. and Gerard, D., 2003. Phenolic diterpenes, flavones, and rosmarinic acid distribution during the development of leaves, flowers, stems, and roots of Rosmarinus officinalis. Antioxidant activity. *Journal of Agricultural and Food Chemistry*, 51(15), pp. 4247–4253.

Elbahnasawy, A.S., Valeeva, E.R., El-Sayed, E.M. and Rakhimov, I.I., 2019. The impact of thyme and rosemary on prevention of osteoporosis in rats. *Journal of Nutrition and Metabolism*, 2019.

Fernández, L.F., Palomino, O.M. and Frutos, G., 2014. Effectiveness of Rosmarinus officinalis essential oil as antihypotensive agent in primary hypotensive patients and its influence on health-related quality of life. *Journal of Ethnopharmacology*, 151(1), pp. 509–516.

Francisqueti-Ferron, F.V., Ferron, A.J.T., Garcia, J.L., Silva, C.C.V.D.A., Costa, M.R., Gregolin, C.S., Moreto, F., Ferreira, A.L.A., Minatel, I.O. and Correa, C.R., 2019. Basic concepts on the role of nuclear factor erythroid-derived 2-like 2 (Nrf2) in age-related diseases. *International Journal of Molecular Sciences*, 20(13), p. 3208.

Fusch, L., 1555. *De Historia Stirpium Commetarii Insignes*. Lugduni Apud Ioan Tornaesium.

Guibourt, N.J.B.G., 1836. *Histoire abrégée des drogues simples*. Méquignon-Marvis Père et fils.

Ozarowski, M., Mikolajczak, P.L., Bogacz, A., Gryszczynska, A., Kujawska, M., Jodynis-Liebert, J., Piasecka, A., Napieczynska, H., Szulc, M., Kujawski, R. and Bartkowiak-Wieczorek, J., 2013. Rosmarinus officinalis L. leaf extract improves memory impairment and affects acetylcholinesterase and butyrylcholinesterase activities in rat brain. *Fitoterapia*, 91, pp. 261–271.

Posadzki, P., Watson, L. and Ernst, E., 2013. Herb–drug interactions: An overview of systematic reviews. *British Journal of Clinical Pharmacology*, 75(3), pp. 603–618.

Pereira, J., 1843. *The Elements of Materia Medica and Therapeutics*. Lea and Blanchard.

Sahu, B.D., Putcha, U.K., Kuncha, M., Rachamalla, S.S. and Sistla, R., 2014. Carnosic acid promotes myocardial antioxidant response and prevents isoproterenol-induced myocardial oxidative stress and apoptosis in mice. *Molecular and Cellular Biochemistry*, *394*(1), pp. 163–176.

Wang, Z.H., Xie, Y.X., Zhang, J.W., Qiu, X.H., Cheng, A.B., Tian, L., Ma, B.Y. and Hou, Y.B., 2016. Carnosol protects against spinal cord injury through Nrf-2 upregulation. *Journal of Receptors and Signal Transduction*, *36*(1), pp. 72–78.

95 Sage
(*Salvia officinalis* L.)

Salvia officinalis L.

Etymology: From the Latin *salvia* = to save and *officinalis* = of medicinal value

Family: Lamiaceae

Common names: Sage; sauge (Fr.); salbei (Ger.); sábio (Port.); sabio (Spa.)

Part used: Leaf

Constituents: Essential oil (thujone) (Mockutë et al., 2003), labdane-type diterpenes (carnosic acid, carnosol, sclareol) (Fischedick et al., 2013; Park et al., 2016).

Medical history: Sage was known to Roman physicians as a diuretic as well as for pruritus and wounds (Dioscorides). The Greek physician Aetius of Amida (6th century AD), in his *Aetii medici graeci ex veteribus medicinae*, states that sage promotes fertility in women. The Medical School of Salerno (10th century) holds sage as able to save (in Latin *salvia*) from death: "*Cur moriatur homo cui Salvia crescit in horto?*" and uses it to strengthen nerves and to treat fever and paralysis. It was considered a hot and dry remedy by German physicians of the 16th century (Fusch, 1555). In 19th-century France, sage was given as a tonic and carminative (Guibourt, 1836).

Medicinal uses: Carminative (1–2 g in 150 mL boiling water) (European Union). Cold, diuretic, respiratory disorders, carminative, sore throat, kidney stones (Turkey); relieve pain during pregnancy, regulate menstrual cycle, diabetes, high cholesterol, flatulence, fever (Iraq)

Plasma lipids and glucose: An extract of sage at 500 mg given to hyperlipidemic patients every 8 h for 2 months decreased total cholesterol, triglyceride, LDL-cholesterol, and VLDL-cholesterol and increased HDL-cholesterol (Kianbakht et al., 2011). In diabetic patients given 150 mg of an extract 3 times a day for 3 months, a decrease was observed after 2 hours in postprandial glycemia and cholesterolemia (Behradmanesh et al., 2013).

Heart: A methanol extract given to rats at 200 mg/kg/day for 28 days attenuated ventricular tachycardia, ventricular fibrillation, and ventricular premature beats induced by the injection of a solution calcium chloride (Radan & Dianat, 2018).

Bones and cartilages: Infusion (2 g in 150 mL of boiling water) mixed with drinking water to aged non-cycling female rats improved the bone mineral density of the left femur and left tibia and increased plasma levels of calcium ions and estradiol while decreasing plasma alkaline phosphatase (Abdallah et al., 2010).

DOI: 10.1201/9781003301455-95

Skin aging and hairs: Applying a topical preparation containing 0.02% of sclareol for 12 weeks to volunteers decreased the development of wrinkles. *In vitro*, sclareol inhibited the secretion of matrix metalloproteinase by human fibroblasts irradiated with ultraviolet B via the upregulation of AP-1 (Park et al., 2016).

Brain: Extract given to Alzheimer's patients for 4 months improved cognitive function and attenuated agitation (Akhondzadeh et al., 2003).

Warnings: Thujone acts on brain GABA receptors and can induce excitation and convulsion (Pelkonen et al., 2013. Not to be taken during pregnancy and lactation (Spiteri, 2011).

Comment: Arrythmia risk increases with aging. Connexin 43 is a transmembrane protein that conveys the electrical impulse from myocytes to myocytes, and with aging, and especially in the absence of physical exercise, the amount of connexin 43 decreases. This results in arrythmia that is magnified as the autonomic cardiac nervous system ages (Chadda et al., 2018; Jones, 2006).

REFERENCES

Abdallah, I.Z., Khattab, H.A., Sawiress, F.A. and El-Banna, R.A., 2010. Effect of Salvia Officinalis L. (sage) herbs on osteoporotic changes in aged non-cycling female rats. *Medical Journal of Cairo University*, 78(Suppl 2), pp. 1–9.

Akhondzadeh, S., Noroozian, M., Mohammadi, M., Ohadinia, S., Jamshidi, A.H. and Khani, M., 2003. Salvia officinalis extract in the treatment of patients with mild to moderate Alzheimer's disease: A double blind, randomized and placebo-controlled trial. *Journal of Clinical Pharmacy and Therapeutics*, 28(1), pp. 53–59.

Behradmanesh, S., Derees, F. and Rafieian-Kopaei, M., 2013. Effect of Salvia officinalis on diabetic patients. *Journal of Renal Injury Prevention*, 2(2), p. 51.

Chadda, K.R., Ajijola, O.A., Vaseghi, M., Shivkumar, K., Huang, C.L.H. and Jeevaratnam, K., 2018. Ageing, the autonomic nervous system and arrhythmia: From brain to heart. *Ageing Research Reviews*, 48, pp. 40–50.

Fischedick, J.T., Standiford, M., Johnson, D.A. and Johnson, J.A., 2013. Structure activity relationship of phenolic diterpenes from Salvia officinalis as activators of the nuclear factor E2-related factor 2 pathway. *Bioorganic & Medicinal Chemistry*, 21(9), pp. 2618–2622.

Fusch, L., 1555. *De Historia Stirpium Commetarii Insignes*. Lugduni Apud Ioan Tornaesium.

Guibourt, N.J.B.G., 1836. *Histoire abrégée des drogues simples*. Méquignon-Marvis Père et fils.

Jones, S.A., 2006. Ageing to arrhythmias: Conundrums of connections in the ageing heart. *Journal of Pharmacy and Pharmacology*, 58(12), pp. 1571–1576.

Kianbakht, S., Abasi, B., Perham, M. and Hashem Dabaghian, F., 2011. Antihyperlipidemic effects of Salvia officinalis L. leaf extract in patients with hyperlipidemia: A randomized double-blind placebo-controlled clinical trial. *Phytotherapy Research*, 25(12), pp. 1849–1853.

Mockutë, D., Nivinskienë, O., Bernotienë, G. and Butkienë, R., 2003. The cis-thujone chemotype of Salvia officinalis L. essential oils. *Chemija*, 14(4), pp. 216–220.

Park, J.E., Lee, K.E., Jung, E., Kang, S. and Kim, Y.J., 2016. Sclareol isolated from Salvia officinalis improves facial wrinkles via an antiphotoaging mechanism. *Journal of Cosmetic Dermatology*, 15(4), pp. 475–483.

Pelkonen, O., Abass, K. and Wiesner, J., 2013. Thujone and thujone-containing herbal medicinal and botanical products: Toxicological assessment. *Regulatory Toxicology and Pharmacology*, *65*(1), pp. 100–107.

Radan, M. and Dianat, M., 2018. Sage (Salvia officinalis L.) Protects against cardiac arrhythmias and electrocardiogram irregularity in rats. *Jundishapur Journal of Physiology*, *1*(1), pp. 1–5.

Spiteri, M., 2011. *Herbal Monographs Including Herbal Medicinal Products and Food Supplements*. Department of Pharmacy, University of Malta.

96 Common Elder (*Sambucus nigra* L.)

Sambucus nigra L.

Etymology: From the Latin *sambuca* = a wind instrument made from the hollowed stems of common elder and *nigra* = black

Family: Viburnaceae

Common names: Common elder, elderberry, boor tree; sureau, suyer (Fr.); holunder (Ger.); sabugueiro (Port.); бузина́ (Rus.); saúco (Spa.)

Pat used: Fruits

Constituents: Anthocyanins (cyanidin-3-sambubioside, cyanidin-3-glucoside) (Bronnum-Hansen & Hansen, 1983), citric acid, phenolics, hydroxycinnamic acid derivatives, flavone glycosides, cyanogenic glycosides (sambunigrin, 19 mg/kg in fresh berries) (Ferreira et al., 2022).

Medical history: Diuretic (Dioscorides). Of no use for the Medical School of Salerno (10th century). German physicians in the 16th century considered common elder hot and dry (Fusch, 1555), while in Italy it was used for syphilis and sciatica and to darken the hair (Matthioli, 1572). In 17th-century France, it was used as a diuretic, for burns, and otalgia (Daléchamps, 1615). In 18th-century France, common elder was a remedy for dysentery (Lémery, 1716) and in Scotland, it was used as a diuretic as well as for dropsy ("*hydropicos juvat*") and for infections, and it was known to be emetic and cathartic. In 19th-century Europe and North America, common elder was used to refresh and to increase appetite and as a diuretic (Pereira, 1843).

Medicinal uses: Asthma, diuretic, fever, laxative, prostate problems (Turkey)

Blood pressure: The systolic blood pressure of L-NAME-induced hypertensive rats decreased from about 170 to 150 mmHg and diastolic blood pressure from about 120 to 100 mmHg with the oral administration of ethanol extract of fruits given at 46 mg/kg body every 2 days for 8 weeks (Ciocoiu et al., 2016).

Heart: L-NAME-induced hypertensive rats given ethanol extract of fruits at 46 mg/kg body orally at every 2 days for 8 weeks had heart rate decreased from about 430 to 275 bpm (Ciocoiu et al., 2016).

Plasma lipids and glucose: Rats fed an extract for 6 weeks were protected against the aortic deposition of cholesterol in aorta (Farrell et al., 2015).

Kidneys: Aqueous extract increased urinary sodium ions excretion in rats (Beaux et al., 1998).

Brain: Common elder berries added to the diet of Alzheimer's disease model rats for 8 weeks improved cognitive function and decreased hippocampal

DOI: 10.1201/9781003301455-96

content of caspase-3 while increasing hippocampal pyramidal neurons (Jahanbakhshi et al., 2022).

Warning: Cyanogenic glycosides decompose into hydrogen cyanide, a violent poison (Senica et al., 2016).

Comment: The fruits are of no therapeutic use and are poisonous. Fractions of anthocyanins, however, might have the potential to slow aging. More experiments are needed.

REFERENCES

Beaux, D., Fleurentin, J. and Mortier, F., 1998. Effect of extracts of Orthosiphon stamineus benth, Hieracium pilosella l., Sambucus nigra l. and Arctostaphylos uva-ursi (l.) spreng. in rats. *Phytotherapy Research: An International Journal Devoted to Pharmacological and Toxicological Evaluation of Natural Product Derivatives, 12*(7), pp. 498–501.

Bronnum-Hansen, K. and Hansen, S.H., 1983. High-performance liquid chromatographic separation of anthocyanins of Sambucus nigra L. *Journal of Chromatography, 262*, pp. 385–392.

Ciocoiu, M., Badescu, M., Badulescu, O. and Badescu, L., 2016. The beneficial effects on blood pressure, dyslipidemia and oxidative stress of Sambucus nigra extract associated with renin inhibitors. *Pharmaceutical Biology, 54*(12), pp. 3063–3067.

Daléchamps, 1615. *De l' histoire generale des plantes simples.* Chez Heritier Guillaume Rouille.

Farrell, N., Norris, G., Lee, S.G., Chun, O.K. and Blesso, C.N., 2015. Anthocyanin-rich black elderberry extract improves markers of HDL function and reduces aortic cholesterol in hyperlipidemic mice. *Food & Function, 6*(4), pp. 1278–1287.

Ferreira, S.S., Silva, A.M. and Nunes, F.M., 2022. Sambucus nigra L. fruits and flowers: Chemical composition and related bioactivities. *Food Reviews International, 38*(6), pp. 1237–1265.

Fusch, L., 1555. *De Historia Stirpium Commetarii Insignes.* Lugduni Apud Ioan Tornaesium.

Jahanbakhshi, H., Moghaddam, M.H., Parvardeh, S., Boroujeni, M.E., Vakili, K., Azimi, H., Mehranpour, M., et al., 2022. The elderberry diet protection against intrahippocampal Aβ-induced memory dysfunction: The abrogated apoptosis and neuroinflammation.

Lémery, N., 1716. *Traité universel des drogues simples, mises en ordre alphabétique. Où l'on trouve leurs différens noms . . . et tout ce qu'il y a de particulier dans les animaux, dans les végétaux, et dans les minéraux.* Au dépend de la Companie.

Matthioli, P.A., 1572. *Commentaires sur les Six Livres de Pedacius Dioscorides Anazarbeen de la matière medicinale.* A l'Escue de Milan.

Pereira, J., 1843. *The Elements of Materia Medica and Therapeutics.* Lea and Blanchard.

Senica, M., Stampar, F., Veberic, R. and Mikulic-Petkovsek, M., 2016. Processed elderberry (Sambucus nigra L.) products: A beneficial or harmful food alternative? *LWT-Food Science and Technology, 72*, pp. 182–188.

97 Summer Savory (*Satureja hortensis* L.)

Satureja hortensis L.

Etymology: From the Latin *satureja* = savory and *hortensis* = garden

Family: Lamiaceae

Synonyms: *Satureja laxiflora* (Hayata) Matsum. & Kudô; *Thymus cunila* (L.) E.H.L. Krause

Common names: Summer savory; sariette (Fr.); bohnenkraut (Ger.); sergurelha (Port.); чабер садовый (Rus.); ajedrea (Spa.)

Part used: Leaf

Constituents: Essential oil (carvacrol, γ-terpinene) (Ghannadi, 2002), flavones, flavone glycosides (Hudz et al., 2020).

Medical history: Summer savory was used as a diuretic by Roman physicians, including for Dioscorides. In 16th-century Netherlands, physicians considered the plant hot and dry to the third degree (Dodoens, 1557). In 18th-century France, summer savory was used as a spice for grilled meat, to ease breathing, and to induce urination (Lémery, 1716). In the lectures of Alston (1770), summer savory is advised for asthma and as an aphrodisiac and carminative. Summer savory was used as a tonic in 19th-century France (Guibourt, 1836).

Medicinal uses: Boosts immune system (Turkey); gout (Iran); aphrodisiac (Fr.)

Blood pressure: Obese patients receiving 450 mg/day of leaf powder before meals daily for 10 weeks had a decrease in diastolic blood pressure from 83.1 to 75.3 mmHg and reduced plasma hs-CRP (Nikaein et al., 2017).

Plasma lipids and glucose: Obese patients receiving 450 mg/day of leaf powder before meals daily for 10 weeks had decreases in total cholesterol from 239.4 to 222.3 mg/dL, LDL-cholesterol from 138.6 to 117.6 mg/dL, and triglycerides from 220 to 187.5 mg/dL (Nikaein et al., 2017).

Comment: Rembert Dodoens (1517–1585) was a Flemish physician and botanist (physicians were all botanists at that time) to the emperor Maximilian II and professor of medicine at the University of Leiden, as well as the author of *Histoire des Plantes* (1557).

DOI: 10.1201/9781003301455-97

REFERENCES

Alston, C., 1770. *Lectures on the Materia Medica: Containing the Natural History of Drugs, their Virtues and Doses: Also Directions for the Study of the Materia Medica; and an Appendix on the Method of Prescribing.* Edward and Charles Dilly.

Dodoens, R., 1557. *Histoire des plantes en laquelle est contenue la description entiere des herbes . . . non seulement de celles qui croissent en ce païs, mais aussi des autres estrangères qui viennent en usage de médecine par Rembert Dodoens . . . traduite de bas aleman en françois par Charles de L'Escluse. (n.p.).* de l'imprimerie de Jean Loë.

Ghannadi, A., 2002. Composition of the essential oil of Satureja hortensis L. seeds from Iran. *Journal of Essential Oil Research, 14*(1), pp. 35–36.

Guibourt, N.J.B.G., 1836. *Histoire abrégée des drogues simples.* Méquignon-Marvis Père et fils.

Hudz, N., Makowicz, E., Shanaida, M., Białoń, M., Jasicka-Misiak, I., Yezerska, O., Svydenko, L. and Wieczorek, P.P., 2020. Phytochemical evaluation of tinctures and essential oil obtained from Satureja montana herb. *Molecules, 25*(20), p. 4763.

Lémery, N., 1716. *Traité universel des drogues simples, mises en ordre alphabétique. Où l'on trouve leurs différens noms . . . et tout ce qu'il y a de particulier dans les animaux, dans les végétaux, et dans les minéraux.* Au dépend de la Companie.

Nikaein, F., Babajafari, S., Mazloomi, S.M., Zibaeenezhad, M. and Zargaran, A., 2017. The effects of Satureja hortensis L. dried leaves on serum sugar, lipid profiles, hs-CRP, and blood pressure in metabolic syndrome patients: A double-blind randomized clinical trial. *Iranian Red Crescent Medical Journal, 19*(1), p. e34931.

98 Chayote (*Sechium edule* (Jacq.) Sw.)

Sechium edule (Jacq.) Sw.

Etymology: From the Greek *sikyos* = cucurbit and *edulis* = edible

Family: Cucurbitaceae

Synonyms: *Chayota edulis* (Jacq.) Jacq., *Sechium americanum* Poir.; *Sicyos edulis* Jacq.

Common names: Chayote; chuchu (Port.); чайот (Rus.)

Part used: Fruit

Constituents: Cucurbitane-type triterpenes (cucurbitacins B and D) (Cadena-Iñiguez et al., 2007).

Medical history: Chayote has been used by natives in Central America (local name: *chocho*) for food since the dawn of time. It was brought to the knowledge of Europeans in the 18th century by Nicolaus Joseph von Jacquin (1760) and later by Olof Peter Schwartz (1797).

Medicinal uses: Hypertension (the Philippines)

Plasma lipids and glucose: Hydroalcoholic extract given orally for 3 weeks at 400 mg/kg to rats on a high-fat diet decreased total cholesterol from 200.4 to 92.3 mg/dL and triglycerides from 137.3 to 82.5 mg/dL (Mohammed et al., 2022).

Blood pressure: Juice expressed from 100 g of fruits given to elderly persons daily for 10 days decreased systolic blood pressure from 156.2 to 152.4 mmHg (Fauziningtyas & Ristanto, 2020).

Heart: Ethanol extract of fruits given orally at 200 mg/kg/day for 28 days protected rats against isoproterenol-induced myocardial infarction (Neeraja et al., 2015). Hydroalcoholic extract given orally to rats for 3 weeks at 200 mg/kg/day on a high-fat diet normalized the prolonged QT or QTc and RR interval as well as heart rate (Mohammed et al., 2022).

Kidneys: Capsules containing 500 mg of fruit powder given before each meal for 6 weeks to obese elderly decreased plasma uric acid and creatinine (Gavia-García et al., 2020).

Warning: No to be consumed during pregnancy as it tends to induce hypokalemia (Jensen & Lai, 1986).

Comments: (i) Nicolaus Joseph von Jacquin (1727–1817) was a Dutch botanist.

(ii) Olof Peter Schwartz (1760–1816) was a Swedish botanist.

DOI: 10.1201/9781003301455-98

REFERENCES

Cadena-Iñiguez, J., Arévalo-Galarza, L., Avendaño-Arrazate, C.H., Soto-Hernández, M., Ruiz-Posadas, L.D.M., Santiago-Osorio, E., Acosta-Ramos, M., Cisneros-Solano, V.M., Aguirre-Medina, J.F. and Ochoa-Martínez, D., 2007. Production, genetics, postharvest management and pharmacological characteristics of Sechium edule (Jacq.) Sw. *Fresh Produce*, *1*(1), pp. 41–53.

Fauziningtyas, R. and Ristanto, A.C.A., 2020. Effectiveness of consumption sechium edule on decreasing blood pressure in elderly with hypertension in coastal area. In *IOP Conference Series: Earth and Environmental Science* (Vol. 519, No. 1, p. 012005). IOP Publishing, June.

Gavia-García, G., Rosado-Pérez, J., Aguiñiga-Sánchez, I., Santiago-Osorio, E. and Mendoza-Núñez, V.M., 2020. Effect of Sechium edule var. nigrum spinosum (Chayote) on telomerase levels and antioxidant capacity in older adults with metabolic syndrome. *Antioxidants*, *9*(7), p. 634.

Jacquin, N.J.F.v., 1760. *Enumeratio systematica plantarum, quas in insulis Caribaeis vicinaque Americes continente detexit novas, aut jam cognitas emendavit.* Haak.

Jensen, L.P. and Lai, A.R., 1986. Chayote (Sechium edule) causing hypokalemia in pregnancy. *American Journal of Obstetrics and Gynecology*, *155*(5), pp. 1048–1049.

Mohammed, F.S., Ghosh, A., Pal, S., Das, C., Alomar, S.Y., Patwekar, M., Patwekar, F., Jeon, B.H. and Islam, F., 2022. Hydroalcoholic extract of sechium edule fruits attenuates QT prolongation in high fat diet-induced hyperlipidemic mice. *Evidence-based Complementary and Alternative Medicine*, *2022*.

Neeraja, K., Debnath, R. and Firdous, S.M., 2015. Cardioprotective activity of fruits of Sechium edule. ||| *Bangladesh Journal of Pharmacology*|||, *10*(1), pp. 125–130.

99 Sesame (*Sesamum indicum* L.)

Sesamum indicum **L.**

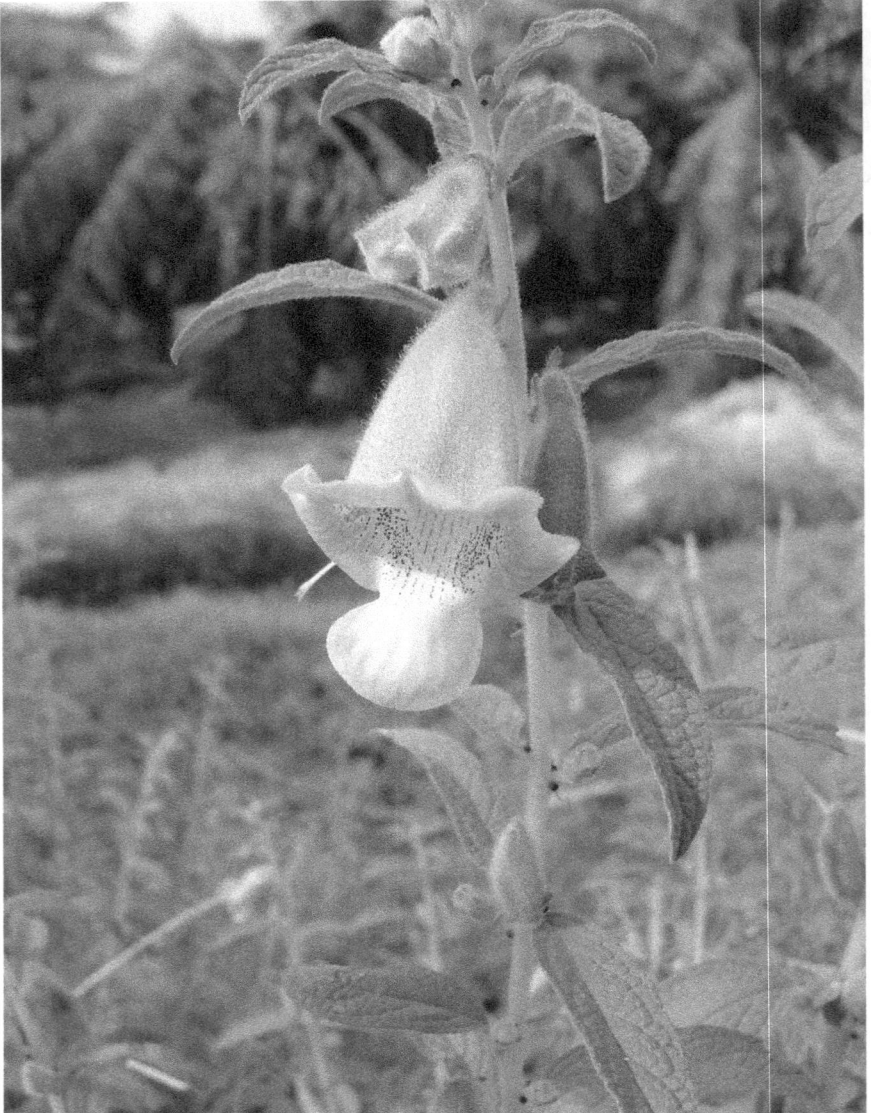

DOI: 10.1201/9781003301455-99

Etymology: From the Latin *sesama* = sesame and *indicum* = from India

Family: Pedaliaceae

Synonym: *Sesamum orientale* L.

Common names: Benne, sesame; sésame (Fr.); sesam (Ger.); sésamo (Port.; Spa.); кунжýт (Ger.)

Part used: Seed

Constituents: Fixed oil (unsaturated fatty acids: linoleic acid, oleic acid), lignans (sesamin, sesamolin, sesamol) (Were et al., 2006).

Medical history: Sesame seeds have been used by Hindus since the dawn of time as a source of oil. Sesame was known to Dioscorides, who tells us that it was used in Ancient Egypt. He also asserts that sesame is anti-inflammatory and good for colic while warning that it is unpleasant for the stomach. Pliny the Elder recommends sesame to stop vomiting, and Galen used to consume it. According to the Persian physician Yuhanna ibn Masawaih (777–857 AD), sesame increases male fertility. The 16th-century Indian physician Bhavmisra, in his book titled "*Bhavaprakasha*", recommends sesame as a tonic and diuretic and for piles, constipation, and dysentery (Dutt, 1877). French physicians in the 18th century used sesame for breathing difficulties, to induce menses, for pain, to enhance male fertility, and to moisten the skin (Lémery, 1716)

Medicinal uses: Emollient (Iran, Iraq); aphrodisiac, cooling, biliousness, boils, diuretic, inflammation, leprosy, skin irritation, syphilitic ulcers, tonic, piles (India); diuretic, laxative (Myanmar); laxative (Korea)

Blood pressure: In hypertensive rats, addition of sesamin to the diet at 1% for 2 weeks caused a decrease in diastolic blood pressure from 199.7 to 153.7 mmHg (Nakano et al., 2003).

Hypertensive diabetic patients using sesame oil for cooking at 35 g/day for 45 days had decreases in systolic blood pressure from 150 to 129.6 mmHg and diastolic blood pressure from 98.3 to 80 mmHg (Sankar et al., 2006).

Volunteers taking 252 mg of sesame seeds each day for 4 weeks had decreases in systolic blood pressure from 129.3 to 121 mmHg and in diastolic blood pressure from 77 to 72.8 mmHg together with decreased plasma MDA (Wichitsranoi et al., 2011).

Intake of sesame oil at 35 g daily for 2 months improved vascular endothelium function, as evidenced by an increase in flow-mediated dilatation (Karatzi et al., 2013).

Sesamin given orally at 60 mg/day to hypertensive patients for 4 weeks decreased systolic blood pressure from 137.6 to 134.1 mmHg and diastolic blood pressure from 87.7 to 85.8 mmHg (Miyawaki et al., 2009).

Plasma lipids and glucose: Rabbits on a high-cholesterol diet given seeds at 10% of diet for 60 days decreased plasma cholesterol and LDL-cholesterol and plasma triglycerides. The same rabbits given oil of sesame as 5% of a high-fat diet experienced a decrease LDL-cholesterol from 1337 to 570 mg/dL (normal: 101 mg/dL) (Asgary et al., 2013). Sesamin given at 160 mg/kg/

day for 7 weeks to rats on a high-fat diet normalized total plasma cholesterol from 2.6 to 1.6 mmol/L and decreased plasma triglycerides from 1.2 to 0.8 mmol/L, and LDL-cholesterol from 1.3 to 1 mmol/L (Zhang et al., 2016).

Hypertensive diabetic patients using sesame oil for cooking at 35 g/day for 45 days had decreases in total cholesterol from 250.7 to 205.8 mg/dL, LDL-cholesterol from 158.9 to 121.3 mg/dL, and triglycerides from 236 to 181.6 mg/dL (Sankar et al., 2006).

Kidneys: In rats poisoned with a cholesterol-enriched diet, supplementation with 5% sesame oil for four weeks decreased plasma urea from 12.2 to 10.9 μmol/L (Olubumi et al., 2014).

Sesamin given at 160 mg/kg/day for 7 weeks to rats on a high-fat diet normalized plasma creatinine from 42.1 to 34.5 mmol/L and prevented renal cytoarchitecture abnormalities (Zhang et al., 2016).

Bones and cartilages: Ethanol extract of seeds given orally to rats at 800 mg/kg/day for 20 days attenuated CFA-induced arthritis with reduced synovial inflammation, cartilage damage, and decreased the expression of interleukin-6 and TNF α (Ruckmani et al., 2018).

Warning: Sesame seeds are bitter and must be avoided in patients with hyperkaliemia.

REFERENCES

Asgary, S., Rafieian-Kopaei, M., Najafi, S., Heidarian, E. and Sahebkar, A., 2013. Antihyperlipidemic effects of Sesamum indicum L. in rabbits fed a high-fat diet. *The Scientific World Journal*, *2013*.

Dutt, U.C., 1877. *The Materia Medica of the Hindus*. Thacker, Spink, & Co.

Karatzi, K., Stamatelopoulos, K., Lykka, M., Mantzouratou, P., Skalidi, S., Zakopoulos, N., Papamichael, C. and Sidossis, L.S., 2013. Sesame oil consumption exerts a beneficial effect on endothelial function in hypertensive men. *European Journal of Preventive Cardiology*, *20*(2), pp. 202–208.

Lémery, N., 1716. *Traité universel des drogues simples, mises en ordre alphabétique. Où l'on trouve leurs différens noms . . . et tout ce qu'il y a de particulier dans les animaux, dans les végétaux, et dans les minéraux.* Au dépend de la Companie.

Miyawaki, T., Aono, H., Toyoda-Ono, Y., Maeda, H., Kiso, Y. and Moriyama, K., 2009. Antihypertensive effects of sesamin in humans. *Journal of Nutritional Science and Vitaminology*, *55*(1), pp. 87–91.

Nakano, D., Itoh, C., Ishii, F., Kawanishi, H., Takaoka, M., Kiso, Y., Tsuruoka, N., Tanaka, T. and Matsumura, Y., 2003. Effects of sesamin on aortic oxidative stress and endothelial dysfunction in deoxycorticosterone acetate-salt hypertensive rats. *Biological and Pharmaceutical Bulletin*, *26*(12), pp. 1701–1705.

Olubumi, A.B., Akomolafe, S.F., Oluwaseyi, M.I. and Adebowale, O.S., 2014. Effect of Sesamum indicum L. seed oil supplementation on the kidney function parameters of hypercholesterolemic rats. *Journal of Nutrition & Food Sciences*, *4*(5), p. 1.

Ruckmani, A., Meti, V., Vijayashree, R., Arunkumar, R., Konda, V.R., Prabhu, L., Madhavi, E. and Devi, S., 2018. Anti-rheumatoid activity of ethanolic extract of Sesamum indicum seed extract in Freund's complete adjuvant induced arthritis in Wistar albino rats. *Journal of Traditional and Complementary Medicine*, 8(3), pp. 377–386.

Sankar, D., Rao, M.R., Sambandam, G. and Pugalendi, K.V., 2006. A pilot study of open label sesame oil in hypertensive diabetics. *Journal of Medicinal Food*, 9(3), pp. 408–412.

Were, B.A., Onkware, A.O., Gudu, S., Welander, M. and Carlsson, A.S., 2006. Seed oil content and fatty acid composition in East African sesame (Sesamum indicum L.) accessions evaluated over 3 years. *Field Crops Research*, 97(2–3), pp. 254–260.

Wichitsranoi, J., Weerapreeyakul, N., Boonsiri, P., Settasatian, C., Settasatian, N., Komanasin, N., Sirijaichingkul, S., Teerajetgul, Y., Rangkadilok, N. and Leelayuwat, N., 2011. Anti-hypertensive and antioxidant effects of dietary black sesame meal in pre-hypertensive humans. *Nutrition Journal*, 10(1), pp. 1–7.

Zhang, R., Yu, Y., Deng, J., Zhang, C., Zhang, J., Cheng, Y., Luo, X., Han, B. and Yang, H., 2016. Sesamin ameliorates high-fat diet–induced dyslipidemia and kidney injury by reducing oxidative stress. *Nutrients*, 8(5), p. 276.

100 Milk Thistle (*Silybum marianum* (L.) Gaetn.)

Silybum marianum (L.) Gaetn

Etymology: From the Greek *sillibos* = an ancient thistle plant and the Latin *marianum* = of Mary

Family: Asteraceae

Synonyms: *Carduus mariae* Crantz; *Carduus marianus* L.; *Carthamus maculatum* (Scop.) Lam.; *Cirsium maculatum* Scop.; *Mariana lactea* Hill; *Silybum maculatum* (Scop.) Moench; *Silybum mariae* (Crantz) Gray

Common names: Milk thistle; chardon argentin, chardon sauvage, chardon de Notre Dame, chardon Marie (Fr.); mariendistel (Ger.); cardo mariao (Port.); расторопша (Rus.); cardo lechoso (Spa.)

Part used: Fruit

Constituents: Flavonolignans (collectively known as sylimarin: silybins A and B, isosilybins A and B, silychristin, isosilychristin, silydianin) (Elwekeel et al., 2013; Kim et al., 2003), amide alkaloids (Qin et al., 2017).

Medical history: Milk thistle was used for sciatica by Dioscorides and for water retention by Galen. French physicians in the 17th century used milk thistle for heart and liver health (Daléchamps, 1615), while in Italy, it was used as a diuretic and for jaundice and painful kidneys. It was also considered to be hot to the third degree and dry to the second (Matthioli, 1572). In 18th-century France, it was used for water retention (Lémery, 1716), as well as for rheumatism and to improve breathing (Alston, 1770).

Medicinal uses: Carminative, liver complaints (3–5 g in 100 mL of boiling water, 2–3 times daily, before meals) (European Union); jaundice, fever (Iran); liver diseases (Pakistan).

Blood pressure: Silymarin given orally at 200 mg 3 times daily to diabetic patients for 4 months attenuated systolic and diastolic blood pressure (Husseini et al., 2006).

Plasma lipids and glucose: Silymarin given orally at 200 mg 3 times per day to diabetic patients for 4 months decreased plasma glucose from 156 to 133 mg/dL, total plasma cholesterol from 225 to 198 mg/dL, LDL-cholesterol from 140 to 123 mg/dL, and triglycerides from 284 to 211 mg/dL (Husseini et al., 2006).

DOI: 10.1201/9781003301455-100

Heart: In rabbits poisoned from a high-cholesterol diet, the ingestion of 200 mg/kg/day for 2 months decreased plasma cholesterol from 1285.8 to 745.9 mg/dL, LDL-cholesterol from 1079.6 to 685.2 mg/dL, and triglycerides from 218.6 to 115.2 mg/dL and decreased atheromatous plaque formation from about 20 to 6% (Radjabian & Fallah, 2010). Sylimarin given orally at 200 mg/kg/day for 9 weeks to alloxan-induced diabetic rats normalized plasma glucose from 23.2 to 8.7 mmol/L and prevented the development of glomerular damages, cellular disruption in the renal in medulla and cortex, as well as vacuolization, and tubules lysis in the renal epithelium zones (Soto et al., 2010). Sylimarin given orally at 500 mg/kg/day for a week protected the hearts of rats against the occlusion and reperfusion of the left anterior descending coronary artery (Rao et al., 2007).

Kidneys: Sylimarin given orally at 200 mg/kg/day for 9 weeks to alloxan-induced diabetic rats prevented the development of glomerular damage, cell disruption (in medulla and cortex), vacuolization, and tubule lysis in epithelium zones in kidneys (Soto et al., 2010).

Bones and cartilages: 420 mg of silymarin given daily to patients with rheumatic arthritis for 3 months decreased join swelling (Gupta et al., 2000).

Brain: An extract given orally and daily at 200 mg/kg/day for 4 weeks improved the cognitive function of mice model of Alzheimer's disease (Hadinia et al., 2010).

Warnings: Anaphylactic shock occurs in subjects allergic to milk thistle (Geier et al., 1990). Interacts with cisplatin, metronidazole, nifedipine, protease inhibitors, pyrazinamide, and warfarin (Spiteri, 2011).

REFERENCES

Alston, C., 1770. *Lectures on the Materia Medica: Containing the Natural History of Drugs, their Virtues and Doses: Also Directions for the Study of the Materia Medica; and an Appendix on the Method of Prescribing.* Edward and Charles Dilly.

Daléchamps, 1615. *De l' histoire generale des plantes simples.* Chez Heritier Guillaume Rouille.

Elwekeel, A., Elfishawy, A. and AbouZid, S., 2013. Silymarin content in Silybum marianum fruits at different maturity stages. *Journal of Medicinal Plants Research, 7*(23), pp. 1665–1669.

Geier, J., Fuchs, T.H. and Wahl, R., 1990. Anaphylactic shock due to an extract of Silybum marianum in a patient with immediate-type allergy to kiwi fruit. *Allergologie, 13*(10), pp. 387–388.

Gupta, O.P., Sing, S., Bani, S., Sharma, N., Malhotra, S., Gupta, B.D., Banerjee, S.K. and Handa, S.S., 2000. Anti-inflammatory and anti-arthritic activities of silymarin acting through inhibition of 5-lipoxygenase. *Phytomedicine, 7*(1), pp. 21–24.

Hadinia, A., Aryanpour, R., Mehdizadeh, M., Mahmodi, R., Mossavizadeh, A., Delaviz, H., Pirhajati, H. and Ghnbari, A., 2010. The effect of silybum marianum on GFAP and spatial memory in a mouse model of alzheimer's disease. *Armaghane Danesh, 14*(4), pp. 65–75.

Huseini, H.F., Larijani, B., Heshmat, R., Fakhrzadeh, H., Radjabipour, B., Toliat, T. and Raza, M., 2006. The efficacy of Silybum marianum (L.) Gaertn. (silymarin) in the treatment of type II diabetes: A randomized, double-blind, placebo-controlled, clinical trial. *Phytotherapy Research: An International Journal Devoted to Pharmacological and Toxicological Evaluation of Natural Product Derivatives*, *20*(12), pp. 1036–1039.

Kim, N.C., Graf, T.N., Sparacino, C.M., Wani, M.C. and Wall, M.E., 2003. Complete isolation and characterization of silybins and isosilybins from milk thistle (Silybum marianum). *Organic & Biomolecular Chemistry*, *1*(10), pp. 1684–1689.

Lémery, N., 1716. *Traité universel des drogues simples, mises en ordre alphabétique. Où l'on trouve leurs différens noms . . . et tout ce qu'il y a de particulier dans les animaux, dans les végétaux, et dans les minéraux*. Au dépend de la Companie.

Matthioli, P.A., 1572. *Commentaires sur les Six Livres de Pedacius Dioscorides Anazarbeen de la matière medicinale*. A l'Escue de Milan.

Qin, N.B., Jia, C.C., Xu, J., Li, D.H., Xu, F.X., Bai, J., Li, Z.L. and Hua, H.M., 2017. New amides from seeds of Silybum marianum with potential antioxidant and antidiabetic activities. *Fitoterapia*, *119*, pp. 83–89.

Radjabian, T. and Fallah, H.H., 2010. Anti-hyperlipidemic and anti-atherosclerotic activities of silymarins from cultivated and wild plants of Silybum marianum L. with different content of flavonolignans. *Iranian Journal of Pharmacology and Therapeutics*, *9*, pp. 63–67.

Rao, B.N., Srinivas, M., Kumar, Y.S. and Rao, Y.M., 2007. Effect of silymarin on the oral bioavailability of ranitidine in healthy human volunteers. *Drug Metabolism and Drug Interactions*, *22*(2–3), pp. 175–186.

Soto, C., Pérez, J., García, V., Uría, E., Vadillo, M. and Raya, L., 2010. Effect of silymarin on kidneys of rats suffering from alloxan-induced diabetes mellitus. *Phytomedicine*, *17*(14), pp. 1090–1094.

Spiteri, M., 2011. *Herbal Monographs Including Herbal Medicinal Products and Food Supplements*. Department of Pharmacy, University of Malta.

101 Spinach (*Spinacia oleracea* L.)

Spinacia oleracea L.

Etymology: From the Latin *spina* = thorn and *oleraceus* = a pot herb

Family: Amaranthaceae

Synonym: *Chenopodium oleraceum* (L.) E.H.L. Krause

Common names: Spinach; épinards (Fr.); espinafre (Port.); spinat (Ger.); шпинат (Rus.); espinaca (Spa.)

Part used: Leaf

Constituents: Nitrates (Barker et al., 1974), flavone glycosides (Aritomi et al., 1985; Ferreres et al., 1997).

Medical history: Physicians in 16th-century Germany used spinach as a cold and humid medicine (Fusch, 1555). In 18th-century France, it was considered good for the stomach and for purifying the blood (Lémery, 1716). The French King Louis Philippe was advised by his personal physician to eat large amounts of spinach. Spinach was used as a laxative, for cooling, for fever, and for diseases of the blood and heart at the time of the First World War (Kirtikar & Basu, 1918).

Medicinal uses: Anemia (Pakistan); cooling, laxative, urinary tract infection, diseases of the blood (India); diuretic, fever, inflammation (Iran, Iraq); thirst, laxative, blood circulation, alcoholism (Korea)

Blood pressure: Consumption of 200 g of spinach daily for 4 weeks increased plasma NO by about threefold, resulting in a decrease of systolic and diastolic blood pressure (Bondonno et al., 2012).

Plasma lipids and cholesterol: Rats poisoned with cholesterol-enriched diet were protected against hypercholesterolemia when consuming spinach for 56 days (Iritani, 1969).

Gallbladder: 100 mg of dry spinach binds 3.2 μmol of bile acid (Kahlon et al., 2007).

Bones and cartilages: An extract given to ovariectomized rats for 12 weeks alleviated symptoms of osteoporosis (Adhikary et al., 2017).

Warnings: Excessive intake of nitrate causes methemoglobinemia, gastric cancer, and other pathologies (Umar et al., 2007). Spinach contains purine and should not be taken by patients with gout (Tang, 2010).

Comment: Dietary nitrates are converted into nitrite by gut bacteria, and nitrite is converted into NO by nitric reductase (Kapil et al., 2020).

DOI: 10.1201/9781003301455-101

REFERENCES

Adhikary, S., Choudhary, D., Ahmad, N., Kumar, S., Dev, K., Mittapelly, N., Pandey, G., Mishra, P.R., Maurya, R. and Trivedi, R., 2017. Dried and free flowing granules of Spinacia oleracea accelerate bone regeneration and alleviate postmenopausal osteoporosis. *Menopause*, 24(6), pp. 686–698.

Aritomi, M., Komori, T. and Kawasaki, T., 1985. Flavonol glycosides in leaves of Spinacia oleracea. *Phytochemistry*, 25(1), pp. 231–234.

Barker, A.V., Maynard, D.N. and Mills, H.A., 1974. Variations in nitrate accumulation among Spinach cultivars. *Journal of the American Society for Horticultural Science*, 99(2), pp. 132–134.

Bondonno, C.P., Yang, X., Croft, K.D., Considine, M.J., Ward, N.C., Rich, L., Puddey, I.B., Swinny, E., Mubarak, A. and Hodgson, J.M., 2012. Flavonoid-rich apples and nitrate-rich spinach augment nitric oxide status and improve endothelial function in healthy men and women: A randomized controlled trial. *Free Radical Biology and Medicine*, 52(1), pp. 95–102.

Ferreres, F., Castañer, M. and Tomás-Barberán, F.A., 1997. Acylated flavonol glycosides from spinach leaves (Spinacia oleracea). *Phytochemistry*, 45(8), pp. 1701–1705.

Fusch, L., 1555. *De Historia Stirpium Commetarii Insignes*. Lugduni Apud Ioan Tornaesium.

Iritani, N., 1969. Influence of spinach and seaweeds on cholesterol metabolism. *Journal of Japanese Society of Food and Nutrition*, 22, pp. 258–261.

Kahlon, T.S., Chapman, M.H. and Smith, G.E., 2007. In vitro binding of bile acids by spinach, kale, brussels sprouts, broccoli, mustard greens, green bell pepper, cabbage and collards. *Food Chemistry*, 100(4), pp. 1531–1536.

Kapil, V., Khambata, R.S., Jones, D.A., Rathod, K., Primus, C., Massimo, G., Fukuto, J.M. and Ahluwalia, A., 2020. The noncanonical pathway for in vivo nitric oxide generation: The nitrate-nitrite-nitric oxide pathway. *Pharmacological Reviews*, 72(3), pp. 692–766.

Kirtikar, K.R. and Basu, B.D., 1918. *Indian Medicinal Plants* (Vol. 2).

Lémery, N., 1716. *Traité universel des drogues simples, mises en ordre alphabétique. Où l'on trouve leurs différens noms . . . et tout ce qu'il y a de particulier dans les animaux, dans les végétaux, et dans les minéraux*. Au dépend de la Companie.

Tang, G., 2010. Spinach and carrots: Vitamin A and health. In *Bioactive Foods in Promoting Health* (pp. 381–392). Academic Press.

Umar, S., Iqbal, M. and Abrol, Y.P., 2007. Are nitrate concentrations in leafy vegetables within safe limits? *Current Science*, pp. 355–360.

102 Dandelion (*Taraxacum officinale* F.H. Wigg.)

Taraxacum officinale F.H. Wigg.

Etymology: From the Greek *taraktikos* = to move and the Latin *officinale* = found in pharmacies

Family: Asteraceae

Synonym: *Leontodon taraxacum* L.

Common names: Dandelion; dent de lion, pissenlit, teste de moine (Fr.); löwenzahn (Ger.); dente de leão (Port.); одуванчик (Rus.); diente de león (Span.)

Parts used: Leaf, root

Constituents: Hydroxycinnamic acid derivatives (chicoric acid, monocaffeyltartaric acid, chlorogenic acid), coumarins (cichoriin, aesculin), flavonol glycosides (Williams et al., 1996), triterpenes (taraxasterol) (Furuno et al., 1993).

Medical history: Dioscorides considered dandelion beneficial for the heart. French physicians in the 17h century used dandelion for dysentery and convulsions (Daléchamps, 1615). In the lectures of Lémery (1716), dandelion purifies the blood, while Alston (1770) advocates its use as diuretic and defines it as cold and dry to the second degree. In 18th-century England, it was used for fever and as a salad in 19th-century France (Guibourt, 1836). It was also a diuretic, a laxative, and a treatment for liver problems in 19th-century North America (Pereira, 1843).

Medicinal uses: 4–10 g of leaves as an infusion 3 times daily to promote urination (European Union); laxative, diuretic (Turkey, Pakistan); jaundice (Pakistan); stomach disorders (Nepal); tonic (India)

Blood pressure: A single oral administration of an ethanol extract of leaves or roots at the dose of 500 mg to rats made hypertensive by L-NAME poisoning was able to reduce both diastolic and systolic blood pressure within 4 hours (Aremu et al., 2019).

Plasma lipids and glucose: Ethanol extract given orally at 300 mg/kg/day for 10 weeks to rats poisoned with a high-fat diet reduced plasma lipids (Raghu et al., 2015). Rabbits on a high-cholesterol diet for 4 weeks were protected against arterial wall thickening by adding 1% of leaf powder in their diet, which decreased plasma triglycerides from 118.7 to 75 mg/dL (Choi et al., 2010). Ingestion of leaves as salad decreases plasma glucose (Goksu et al., 2010).

DOI: 10.1201/9781003301455-102

Adipose tissue: Ethanol extract given orally at 300 mg/kg/day for 10 weeks to rats poisoned with a high-fat diet reduced fat pad weights (Raghu et al., 2015).

Kidneys: Ethanol extract of leaves (1 g/1mL) give to human volunteers at the volume of 8 mL caused diuretic effects (Clare et al., 2009).

Bones and cartilages: Taraxasterol given orally to rats at 8 mg/kg/day for 26 days decreased synovial hyperplasia, bone and cartilage damages, and inflammatory cells infiltration together with a decreased expression in TNF α and interleukin-1β (Chen et al., 2019).

Warnings: A diabetic patient taking insulin developed severe hypoglycemia after eating salads of dandelion for 2 weeks (Goksu et al., 2010). Patients using antihypertensive and/or antidiabetic agents or insulin should be monitored closely while using dandelion (Posadzki et al., 2013).

Comment: Misidentification of plants as dandelion resulted in poisoning (Colombo et al., 2010), exemplifying the fact that teaching of botany and medicinal plants is a must for pharmacy students, even if it displeases the big pharma lobbies and the accreditation boards working for them. Materia medica and the teaching of medicinal plants are being ridiculed by accreditation boards visiting third-world universities. How many fatalities owed to the botanical incompetence of pharmacy students? We need to understand that providing proper medicinal plant teaching in schools of pharmacy saves lives. It is also time perhaps to stop disturbing lecturer and students with continuous modifications in teaching techniques with emphasis on "soft skills" and "communication" and other strange "group teaching" and "personality assessment", etc...(?). Knowledge acquired by effort and discipline and graduation of students solely made upon satisfactory acquisition of skills by professional bodies should be followed as in most US top universities.

REFERENCES

Alston, C., 1770. *Lectures on the Materia Medica: Containing the Natural History of Drugs, their Virtues and Doses: Also Directions for the Study of the Materia Medica; and an Appendix on the Method of Prescribing.* Edward and Charles Dilly.

Aremu, O.O., Tata, C.M., Sewani-Rusike, C.R., Oyedeji, A.O., Oyedeji, O.O., Gwebu, E.T. and Nkeh-Chungag, B.N., 2019. Acute and sub-chronic antihypertensive properties of Taraxacum officinale leaf (TOL) and root (TOR). *Transactions of the Royal Society of South Africa*, 74(2), pp. 132–138.

Chen, J., Wu, W., Zhang, M. and Chen, C., 2019. Taraxasterol suppresses inflammation in IL-1β-induced rheumatoid arthritis fibroblast-like synoviocytes and rheumatoid arthritis progression in mice. *International Immunopharmacology*, 70, pp. 274–283.

Choi, U.K., Lee, O.H., Yim, J.H., Cho, C.W., Rhee, Y.K., Lim, S.I. and Kim, Y.C., 2010. Hypolipidemic and antioxidant effects of dandelion (taraxacum officinale) root and leaf on cholesterol-fed rabbits. *International Journal of Molecular Sciences*, 11(1), pp. 67–78.

Clare, B.A., Conroy, R.S. and Spelman, K., 2009. The diuretic effect in human subjects of an extract of Taraxacum officinale folium over a single day. *The Journal of Alternative and Complementary Medicine*, 15(8), pp. 929–934.

Colombo, M.L., Assisi, F., Della Puppa, T., Moro, P., Sesana, F.M., Bissoli, M., Borghini, R., Perego, S., Galasso, G., Banfi, E. and Davanzo, F., 2010. Most commonly plant exposures and intoxications from outdoor toxic plants. *Journal of Pharmaceutical Sciences and Research*, 2(7), p. 417.

Daléchamps, 1615. *De l' histoire generale des plantes simples*. Chez Heritier Guillaume Rouille.

Furuno, T., Kamiyama, A., Akashi, T., Usui, M., Takahashi, T. and Ayabe, S.I., 1993. Triterpenoid constituents of tissue cultures and regenerated organs of Taraxacum officinale. *Plant Tissue Culture Letters*, 10(3), pp. 275–280.

Goksu, E., Eken, C., Karadeniz, O. and Kucukyilmaz, O., 2010. First report of hypoglycemia secondary to dandelion (Taraxacum officinale) ingestion. *American Journal of Emergency Medicine*, 28(1).

Guibourt, N.J.B.G., 1836. *Histoire abrégée des drogues simples*. Méquignon-Marvis Père et fils.

Lémery, N., 1716. *Traité universel des drogues simples, mises en ordre alphabétique. Où l'on trouve leurs différens noms . . . et tout ce qu'il y a de particulier dans les animaux, dans les végétaux, et dans les minéraux*. Au dépend de la Companie.

Pereira, J., 1843. *The Elements of Materia Medica and Therapeutics*. Lea and Blanchard.

Posadzki, P., Watson, L. and Ernst, E., 2013. Contamination and adulteration of herbal medicinal products (HMPs): An overview of systematic reviews. *European Journal of Clinical Pharmacology*, 69, pp. 295–307.

Raghu, M.R.P., Jyothi, Y. and Rabban, S.I., 2015. Anti-obesity activity of Taraxacum officinale in high fat diet induced obese rats. *Journal of Chemical and Pharmaceutical Research*, 7(4), pp. 244–248.

Williams, C.A., Goldstone, F. and Greenham, J., 1996. Flavonoids, cinnamic acids and coumarins from the different tissues and medicinal preparations of Taraxacum officinale. *Phytochemistry*, 42(1), pp. 121–127.

103 Thyme (*Thymus vulgaris* L.)

Thymus vulgaris L.

Etymology: From the Greek *thumos* = a warty excrescence and the Latin *vulgaris* = common

Family: Lamiaceae

Synonym: *Origanum thymus* (L.) Kuntze

Common names: Thyme; thym (Fr.); thymian (Ger.); tomilho (Port.); tomillo (Spa.)

Part used: Leaf

Constituents: Essential oil (thymol) (Borugă et al., 2014), hydroxycinnamic acid derivatives (rosmarinic acid) (Shekarchi et al., 2012).

Medical history: Dioscorides recommends thyme for asthma and to induce urination, dissolve blood clots, clean the eyes, and remove warts. Galen asserts that thyme is good for the lungs. German physicians in the 16th century defined thyme as hot and dry to the third degree. It was used for arthritis in 18th-century Scotland (Alston, 1770).

Medicinal uses: Cough (1–2 g in 150 mL of boiling water, 3–4 times daily, European Union). Gingivitis, dyspepsia, abdominal cramps, cough, tonic, immune system, cystitis, nephritis (Iraq); hypoglycemic, hypolipidemic, gastric tonic, antifungal, analgesic (Iran)

Blood pressure: Aqueous extract (1 g boiled in 100 mL water for 30 mins) given to hypertensive rats orally at 100 mg/kg/day for 8 days decreased systolic blood pressure from 186 to 138 mmHg (Osama et al., 2013) via a mechanism involving, at least in part, ACE inhibition (Shimada et al., 2014).

Thymol given to rats at doses ranging from 1 to 10 mg/kg evoked dose-dependent decreases in blood pressure and heart rate by antagonizing calcium ions channels (Aftab et al., 1995).

Heart: Thymol given to rats at doses ranging from 1 to 10 mg/kg evoked a dose-dependent decrease in heart rate (Aftab et al., 1995). Cardiac SOD activity and antioxidant capacity increased in rat treated with essential oil (Youdim & Deans, 1999). The heart function of diabetic and hyperlipidemic rats orally given an ethanol extract at 2 g/kg/day for 2 weeks was improved (Koohi-Hosseinabadi et al., 2015).

Plasma lipids and glucose: Broilers given in their diet an alcoholic extract (0.6%) experienced decreased levels of plasma triglycerides, total cholesterol, and LDL-cholesterol (Abdulkarimi et al., 2011).

DOI: 10.1201/9781003301455-103

Kidneys: The kidney function of diabetic and hyperlipidemic rats orally given at 2 g/kg/day for 2 weeks was protected.

Bones and cartilages: Thyme added as 5% of a low-calcium diet for 8 weeks increased femur bone mineral density in rats (Elbahnasawy et al., 2019).

Brain: An extract given to rats at 100 mg/kg/day for 15 days attenuated scopolamine-induced dementia in rats (Rabiei et al., 2015).

Warnings: Thymol has inotropic properties and can induce bradycardia (Szentandrássy et al., 2004). Thyme may decrease concentrations of thyroid hormone (Ulbricht et al., 2008). Nausea, vomiting, headache, dizziness, convulsions, and cardiac or respiratory arrest can occur if taken at more than 2 g daily (Spiteri, 2011). In general, infusions are much safer than essential oils, and essential oils should be used externally, not ingested (Bronstein et al., 2007; Yürüktümen et al., 2011).

REFERENCES

Abdulkarimi, R., Daneshyar, M. and Aghazadeh, A., 2011. Thyme (Thymus vulgaris) extract consumption darkens liver, lowers blood cholesterol, proportional liver and abdominal fat weights in broiler chickens. *Italian Journal of Animal Science*, 10(2), p. e20.

Aftab, K., Atta-U-Rahman, and Usmanghani, K., 1995. Blood pressure lowering action of active principle from Trachyspermum ammi (L.) Sprague. *Phytomedicine*, 2(1), pp. 35–40.

Alston, C., 1770. *Lectures on the Materia Medica: Containing the Natural History of Drugs, their Virtues and Doses: Also Directions for the Study of the Materia Medica; and an Appendix on the Method of Prescribing.* Edward and Charles Dilly.

Borugă, O., Jianu, C., Mişcă, C., Goleţ, I., Gruia, A.T. and Horhat, F.G., 2014. Thymus vulgaris essential oil: Chemical composition and antimicrobial activity. *Journal of Medicine and Life*, 7(Special Issue 3), p. 56.

Bronstein, A.C., Spyker, D.A., Cantilena, L.R. Jr, Green, J.L., Rumack, B.H., Heard, S.E. and American Association of Poison Control Centers, 2007. Annual report of the American association of poison control centers' national poison data system (NPDS): 25th annual report. *Clinical Toxicology (Phila) 2008*, 46, pp. 927–1057.

Elbahnasawy, A.S., Valeeva, E.R., El-Sayed, E.M. and Rakhimov, I.I., 2019. The impact of thyme and rosemary on prevention of osteoporosis in rats. *Journal of Nutrition and Metabolism, 2019.*

Koohi-Hosseinabadi, O., Moini, M., Safarpoor, A., Derakhshanfar, A. and Sepehrimanesh, M., 2015. Effects of dietary thymus vulgaris extract alone or with atorvastatin on the liver, kidney, heart, and brain histopathological features in diabetic and hyperlipidemic male rats. *Comparative Clinical Pathology*, 24(6), pp. 1311–1315.

Osama, A.K., Naser, A.E., Adel, G.E.S. and Eslam, A.H., 2013. Thymus vulgaris supplementation attenuates blood pressure and aorta damage in hypertensive rats. *Journal of Medicinal Plants Research*, 7(11), pp. 669–676.

Rabiei, Z., Mokhtari, S., Asgharzade, S., Gholami, M., Rahnama, S. and Rafieian-kopaei, M., 2015. Inhibitory effect of thymus vulgaris extract on memory impairment induced by scopolamine in rat. *Asian Pacific Journal of Tropical Biomedicine*, 5(10), pp. 845–851.

Shekarchi, M., Hajimehdipoor, H., Saeidnia, S., Gohari, A.R. and Hamedani, M.P., 2012. Comparative study of rosmarinic acid content in some plants of Labiatae Family. *Pharmacognosy Magazine*, 8(29), p. 37.

Shimada, A. and Inagaki, M., 2014. Angiotensin I-converting enzyme (ACE) inhibitory activity of ursolic acid isolated from thymus vulgaris, L. *Food Science and Technology Research*, *20*(3), pp. 711–714.

Spiteri, M., 2011. *Herbal Monographs Including Herbal Medicinal Products and Food Supplements*. Department of Pharmacy, University of Malta.

Szentandrássy, N., Szigeti, G., Szegedi, C., Sárközi, S., Magyar, J., Bányász, T., Csernoch, L., Kovács, L., Nánási, P.P. and Jóna, I., 2004. Effect of thymol on calcium handling in mammalian ventricular myocardium. *Life Sciences*, *74*(7), pp. 909–921.

Ulbricht, C., Chao, W., Costa, D., Rusie-Seamon, E., Weissner, W. and Woods, J., 2008. Clinical evidence of herb-drug interactions: A systematic review by the natural standard research collaboration. *Current Drug Metabolism*, *9*(10), pp. 1063–1120.

Youdim, K.A. and Deans, S.G., 1999. Dietary supplementation of thyme (Thymus vulgaris L.) essential oil during the lifetime of the rat: Its effects on the antioxidant status in liver, kidney and heart tissues. *Mechanisms of Ageing and Development*, *109*(3), pp. 163–175.

Yürüktümen, A., Hocaoglu, N.İ.L., Ersel, M., Özsaraç, M. and Kiyan, S., 2011. Acute hepatitis associated with Thymus Vulgaris oil ingestion: Case report. *Turkish Journal of Emergency Medicine*, *11*(2).

104 Ajowan (*Trachyspermum ammi* (L.) Sprague)

Trachyspermum ammi (L.) **Sprague**

Etymology: From the Greek *trachys* = rough and *spermum* = seeds and the Latin name of the plant, *ammi*

Family: Apiaceae

Synonyms: *Ammi copticum* L.; *Bunium copticum* (L.) Spreng.; *Carum copticum* (L.) Benth. & Hook. f.; *Daucus copticus* (L.) Pers.; *Ptychotis ajowan* DC.; *Ptychotis coptica* (L.) DC.; *Sison ammi* L.; *Trachyspermum copticum* (L.) Link

Common names: Ajowan, bishop's weed, carom seeds; ammi de l'Inde (Fr.); ajowan-kümmel (Ger.)

Part used: Seed

Constituents: Essential oil (thymol), phenolic glycosides (Ishikawa et al., 2001; Bairwa et al., 2012).

Medical history: The plant was known to ancient Ayurvedic physicians including Sushruta and used as a carminative (Dutt, 1877). It was used for colic, cholera, and alcoholism and as a carminative in 19th-century North America and Europe (Pereira, 1843).

Medicinal uses: Antispasmodic, diarrhea, diuretic, carminative, appetite stimulant, galactagogue (Pakistan); stimulant, carminative, antispasmodic; atonic dyspepsia, diarrhea (India); carminative, stomach aches, dysentery, burns, vomiting (Myanmar)

Blood pressure: Thymol given to rats at doses ranging from 1 to 10 mg/kg evoked dose-dependent decreases in blood pressure and heart rate by blocking calcium ions channels (Aftab et al., 1995).

Heart: Thymol given to rats at doses ranging from 1 to 10 mg/kg evoked a dose-dependent decrease in heart rate (Aftab et al., 1995).

Plasma lipids and glucose: In rabbits poisoned with butter and a high-cholesterol diet for 90 days, oral intake of petroleum ether extract of seeds at 2 g/kg/day for the following 46 days decreased plasma cholesterol from 615 to 335 mg/dL (normal: 334 mg/dL) (Javed et al., 2006).

Methanol extract given at 500 mg/kg/day for 60 days to streptozotocin-induced diabetic rats caused a decrease in glycemia from about 370 to 100 mg/dL and decreased hepatic interleukin-1β and TNF α (Zolfaghari et al., 2023).

DOI: 10.1201/9781003301455-104

Thymol given orally to streptozotocin-induced diabetic rats corrected glycemia from 308.7 to 117.1 mg/dL (normal: 87 mg/dL) (Sachan et al., 2022).

Bones and cartilages: An extract given at 100 mg/kg/day for 21 days attenuated collagen-induced arthritis in rats (Umar et al., 2012).

Brain: Thymol attenuated scopolamine-induced dementia in mice (Timalsina et al., 2023). Seed added as 2% of diet protected rats against scopolamine-induced loss of cognitive function (Soni & Parle, 2017).

Warnings: A 34-year-old man was admitted for heart block after consuming high amounts of ajowan (Hosseinjani et al., 2016). Thymol has inotropic negative properties and at non-culinary doses can induce bradycardia (Szentandrássy et al., 2004).

REFERENCES

Aftab, K., Atta-U-Rahman, and Usmanghani, K., 1995. Blood pressure lowering action of active principle from Trachyspermum ammi (L.) Sprague. *Phytomedicine*, *2*(1), pp. 35–40.

Bairwa, R., Sodha, R.S. and Rajawat, B.S., 2012. Trachyspermum ammi. *Pharmacognosy Reviews*, *6*(11), p. 56.

Dutt, U.C., 1877. *The Materia Medica of the Hindus*. Thacker, Spink, & Co.

Hosseinjani, H., Sadeghian, S. and Talasaz, A.H., 2016. Complete Heart Block in a Patient Taking Trachyspermum ammi and Zingiber officinale. *Journal of Pharmaceutical Care*, pp. 91–92.

Ishikawa, T., Sega, Y. and Kitajima, J., 2001. Water-soluble constituents of ajowan. *Chemical and Pharmaceutical Bulletin*, *49*(7), pp. 840–844.

Javed, I., Iqbal, Z., Rahman, Z.U., Khan, F.H., Muhammad, F., Aslam, B. and Ali, L., 2006. Comparative antihyperlipidaemic efficacy of Trachyspermum ammi extracts in albino rabbits. *Pakistan Veterinary Journal*, *26*(1), p. 23.

Pereira, J., 1843. *The Elements of Materia Medica and Therapeutics*. Lea and Blanchard.

Sachan, N., Saraswat, N., Chandra, P., Khalid, M. and Kabra, A., 2022. Isolation of thymol from Trachyspermum ammi fruits for treatment of diabetes and diabetic neuropathy in STZ-induced rats. *BioMed Research International*, *2022*.

Soni, K. and Parle, M., 2017. Trachyspermum Ammi seeds supplementation helps reverse scopolamine, alprazolam and electroshock induced amnesia. *Neurochemical Research*, *42*, pp. 1333–1344.

Szentandrássy, N., Szigeti, G., Szegedi, C., Sárközi, S., Magyar, J., Bányász, T., Csernoch, L., Kovács, L., Nánási, P.P. and Jóna, I., 2004. Effect of thymol on calcium handling in mammalian ventricular myocardium. *Life Sciences*, *74*(7), pp. 909–921.

Timalsina, B., Haque, M.N., Choi, H.J., Dash, R. and Moon, I.S., 2023. Thymol in Trachyspermum Ammi seed extract exhibits neuroprotection, learning, and memory enhancement in scopolamine_induced Alzheimer's disease mouse model. *Phytotherapy Research*. *37*(7): 2811–2826.

Umar, S., Asif, M., Sajad, M., Ansari, M.M., Hussain, U., Ahmad, W., Siddiqui, S.A., Ahmad, S. and Khan, H.A., 2012. Anti-inflammatory and antioxidant activity of Trachyspermum ammi seeds in collagen induced arthritis in rats. *International Journal of Drug Development and Research*, *4*(1), pp. 210–219.

Zolfaghari, N., Monajemi, R., ShahaniPour, K. and Ahadi, A.M., 2023. TNF-α and IL-1β in diabetes-induced liver damage: The relationship between Trachyspermum ammi Seeds methanol extract and inflammatory cytokine inhibition. *Journal of Food Biochemistry*, *2023*.

105 Stinging Needle (*Urtica dioica* L.)

Urtica dioica L.

Etymology: From the Latin *urere* = to burn and *dioica* = dioecious
Family: Urticaceae
Synonym: *Urtica galeopsifolia* Wierzb. ex Opiz
Common names: Stinging nettle; ortie (Fr.); brennnessel (Ger.); urtiga (Port.); крапива (Russian); ortiga (Spa.)
Part used: Root
Constituents: Flavonols, coumarins, acetylcholine, 5-hydroxytryptamine (Asgarpanah & Mohajerani, 2012).
Medical history: Dioscorides recommends stinging nettle for sexual impotence, for those who cannot breathe if not standing up, for inflammation, as a diuretic, and to clean secretions from the lungs. Pliny the Elder asserts that it is of value for cancer. German physicians in the 16th century consider stinging nettle hot and dry (Fusch, 1555), while in17th century France, it was used as a diuretic and eaten as a vegetable to prevent illness (Daléchamps, 1615). During the 18th-century France, it was prescribed for kidney stones, asthma, gout, hemorrhages, kidney stones, fever, and cleaning the blood (Lémery, 1716). In 19th-century France, it was taken as a diuretic and a tonic and to induce menses (Guibourt, 1836).
Medicinal uses: Diuretic (1.5 g of leaves as a decoction 3–4 times daily (European Union). Anemia cancer, emmenagogue, tonic, hypertension, kidney stones, rheumatism (Turkey); diabetes (Pakistan); bone fracture (India); diuretic (Myanmar)

Blood pressure: Methanol extract of leaves given orally at 200 mg/kg/day for 4 weeks to hypertensive rats reduced systolic and diastolic blood pressure (Vajic et al., 2018). Ethanol extract of leaves given at 500 mg every 8 hours for 3 months to type-2 diabetic patients had no effect on blood pressure (Kianbakht et al., 2013).
Plasma glucose and lipids: Aqueous extract of aerial parts given orally at the single dose of 250 mg/kg to rats before glucose oral load decreased glycemia by 33% after 1 hour (Bnouham et al., 2003). Ethanol extract of leaves given at 500 mg every 8 hours for 3 months to type-2 diabetic patients decreased glycemia by 34.2% (Kianbakht et al., 2013).
Kidneys: Ethanol extract given orally at 700 mg/kg/day for 8 weeks to rats prevented the formation of kidney stones induced by ethylene glycol (Keleş et al., 2020).

DOI: 10.1201/9781003301455-105

Bones and cartilages: Ethanol extract given orally to ovariectomized rats at 200 mg/kg/day for 8 weeks increased femur bone weight and increased bone mineral density (Gupta et al., 2014).

Brain: An extract given orally to rats attenuated scopolamine-induced dementia (Abu Almaaty et al., 2021).

Warnings: Nettle is contraindicated with diuretics, anticoagulants, antihypertensive medicines, finasteride, anticonvulsants, and hypoglycemic drugs as well as for patients with tachycardia (Spiteri, 2011).

REFERENCES

Abu Almaaty, A.H., Mosaad, R.M., Hassan, M.K., Ali, E.H., Mahmoud, G.A., Ahmed, H., Anber, N., Alkahtani, S., Abdel-Daim, M.M., Aleya, L. and Hammad, S., 2021. Urtica dioica extracts abolish scopolamine-induced neuropathies in rats. *Environmental Science and Pollution Research*, 28, pp. 18134–18145.

Asgarpanah, J. and Mohajerani, R., 2012. Phytochemistry and pharmacologic properties of Urtica dioica L. *Journal of Medicinal Plants Research*, 6(46), pp. 5714–5719.

Bnouham, M., Merhfour, F.Z., Ziyyat, A., Mekhfi, H., Aziz, M. and Legssyer, A., 2003. Antihyperglycemic activity of the aqueous extract of Urtica dioica. *Fitoterapia*, 74(7–8), pp. 677–681.

Daléchamps, 1615. *De l' histoire generale des plantes simples*. Chez Heritier Guillaume Rouille.

Fusch, L., 1555. *De Historia Stirpium Commetarii Insignes*. Lugduni Apud Ioan Tornaesium.

Guibourt, N.J.B.G., 1836. *Histoire abrégée des drogues simples*. Méquignon-Marvis Père et fils.

Gupta, R., Singh, M., Kumar, M., Kumar, S. and Singh, S.P., 2014. Anti-osteoporotic effect of Urtica dioica on ovariectomised rat. *Indian Journal of Research in Pharmacy and Biotechnology*, 2(1), p. 1015.

Keleş, R., Şen, A., Ertaş, B., Kayali, D., Eker, P., Şener, T.E., Doğan, A.H.M.E.T., Çetinel, Ş.U.L.E. and Şener, G.Ö.K.S.E.L., 2020. The effects of Urtica dioica L. ethanolic extract against urinary calculi in rats. *Journal of Research in Pharmacy*, 24(2), pp. 205–217.

Kianbakht, S., Khalighi-Sigaroodi, F. and Dabaghian, F.H., 2013. Improved glycemic control in patients with advanced type 2 diabetes mellitus taking Urtica dioica leaf extract: A randomized double-blind placebo-controlled clinical trial. *Clinical Laboratory*, 59(9–10), pp. 1071–1076.

Lémery, N., 1716. *Traité universel des drogues simples, mises en ordre alphabétique. Où l'on trouve leurs différens noms . . . et tout ce qu'il y a de particulier dans les animaux, dans les végétaux, et dans les minéraux*. Au dépend de la Companie.

Spiteri, M., 2011. *Herbal Monographs Including Herbal Medicinal Products and Food Supplements*. Department of Pharmacy, University of Malta.

Vajic, U.J., Grujic-Milanovic, J., Miloradovic, Z., Jovovic, D., Ivanov, M., Karanovic, D., Savikin, K., Bugarski, B. and Mihailovic-Stanojevic, N., 2018. Urtica dioica L. leaf extract modulates blood pressure and oxidative stress in spontaneously hypertensive rats. *Phytomedicine*, 46, pp. 39–45.

106 Bilberry (*Vaccinium myrtillus* L.)

Vaccinium myrtillus L.

Etymology: From the Latin *vacca* = cow and *myrtillus* = little myrtle

Family: Ericaceae

Synonym: *Vaccinium oreophilum* Rydb

Common names: Bilberry; black-wort, whortle-berry, European blueberry; airelles, myrtilles (Fr.); blaubeere (Ger.); mirtilo (Port.); черника (Rus.); arándano (Spa.);

Part used: Fruit

Constituents: Anthocyanins (delphinidin 3-galactoside) (Burdulis et al., 2007).

Medical history: According to the English physician and botanist John Gerarde (1545–1612) in his *The herball of generall historie of plantes* (1597), bilberry is cold to the second degree and dry, it quenches thirst, and it is astringent for the stomach "binde the belly", antiemetic, and good for fever. It was used for diarrhea and dysentery in 19th-century North America (Pereira, 1843) and given for fever in France (Guibourt, 1836).

Medicinal uses: Diabetes, laxative (Turkey)

Blood pressure: Adding 10% of fruits to a mouse high-fat diet for 3 months decreased systolic blood pressure by 6 mmHg (Mykkänen et al., 2014).

Plasma lipids and glucose: Fruit powder given orally at 2g/day for 4 weeks to alloxan-induced diabetic rats decreased plasma glucose, total cholesterol, LDL-cholesterol, and triglycerides (Asgary et al., 2016). Adding 10% of fruits to a high-fat mouse diet for 3 months had no effects on insulin resistance but prevented plasma TNF α elevation (Mykkänen et al., 2014).

Healthy volunteers consuming 150 g of frozen fruits 3 times a week for 6 weeks had no modification of blood pressure but experienced decreases in LDL-cholesterol (3 to 2.7 mmol/L), total cholesterol (5.3 to 4.9 mmol/L), and glucose (5.9 to 5 mmol/L) (Habanova et al., 2016). An extract corresponding to 50 g of fresh fruits given to type-2 diabetic patients improved post-prandial glycemia (Hoggard et al., 2013).

Kidneys: Aqueous extract of fruits caused diuretic effects (Ryazanova & Zaitceva, 2014).

Brain: Anthocyanins added to the diet of Alzheimer's mice models improved cognitive function (Yamakawa et al., 2016).

DOI: 10.1201/9781003301455-106

Comments: (i) Cold medicinal plants (Hippocratic system) are often anti-inflammatory.

(ii) It is clear that regular intake of bilberry is beneficial for aging.

REFERENCES

Asgary, S., RafieianKopaei, M., Sahebkar, A., Shamsi, F. and Goli-malekabadi, N., 2016. Anti-hyperglycemic and anti-hyperlipidemic effects of Vaccinium myrtillus fruit in experimentally induced diabetes (antidiabetic effect of Vaccinium myrtillus fruit). *Journal of the Science of Food and Agriculture*, 96(3), pp. 764–768.

Burdulis, D., Ivanauskas, L., Dirsė, V., Kazlauskas, S. and Ražukas, A., 2007. Study of diversity of anthocyanin composition in bilberry (Vaccinium myrtillus L.) fruits. *Medicina*, 43(12), p. 971.

Guibourt, N.J.B.G., 1836. *Histoire abrégée des drogues simples*. Méquignon-Marvis Père et fils.

Habanova, M., Saraiva, J.A., Haban, M., Schwarzova, M., Chlebo, P., Predna, L., Gažo, J. and Wyka, J., 2016. Intake of bilberries (Vaccinium myrtillus L.) reduced risk factors for cardiovascular disease by inducing favorable changes in lipoprotein profiles. *Nutrition Research*, 36(12), pp. 1415–1422.

Hoggard, N., Cruickshank, M., Moar, K.M., Bestwick, C., Holst, J.J., Russell, W. and Horgan, G., 2013. A single supplement of a standardised bilberry (Vaccinium myrtillus L.) extract (36% wet weight anthocyanins) modifies glycaemic response in individuals with type 2 diabetes controlled by diet and lifestyle. *Journal of Nutritional Science*, 2, p. e22.

Mykkänen, O.T., Huotari, A., Herzig, K.H., Dunlop, T.W., Mykkänen, H. and Kirjavainen, P.V., 2014. Wild blueberries (Vaccinium myrtillus) alleviate inflammation and hypertension associated with developing obesity in mice fed with a high-fat diet. *PLoS One*, 9(12), p. e114790.

Pereira, J., 1843. *The Elements of Materia Medica and Therapeutics*. Lea and Blanchard.

Ryazanova, T.K. and Zaitceva, H.N., 2014. The study of diuretic activity of bilberry drugs. *Aspirantskiy Vestnik Povolzhiya*, 14(1–2), pp. 249–251.

Yamakawa, M.Y., Uchino, K., Watanabe, Y., Adachi, T., Nakanishi, M., Ichino, H., Hongo, K., Mizobata, T., Kobayashi, S., Nakashima, K. and Kawata, Y., 2016. Anthocyanin suppresses the toxicity of Aβ deposits through diversion of molecular forms in in vitro and in vivo models of Alzheimer's disease. *Nutritional Neuroscience*, 19(1), pp. 32–42.

107 Cassumunar Ginger (*Zingiber cassumunar* Roxb.)

Zingiber cassumunar **Roxb.**

DOI: 10.1201/9781003301455-107

Etymology: From the Greek *ziggiberis* = ancient spice known to Arabs and the Sanskrit *cassumunar* = ginger
Family: Zingiberaceae
Synonym: *Zingiber purpureum* Roscoe
Common names: Cassumunar ginger, phlai; gingembre marron (Fr.)
Part used: Rhizome
Constituents: Essential oil, sesquiterpenes (zerumbone), arylheptanoids, hydroxy-cinnamic acid derivatives ((*E*)-4-(3′,4′-dimethoxyphenyl)but-3-en-2-ol) (Han et al., 2021).
Medical history: Cassumunar ginger was unknown to ancient Romans and Middle Ages European physicians. It was used in Ayurvedic medicine from dawn of time, and Dymock (1884) notes the use of cassumunar ginger for diarrhea and colic, while Gosh in the Calcutta Journal of Medicine, 1873 describes its uses for piles, increased bile secretion, and spleen enlargement in 19th-century India. In 19th-century Europe, it was a remedy for convulsions, epilepsy, and hysteria (Pereira, 1843).
Medicinal uses: Anorexia (India); stomachache (Malaysia)

Blood pressure: Ethanol extract given to healthy volunteers for 4 weeks at 170 mg/day decreased systolic blood pressure from 121 to 109 mmHg, total cholesterol from 193 to 183 mg/dL, and triglycerides from 126 to 106 mg/dL (Kato et al., 2018).
Plasma lipids and glucose: Ethanol extract give orally at 100 mg/kg/day to fructose-induced hyperlipidemic rats for 21 days decreased total cholesterol, LDL-cholesterol, and triglycerides (Hasimun et al., 2019), Ethanol extract (devoid of essential oil) given orally at 3 mg/kg/day for 270 days decreased the glycemia of rats (Koontongkaew et al., 2014).
Bones and cartilages: (*E*)-4-(3′,4′-dimethoxyphenyl)but-3-en-2-ol evoked some levels of protection against CFA-induced arthritis (Panthong et al., 1997).

Warning: To be avoided by patients on anticoagulant medicines (Sukati et al., 2019).
In rats, an oral dose of 5 g/kg of ethanol extract was not toxic. One gram given daily for 270 days was also not toxic (Koontongkaew et al., 2014).

REFERENCES

Dymock, W., 1884. *The Vegetable Materia Medica of Western India*. Education Society Press.
Han, A.R., Kim, H., Piao, D., Jung, C.H. and Seo, E.K., 2021. Phytochemicals and bioactivities of Zingiber cassumunar roxb. *Molecules*, 26(8), p. 2377.
Hasimun, P., Sulaeman, A., Mulyani, Y., Islami, W.N. and Lubis, F.A.T., 2019. Antihyperlipidemic activity and HMG CoA reductase inhibition of ethanolic extract of zingiber cassumunar roxb in fructose-induced hyperlipidemic wistar rats. *Journal of Pharmaceutical Sciences and Research*, 11(5), pp. 1897–1901.

Kato, E., Kubo, M., Okamoto, Y., Matsunaga, Y., Kyo, H., Suzuki, N., Uebaba, K. and Fuku-yama, Y., 2018. Safety assessment of bangle (Zingiber purpureum Rosc.) rhizome extract: Acute and chronic studies in rats and clinical studies in human. *ACS Omega, 3*(11), pp. 15879–15889.

Koontongkaew, S., Poachanukoon, O., Sireeratawong, S., Dechatiwongse Na Ayudhya, T., Khonsung, P., Jaijoy, K., Soawakontha, R. and Chanchai, M., 2014. Safety evaluation of Zingiber cassumunar Roxb. rhizome extract: Acute and chronic toxicity studies in rats. *International Scholarly Research Notices, 2014*.

Panthong, A., Kanjanapothi, D., Niwatananant, W., Tuntiwachwuttikul, P. and Reutrakul, V., 1997. Anti-inflammatory activity of compound D {(E)-4-(3′, 4′-dimethoxyphenyl) but-3-en-2-ol} isolated from Zingiber cassumunar Roxb. *Phytomedicine, 4*(3), pp. 207–212.

Pereira, J., 1843. *The Elements of Materia Medica and Therapeutics*. Lea and Blanchard.

Sukati, S., Jarmkom, K., Techaoei, S., Wisidsri, N. and Khobjai, W., 2019. In vitro antico-agulant and antioxidant activities of prasaplai recipe and Zingiber cassumunar Roxb. extracts. *International Journal of Applied Pharmaceutics, 11*, pp. 26–30.

Latin Names of Plants

Common Names of Plants

Melon tree 98
Milk thistle 162
Moringa 105
Mother's heart 94
Mugwort 67
Nigella 211
Okra 173
Olives 217
Oregano 220
Papaya 98
Paprika pepper 96
Parsley 226
Pear 240
Philippine waxeflower 157
Phlai 250
Pie cherry 234
Piss-a-bed 239
Pomegranate 236
Pride of India 187
Pumpkin 136
Queen's crape myrtle 187
Quince 147
Radish 160
Rapeseed 80
Saffron 128
Sage 249
Sesame 258
Spinach 265
Rosemary 246
Schwarze johannisbeere 244
Squash 136
Shallot 15
Shepherd's purse 94

Sour cherry 234
Soothing wort 61
Southernwood 61
Sponge gourd 193
Stinging nettle 275
Summer savory 254
Swamp cabbage 175
Sweet almonds 40
Sweet potato 177
Stone apple 7
Tarragon 64
Tart cherry 234
Tea 88
Thyme 270
Toothpickweed 38
Torch ginger 158
Tournefort's gundelia 171
True cinnamon 110
Turnip 86
Turmeric 142
Venus's hair fern 4
Vietnamese coriander 232
Walnut 180
Wasabi 160
Watermelon 113
Water spinach 175
White mustard 76
White thorn 121
Whortle-berry 277
Wild Marjoram 220
Wormwood 62
Yarrow 1
Yellow mustard 76

FRENCH

Absinthe amere
Achillée millefeuille 1
Aigremoine eupatoire 12
Aigriottes 234
Aloès 25
Ail 20
Airelles 277
Ajowan 273
Aluyne 62
Amandes douces 40
Ammi de l' Inde 273
Aneth 45
Angélique officinale 47
Armoise amere 62
Armoise vulgaire 67
Artichaut 150
Ase fétide 162
Avocat 229
Aubepine 121
Aubespin 121

Aurone citronelle 61
Aurone des jardins 61
Bardane a grosse tete 55
Basilic 214
Betterave 72
Bon riblet 196
Bourache
Bourse a Pasteur 94
Buglosse azurée 43
Buglosse d' Italie 43
Busserole 54
Calebasse 185
Canelle de Ceylan 110
Canellier 110
Capillaire de Montpellier 4
Cardamome 153
Cassis 244
Carvi 101
Céreli 52
Cerisier aigre 234

GERMAN

MALAY

Asam gelugur 169

Laksa 232

FILIPINO

Banaba 187

Moringa 205

PORTUGUESE

Abacate 229
Abóbora 136
Abrótono 61
Abstinto 62
Açafrão 128
Agrião comum 208
Agrião-do-jardim 190
Agrimônia comum 12
Alcaravia 101
Alecrim 246
Alface 183
Alho 20
Alho-poró 17
Amêndoa doce 40
Aneto 45
Angélica-dos-jardins
Alcachofra 150
Artemísia 67
Assa-fétida 162
Azeitonas 217
Batata doce 177
Babosa 25
Bardana 55
Bastão-do-imperador 158
Beterraba 72
Bolsa do pastor 94
Borragem 74
Cabaça 185
Cabaça esponja 194
Cabaco amargo 202
Canela verdadeira 110
Capilária 4
Cardamomo 153
Cardo Mariao 262
Centela 104
Cavalinha 155
Cerefolion 50
Cereja ácida 211
Chalota 15
Chá 88
Chicória 107
Chuchu 256
Coentro 124
Colza 80

Coentro 124
Coentro vietnamita 232
Cominho 139
Cúrcuma 142
Espinafre 265
Dente-de-leão 267
Espinafre de água 175
Espinheiro 121
Estragão 64
Figos 164
Funcho 166
Galanga maior 30
Galanga menor 33
Grãos-do-paraíso 10
Groselha preta 244
Limão 117
Marmelo de bengala 7
Melão 131
Melancia 113
Melissa 199
Mostarda preta 82
Lingua-de-vaca 43
Mamão 98
Manjericão 214
Moringa 205
Marrohlo 205
Marmelo 147
Mirtilo 277
Mostarda branca 76
Mostarda Indiana 78
Nabo 86
Nigela 211
Noz 180
Orégano 220
Ortossifão 223
Papaya 98
Pepino 133
Pera 158
Pimentão 96
Planta milefólio 1
Quiabo 173
Rabanete 242
Rábano 59
Repolho 84

Russian

SPANISH

Abrótano 61
Aceituna 217
Alcachofera 150
Azafran 128
Absintio 62
Agrimonia común 12
Ajedrea 254
Ajo 20
Albahaca 214
Alcaravea 101
Almendra dulce
Antorcha de jengibre 158
Áloe 25
Angélica archangelica
Apio 52
Arándano 277
Artemisia 67
Asa fétida 162
Banabá de Filipinas 187
Bardana 55
Batata 177
Berro 208
Berro de jardín 190
Bolsa de pastor 94
Borraja 74
Calabaza 185
Calabaza amarga 202
Calbaza 136
Camarruego 196
Cardamomo 153
Cardo lechoso 162
Centella 104
Cerezo ácido 234
Chalote
Chayote 256
Chicória 107
Cilantro 124
Cilantro Vietnamita 232
Cola de caballo 155
Colza 80
Comino 139
Culantrillo de pozo 4
Cúrcuma 142
Diente de león 167
Eneldo 45
Espino 121
Espinaca 265
Espinaca de agua 175
Esponja vegetal 194
Estragón 64
Gayuba 54

Granata 236
Granos del Paraíso 10
Galanga mayor
Galanga menor 33
Granata 236
Grosella nigra 244
Hierba luisa 28
Higos 164
Hinojo 166
Lechuga 183
Lengua de buey 43
Limón 101
Malvavisco 36
Melaon 131
Melisa 199
Marmellos 147
Menbrillo 147
Moringa 205
Mostaza blanca 76
Mostaza negra 82
Mostaza India 78
Membrillo de Bengala 7
Nigella 211
Perifollo 50
Nabo 86
Nuez 180
Orégano 220
Ortiga 275
Ortosifón 233
Palta 229
Pera 240
Perejil 226
Pimiento 96
Planta de milenrama 1
Puerro 17
Quimbombó 173
Rábano 242
Rábano picante 59
Raíz de remolacha 72
Reina de las flores 187
Repollo 84
Romero 246
Sabio 249
Sandía 113
Saúco 252
Sésamo 259
Té 85
Tomillo 270
Verdadera canela 110
Wasabi 160

Index of Natural Products

For Product Safety Concerns and Information please contact our EU
representative GPSR@taylorandfrancis.com
Taylor & Francis Verlag GmbH, Kaufingerstraße 24, 80331 München, Germany

*9 7 8 1 0 3 2 2 9 3 9 8 1 *